Urban Sustainability

The Urban Sustainability Book Series is a valuable resource for sustainability and urban-related education and research. It offers an inter-disciplinary platform covering all four areas of practice, policy, education, research, and their nexus. The publications in this series are related to critical areas of sustainability, urban studies, planning, and urban geography.

This book series aims to put together cutting-edge research findings linked to the overarching field of urban sustainability. The scope and nature of the topic are broad and interdisciplinary and bring together various associated disciplines from sustainable development, environmental sciences, urbanism, etc. With many advanced research findings in the field, there is a need to put together various discussions and contributions on specific sustainability fields, covering a good range of topics on sustainable development, sustainable urbanism, and urban sustainability. Despite the broad range of issues, we note the importance of practical and policy-oriented directions, extending the literature and directions and pathways towards achieving urban sustainability.

The series will appeal to urbanists, geographers, planners, engineers, architects, governmental authorities, policymakers, researchers of all levels, and to all of those interested in a wide-ranging overview of urban sustainability and its associated fields. The series includes monographs and edited volumes, covering a range of topics under the urban sustainability topic, which can also be used for teaching materials.

Zhenwei Peng
Editor

Theories and Frontiers in Territorial Spatial Planning

同濟大學出版社
TONGJI UNIVERSITY PRESS

Springer

Editor
Zhenwei Peng
Tongji University
Shanghai, China

Translated by
Zhang Qian
Xi'an University of Posts & Telecommunications
Xi'an, China

Xu Lihua
Xi'an University of Posts & Telecommunications
Xi'an, China

ISSN 2731-6483 ISSN 2731-6491 (electronic)
Urban Sustainability
ISBN 978-981-96-9824-0 ISBN 978-981-96-9825-7 (eBook)
https://doi.org/10.1007/978-981-96-9825-7

Jointly published with Tongji University Press
The print edition is not for sale in China (Mainland). Customers from China (Mainland) please order the print book from: Tongji University Press.

Translation from the Chinese Simplified language edition: "《国土空间规划理论与前沿》" by Zhenwei Peng, © Tongji University Press 2023. Published by Tongji University Press. All Rights Reserved.

This Springer imprint is published by the registered company Springer Nature Singapore Pte Ltd.
The registered company address is: 152 Beach Road, #21-01/04 Gateway East, Singapore 189721, Singapore

Foreword by Dong Liming

China's vast territorial land and maritime areas constitute invaluable assets. Its important mission in territorial spatial planning is to protect and maximize the abundant territorial space resources, to build a prosperous and beautiful homeland and to achieve the goal of strengthening the socialist country with modernization.

With the gradual deepening of reform and opening up, spatial planning in China has undergone significant transformations in comparison to traditional territorial and regional planning. Notably, the 2019 issuance of "*Several Opinions on Establishing the National Territorial Spatial Planning System and Supervising Its Implementation*" by the CPC Central Committee and the State Council, which not only provides clear guiding ideology, objectives and corresponding strategies for territorial space development and protection, but also builds a relatively comprehensive planning system. This system has facilitated significant progress and breakthroughs in the planning process, reflecting a range of pivotal ideas and national policies proposed by the Party and the State, such as "integrating various types of plans into a single plan", "optimizing the layout of territorial space", and "coordinated development of the region". These aspects highlight the strategic nature of the planning efforts as well as the comprehensive and localized features driven by the Party and the State's national policies.

The core of planning is the "five levels and three categories" planning approval system. The "five levels" are spatial scales, from the national, provincial, municipal and county levels all the way down to the grass-roots units of townships, and for each spatial level, the plan sets out different requirements. The "three categories" encompass richer connotations, which not only embodies the "combination of three plans with one plan" of main functional zone planning, urban and rural planning and integrated land use planning, but also includes the contents of master planning, detailed planning and a number of related special planning, involving multiple disciplines. Confronted with these transformative changes, it is difficult to accomplish the task without the necessary adjustment and enrichment of the existing planning majors in China's colleges and universities, as well as all kinds of planning and design teams and management.

Fortunately, in order to keep pace with the development in the new era and promote the dissemination of fundamental knowledge and technology in territorial spatial planning, Tongji University organized forces to compile five volumes of "Training Series of Territorial Spatial Planning" comprising over one million words in a short period of time. Looking at the contents of the five fascicles, it not only includes strategic and overall issues such as the exploration of the theoretical system of territorial spatial planning, the framework of planning core technologies, the compilation and transmission of plans at different spatial levels, but also provides a large amount of space to provide experiences and cases in the compilation process of special planning related to detailed planning and territorial spatial planning. The book basically covers all aspects of territorial spatial planning, and its content is very rich. The publication of "Territorial Spatial Planning Training Series" is beneficial not only for professionals seeking to acquire knowledge and skills in this field, but also holds invaluable reference significance for universities and colleges that are exploring the establishment of territorial spatial planning majors.

As an educator who has been engaged in urban and rural planning and integrated land use planning for a long time, facing the changes of territorial spatial planning, we deeply feels the importance of keeping up with the pace of the times. Thus, we must broaden our horizons, constantly absorb new things, expand our original professional fields, keep pace with the times, and never rest on our laurels. Back in the late 1970s, the Department of Geography of Peking University established the major of urban planning in response to the national requirements. In order to make up for the shortcomings of the original major, I went to Tongji University twice to learn lessons, and gained a lot of useful experience in running schools in personnel training, teaching plan, curriculum setting and production practice. Through the continuous communication between colleges and universities, Tongji University has also absorbed some theoretical methods in geography to further enrich the teaching content of urban planning major. This method of complementing science and engineering and learning from other's strong points to offset its weaknesses is also an innovation in the field of urban and rural planning education in China. In June 1998, the Ministry of Construction organized professionals to evaluate the urban planning majors of six engineering colleges. As a member of the evaluation expert group, I have the opportunity to fully understand the development and change process of Tongji University since the reform and opening up. In order to meet the national requirements for urban planning talents, the university not only maintains its original professional characteristics, but also adds some new courses according to the development needs, such as urban geography, urban economy, regional planning, urban ecology, etc. At the same time, it also absorbs some personnel from geography colleges to enrich the teaching staff, and continuously optimizes the urban planning major through inter-school exchanges. University not only shows its advantages in traditional planning fields, such as overall urban planning, detailed planning, road traffic planning, urban design, etc., but also opens up new fields, and is recognized by academic circles as one of the universities with the most comprehensive and powerful urban and rural planning in China. From this point of view, it is no accident that Tongji University compiled and published the Training Series of Territorial Spatial Planning.

Without solid professional foundation, long-term exploration and accumulation, it is impossible to undertake the training task of territorial spatial planning.

As a set of training series to popularize territorial spatial planning, there is certainly room for improvement and perfection. In general, the content of the anthology is rich, but there is still room for systematic improvement. The articles collected in the series have different styles, and there may still be collision of viewpoints, which is the characteristic of the times in the current reform of territorial spatial planning.

It should also be noted that although the current territorial spatial planning has formed a relatively comprehensive planning examination and approval system, variables and problems will be bound to arise in the process of planning and implementation due to the lack of practical testing and the absence of promulgated laws and regulations pertaining to territorial spatial planning. For example, at a macro level, although the territorial spatial planning has achieved the "combination of three plans with one plan", how to integrate spatial planning and development planning into a planning system with division of labor, complementary and interdependence, and how to integrate the development planning of town clusters and metropolitan areas with the master planning of central cities into an organic whole still need further exploration. On a micro level, there is a lack of clear division of labor among functional departments for urban design, urban renewal, historical and cultural heritage protection and infrastructure construction planning. This leads to avoidable overlaps of responsibilities. It is crucial to address this issue and strengthen the division of labor and cooperation among functional departments. These efforts are of great significance in improving the compilation system of spatial planning.

To summarize, China's territorial spatial planning is still at an early stage, encountering different opportunities and challenges, with significant progress yet to be made. I hold a strong belief that with the continuous exploration and innovation of more related disciplines and planners in the future, China's territorial spatial planning will reach a series of milestones.

Beijing, China
July 2022

Dong Liming
Professor, College of Urban
and Environmental Sciences,
Peking University

Foreword by Chen Bingzhao

Following the "multi-plan integration" reform and pilot projects, territorial spatial planning has gradually evolved into a new planning system built upon existing spatial plans such as main functional zoning, land use planning, and urban-rural planning. Territorial spatial planning is "the guide of national spatial development, the spatial blueprint of sustainable development, and the basic basis of various development, protection and construction activities".[1] It is not the combination and superposition of the original spatial planning, but a qualitative leap after "integration" and "combination". It has distinct essential characteristics and attributes of "three completeness" and "three highs".

The term "three completeness" refers to "the object of spatial planning is 'the whole region, all factors and general action strategies', involving the protection, development, utilization, restoration and governance of natural spatial elements such as mountains, water, fields, forests, lakes, grass, sand and islands, as well as artificial environments such as cities, towns and villages".[2] The term "three highs" refer to "achieving high-quality development, high-quality life and high-efficiency governance". With the establishment of the Ministry of Natural Resources (MNR) by the State Council, the central government defined its dual mandates as: To carry out the unified exercise of the duties of owners of natural resources assets owned by the whole nation, and the unified exercise of all duties of regulating the use of territorial space and ecological protection and restoration. The requirement of "two unifications" means that it is not enough for the Ministry of Natural Resources to be the guardian of land, ecology and other resources, but also to be transformed into an efficient asset operator.... For territorial space plannng, it is not just simply a matter of solving the problems of spatial arrangement and control of use, but also to study how to promote the preservation and appreciation in natural resources through

[1] *Several opinions on building a national territorial spatial planning system and supervising the implementation by CPC Central Committee and State Council* (2019-05-23). http://www.gov.cn/zhengce/2019-05/23/content5394187.htm.

[2] Sun Shiwen. (2020). *From Urban and Rural Planning to Territorial Spatial Planning*. Journal of Urban and Rural Planning, (4): 11–17.

spatial planning."[3] Since spatial allocation and utilization inherently generate both positive and negative externalities, the distribution of spatial resources must carefully balance competing usage demands. Territorial spatial planning should not only "keep the bottom line", but also "seek development". On the premise of "ecological priority" (rigid management and control), local "green development" should be planned, and the planning should be usable, effective and easy to use. In a word, territorial spatial planning is a national spatial governance behavior and an important means to realize the modernization of national governance. It faces many new challenges and needs to be explored.

While valuable lessons can be drawn from conventional planning practices, the new era demands continuous exploration and knowledge accumulation. "Territorial Space Training Series" is the stage achievement on the road of exploration. From theory to practice, there are five fascicles, including *Theories and Frontiers of Territorial Spatial Planning, Compilation Technology of Territorial Spatial Planning, Special Planning of Territorial Spatial Planning, Urban Renewal and Urban Design and Fine Governance of Territorial Spatial Planning*. The 75 chapters can be described as Expo, which is a great review of theory and practice.

Notwithstanding its comprehensive coverage, the series has yet to delve deeply into critical components such as "double evaluation", "three-line delineation", "a base map", "ocean and coastline", "nature reserve" and "national park". To complete the territorial spatial planning, it is necessary to expand the related professional knowledge of pedology, agronomy, forestry, oceanography, water conservancy and biology on the basis of the original urban and rural planning. Obviously, the compilation of territorial spatial planning can never be completed by single major, but requires collective effort of multi-professional teams. However, from the actual situation of compiling territorial spatial planning in various parts of China, the main force of compiling territorial spatial planning is still the professionals engaged in urban and rural planning. Perhaps, the root cause lies not only in the knowledge structure, but also in: (1) The original urban and rural planners have strong comprehensive thinking ability, and have developed a network thinking habit, that is, once they touch a point (problem) on the Internet, they will immediately make comprehensive analysis, comparison, trade-off and judgment on many related problems. Dealing with the spatial problems of urban and rural planning requires considering not only two dimensional—fixed-locations, but also three-dimensional—altitude and plot-ratio. Further, it is necessary to consider the time sequence of four-dimensional. Some people even say that there are "five dimensions"—people should be thought of, because people are the core of the problem and we should put people at the first place. This three-dimensional, multi-dimensional and dynamic thinking habit can only be formed through long-term cultivation. (2) Urban and rural planning staff have a strong policy concept, because planning is not a simple technical problem,

[3] Zhang Jingxiang. (2021) *Academic Talk on the Relationship between Strategic Guidance and Rigid Management and Control of Territorial Spatial Planning System.* Journal of Urban and Rural Planning, (2): 6–14.

but more importantly, it serves the current policy and is the embodiment of public policy.

As previously mentioned, the compilation of territorial spatial planning requires a broadening of knowledge. It is therefore relevant to discuss a historical experience in this context. In 1980, shortly after Tongji University resumed its major in urban planning, Hua Changyi, a Chinese-American professor from Boston University, was invited to lecture at Tongji University with the support of UNESCO. We took the initiative to invite him to the teaching site (classroom) to observe, and the students were working out the overall urban planning of Pingliang City, Gansu Province. After reading it, he remarked insightfully: This is the same as when I was in America more than 10 years ago. Teachers are like commanders, commanding various arms. You are responsible for land use, and he is responsible for transportation and greening. However, this is no longer applicable in the United States now,

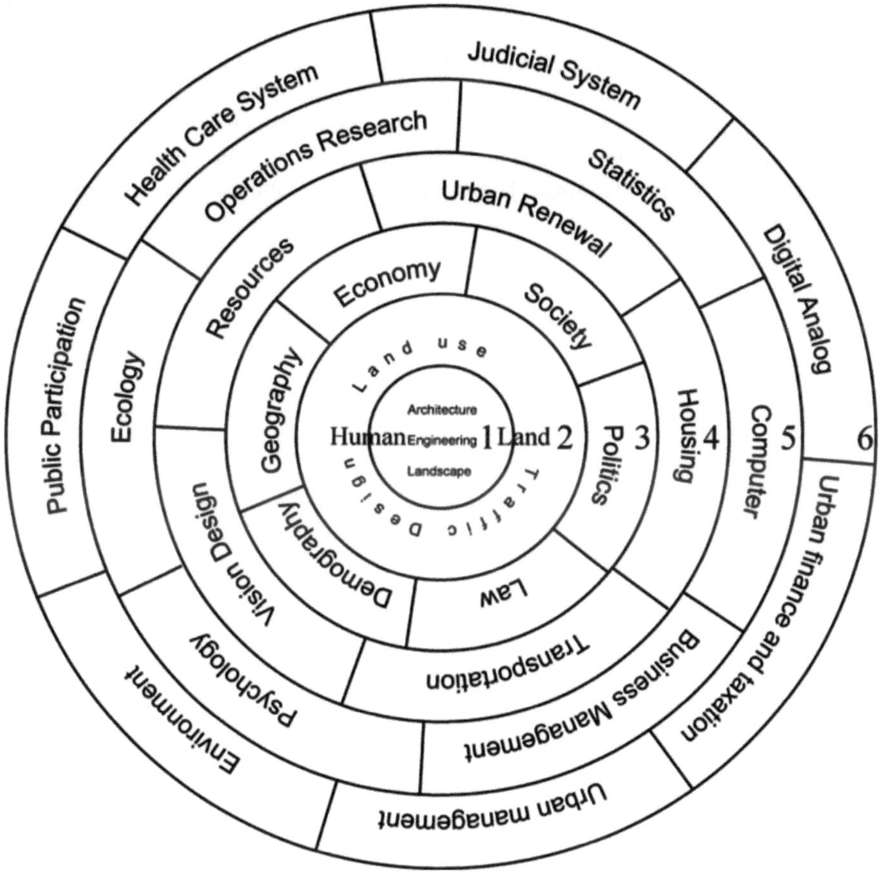

Fig. 1 Expansion diagram of urban planning knowledge used by Professor Hua Changyi in his class

and the overall urban planning is no longer carried out. In his lecture, he used an expansion diagram of urban planning knowledge (Fig. 1): The first circle (core circle) includes architecture, engineering and landscape; The second circle includes land use and traffic design; The third circle includes economy, society, politics, law, population and geography; In addition, there are laps 4, 5 and 6. Due to the expansion of urban planning knowledge, many people who call themselves "urban planning experts" experience communication barriers with each other, because the contents of the core circle are hollowed out. This experience has some enlightenment for our territorial spatial planning today: Although the territorial spatial planning involves rich contents, including marine knowledge and agricultural knowledge in addition to information such as mountains, water, fields, forests, lakes, grasses and sands, the common core-people and land cannot be lost.

In the context of knowledge expansion, the significance lies not in the amount of knowledge you possess, but in your familiarity with relevant knowledge and your ability to locate it. This series could be a useful source of information on territorial spatial planning.

We have deeply felt that great changes have occurred in urban and rural spatial structure and people's lifestyle from the era of carriages to that of automobiles. With the rapid development of science and technology, what problems will the concepts of information technology, virtual space composed of big data and "metaverse" bring to the territorial spatial planning? These issues deserve attention as we continue to explore and gather knowledge.

Shanghai, China
July 2022

Chen Bingzhao
Professor, College of Architecture
and Urban Planning, Tongji University

Preface to the Collection

On 23 May 2019, the CPC Central Committee and the State Council jointly issued the "*Several Opinions on Establishing the National Territorial Spatial Planning System and Supervising Its Implementation*" (hereinafter referred to as the "*Several Opinions*"), which officially inaugurated the establishment of China's territorial spatial planning system and associated reforms. On 30 May of the same year, the Ministry of Natural Resources issued the Notice on Carrying out Land Spatial Planning in an All-round Way, requiring that the competent departments of natural resources at all levels should unify their thoughts and actions to the decision-making arrangements of the CPC Central Committee, take the initiative to perform their duties and fulfill their duties in accordance with the requirements of *Several Opinions*, and establish a multiple compliance territorial spatial planning system and supervise its implementation". Under the vigorous promotion of the Ministry of Natural Resources, all parts of the country continue to promote the research and practice of territorial spatial planning according to the requirements of the documents of the central government and ministries and commissions.

As a world-renowned university renowned for its excellence in engineering education, Tongji University boasts a century-long tradition in urban-rural planning discipline development and talent cultivation. In the process of development, Tongji University has continuously expanded its intersection and integration with related disciplines such as architecture, landscape architecture, land planning, environmental planning, ecology, natural resources development and utilization, ocean engineering and infrastructure. The development system of teaching, research and practice of urban and rural planning has been constructed. Tongji University actively serves the national economic and social development, and a batch of professional leading cadres for the national planning and construction system. Since the establishment of the Urban Construction Cadre Center in Tongji University in the early 1980s, more than 30,000 professional leading cadres have been for the whole country, who are active in the field of urban and rural planning and construction all over the country and have made great contributions to the healthy development of urbanization in China.

The year 2020 marked a critical juncture for advancing China's territorial spatial planning system, yet the outbreak of the world-wide health crisis severely disrupted socioeconomic operations, preventing planning professionals from conducting field work. In order to express condolences to the planners who fought in the front line of anti-epidemic in Wuhan, the Urban Construction Cadre Training Center of Tongji University organized the first live broadcast of online public welfare training on territorial spatial planning in March 2020, and set up a special training course on territorial spatial planning and construction leading cadres, professional and technical personnel, teachers and scientific research groups in Wuhan, which was then extended to Hubei Province and then gradually extended to other provinces in China. In just three months, well-known domestic planning experts including academicians of Chinese Academy of Engineering and Professor Wu Zhiqiang of Tongji University were invited, and 36 online public welfare live training sessions were conducted, with cumulative participation exceeding 200,000 professionals nationwide. According to the feedback from trainees, these public welfare training courses have greatly promoted the study of territorial spatial planning theory and planning techniques, the popularization of laws and knowledge related to land spatial management, and made contributions to further improving the construction level of China's territorial spatial planning system.

Through these public welfare trainings, we have accumulated a lot of valuable teaching resources for territorial spatial planning. Combining the courseware and audio-visual materials of these training courses with the latest concepts of land spatial planning development and changes in circumstances, the lecturers have carefully polished the topics. At the same time, the content structure of territorial spatial planning, a small number of academic papers published in journals were added to form a 75-chapter "National Spatial Planning Training Series".

The themes and sections of the five volumes of the series are:

1. Territorial spatial planning theory and frontier (theoretical exploration and technical frontier);
2. Compilation technology of territorial spatial planning (technical framework, overall planning and detailed planning);
3. Territorial space special planning (transportation and infrastructure, ecosystem, historical protection, special research);
4. Urban renewal and urban design (urban renewal, urban design, practical cases);
5. Fine of territorial space (big cities, rural areas, urban communities, built environment, legal basis and support).

The publication of "Territorial Spatial Planning Training Series" can strengthen the in-depth communication of China's planning industry, promote the improvement of planning practitioners' professional ability and level in the critical period of the current territorial spatial planning system construction, form a good development ecology of territorial spatial planning industry.

Shanghai, China
July 2022

Zhenwei Peng
Deputy Secretary of Tongji University
Party Committee

Preface

The continuous promotion of "multi-regulation integration" since the 18th National Congress of the Communist Party of China is an important countermeasure to solve the contradiction between the spatial layout and scale of cities and towns and the carrying capacity of resources and environment in the process of urbanization in China. The establishment of China's territorial spatial planning system in 2019 provides institutional guarantees for fundamentally resolving spatial disorder in socioeconomic development and safeguarding ecological civilization construction. However, whether it is the continuous exploration since the Central Urbanization Work Conference put forward the "multi-regulation integration" in December 2013, or the "*Several opinions on building a national territorial spatial planning system and supervising the implementation by CPC Central Committee and State Council*" (Zhongfa [2019] No.18 Document). The practice of compiling the overall territorial spatial planning since the establishment of the land spatial planning system shows that it is an extremely arduous task and necessary to integrate the development priority zone, land use planning, urban and rural planning and other spatial planning into a unified national, clear responsibilities and rights.

Structurally, the territorial spatial planning system comprises four major components: planning formulation and approval, implementation and supervision, legal and policy frameworks, and technical standards - each requiring multidimensional interpretation to fully grasp its profound implications. Each part has a complete setup, and its profound connotation needs to be interpreted from different dimensions and perspectives. Secondly, from the theoretical point of view behind the territorial spatial planning, all kinds of spatial planning that constitute the territorial spatial planning system are organized according to certain theories and form their own relatively independent knowledge systems, which are irreplaceable to each other. For example, urban and rural planning is based on the knowledge system of the discipline of urban and rural planning, while land use planning is based on the knowledge system of land administration under the first-level discipline of public management. This also shows that territorial spatial planning is neither urban and rural planning or land use planning, nor the sum of them, but a brand-new one. Therefore, in order to complete the arduous task of establishing the territorial spatial planning system, it is necessary to break the original pattern of all kinds of spatial planning, establish a comprehensive and overall conceptual system and theoretical system, and establish a unified logical framework.

Based on this, the Urban Construction Cadre Training Center of Tongji University has set up a series of courses focusing on the theoretical system and internal logic of territorial spatial planning since 2020, trying to clarify the connotation of this brand-new system of territorial spatial planning and the development direction of the territorial spatial planning system. The contents of these courses have also become the main content of the *National Spatial Planning Theory and Frontier* fascicle of *National Spatial Planning Training Series.*

This fascicle is divided into two sections with 13 papers, among which the first section is the first six papers, which mainly focus on the overall direction and strategic thinking of territorial spatial planning reform, the knowledge base of territorial spatial planning, and the construction logic and theoretical system. The second section is the last seven papers, which mainly focus on the compilation system of territorial spatial planning system, including the elastic mechanism and transmission mechanism of city and county-level land spatial master planning (Articles 7–8), the rural planning in territorial spatial detailed planning (Article 9), and the sector planning research in land spatial planning (Articles 10–13), mainly including land ecological spatial planning, healthy city planning, urban renewal and urban design.

Among them, it should be mentioned particularly, the practice of compiling national territorial spatial planning mainly stays at the level of overall territorial spatial planning, and the planning exploration of urban and rural built environment under the territorial spatial planning system is far from enough. In order to reflect the characteristics and sustainable health of urban life in the compilation of the overall territorial spatial planning, and to create a beautiful living environment in urban and rural areas, papers on urban renewal and urban design are included in this fascicle.

In order to reflect the completeness of the contents of this volumes, five papers "*Discussion on the logic and operation strategy of territorial spatial planning system construction*", "*Knowledge base and structure of land spatial planning*", "*Issues and Prospects of Spatial Planning Reform*: Five papers", "*Thoughts on Planning Compilation and Discipline Development*", "*Constructing Rural Planning System* and *Urban Organic Renewal and Fine Management*", have been published in relevant journals, and are included in this fascicle with the consent of the author and relevant journals.

Shanghai, China

Zhenwei Peng
Deputy Secretary and Professor
of Tongji University Party Committee

Contents

About the Translators

Zhang Qian, (1977–) was born in Shanxi Province. She graduated with a bachelor's degree in English linguistics and culture from Xi'an International Studies University in 2001, a master's degree in educational economy and management from Shaanxi Normal University, and is currently pursuing her doctorate in translation studies at the School of Foreign Languages at Xi'an Jiaotong University. Her research focuses on translation studies. She is an associate professor and supervisor for master's degree candidates, a member of the China Translation Association and the Shaanxi Translation Association. She has published over 10 research papers and has authored 5 books. She has led 4 provincial research projects, 7 university-level educational reform projects, and over 10 corporate research projects. Some of the provincial research projects she has led include "the study of the translation of cultural-loaded terms in folk culture in the Guanzhong region of Shaanxi", "the construction of cultural linguistic context for the translation and introduction of intangible cultural heritage in Shaanxi", and "the intertextual tracing and translation of the Banpo and Yangshao culture". Her publications include *Academic English Translation, Scientific and Technological English The Modern Humanities Perspective of English and American Literature*, and *Language Expression Techniques in English Teaching etc.*

Xu Lihua, (1977–) was born in Jiangsu Province. She graduated with a bachelor's degree in English education from Xi'an International Studies University, a master's degree in educational economy and management from Shaanxi Normal University, and is currently pursuing her doctorate at the School of Humanities and Social Science at Xi'an Jiaotong University. She is a professor and supervisor for master's degree candidates, a member of the TESOL Expert Committee, a director of the Shaanxi Association, an executive of the Cross-border E-commerce Working Committee of the China Association of Enterprises with Foreign Economic Cooperation, and an international business English. Her research focuses on foreign language teaching and research, cross-cultural adaptation, and translation studies. She has received the first prize for the Shaanxi Provincial Graduate Foreign Language Teaching Achievement Award. She has participated in 4 national social science fund projects and has led nearly 10 provincial projects. She has also led 11 university-level educational reform projects. She has edited or participated in the writing of 6 textbooks and books. She has published nearly 20 academic papers, with her main achievements published in CSSCI academic journals such as *Foreign Language Teaching and Research, Foreign Language Electronic Teaching, and Journal of Northwest University.*

Theory

Reform and Prospect of Territorial Spatial Planning

Peng Zhenwei

China's urban and rural development has ushered in a new era. Adhering to a people-oriented approach, scientific development, reform and innovation, and governance based on the rule of law, we are transforming the mode of urban and rural development, perfecting the urban governance system, and enhancing governance capacity. With a focus on addressing pressing issues such as urban problems. We are continuously improving the quality of the urban environment, the standard of living for our people, and the city's competitiveness. Our goal is to build a modern city that is harmonious, livable, vibrant, and unique in its own way. This is not only the overall requirement for advancing the development of new-type urbanization in our country, but also the fundamental backdrop for understanding the reform of the current territorial spatial planning system.

Since the 1980s, China has successively developed various types of spatial planning, which have played a positive role in serving the social-economic development and urbanization since the country's reform and opening up, as well as promoting the rational utilization and effective protection of territorial resources. However, they also face numerous contradictions and conflicts. Since the 18th National Congress of the Communist Party of China (CPC), with the continuous advancement of ecological civilization construction, the central government has been actively promoting institutional reforms in related fields. The establishment of the national institutional reform plan in 2018 laid the foundation for the spatial planning system construction. In 2019, policy documents such as the *"Guidelines on establishing a national territorial spatial planning system and supervising the implementation by CPC Central Committee and State Council"* clarified the requirements for constructing a unified territorial spatial planning system. From the perspective of in-depth institutional reform, the spatial planning system is a crucial lever for advancing the modernization

P. Zhenwei (✉)
Tongji University, Shanghai, China
e-mail: pengzhw@tongji.edu.cn

Z. Peng (ed.), *Theories and Frontiers in Territorial Spatial Planning*, Urban Sustainability, https://doi.org/10.1007/978-981-96-9825-7_1

of the national governance system and governance capacity, as well as an important guarantee for the sustainable and healthy development of China's economic society.[1]

1 Background of Spatial Planning Reform

1.1 Process and Characteristics of Urbanization in China

Since the reform and opening up, China has undergone the largest and fastest urbanization process in world history, with unprecedented urban expansion and remarkable achievements that have captured worldwide attention. Since 1949, China's urbanization has evolved through distinct phases, each marked by unique characteristics. From the perspective of urbanization growth rate, China's urbanization development can be roughly divided into four stages. The first stage (1949–1957) marked a period of normal urbanization development, during which the urbanization rate increased from 10.64 to 15.39%. The second stage began after 1958, during which China experienced a period of fluctuating development. Initially, from 1958 to 1960, the urbanization rate rose from 15.39% to 19.75%, a rise of 4.36 percentage points. However, it then declined from 19.75% to 17.98% between 1960 and 1964, a decline of approximately 2 percentage points. The third stage, from the mid-1960s to 1978, was a period of stagnant urbanization development, with the urbanization rate fluctuating around 17% for many years. The fourth stage, from 1978 after the reform and opening up, witnessed the rapid development of urbanization in China. From 1978 to 1990, the urbanization rate increased from 17.92% to 26.41%, with an average annual growth rate of 0.71 percentage points. According to the data from the seventh national census, China's urbanization rate reached 63.89% in 2020, indicating that during the 30-year period from 1990 to 2020, China's urbanization rate experienced rapid growth, with an average annual increase of 1.25 percentage points, significantly higher than all previous stages. As of 2021, the urbanization rate stands at 64.72% (Fig. 1).

1.2 Key Challenges in Urbanization

The development of urbanization has propelled the growth of cities in China, subsequently driving the overall economic and social development. Urban construction has become a vital catalyst for modernization and has played a significant role in China's economic and social development as well as in improving people's livelihoods. However, it is important to recognize the challenges that arise from the

[1] Peng Zhenwei (2021).

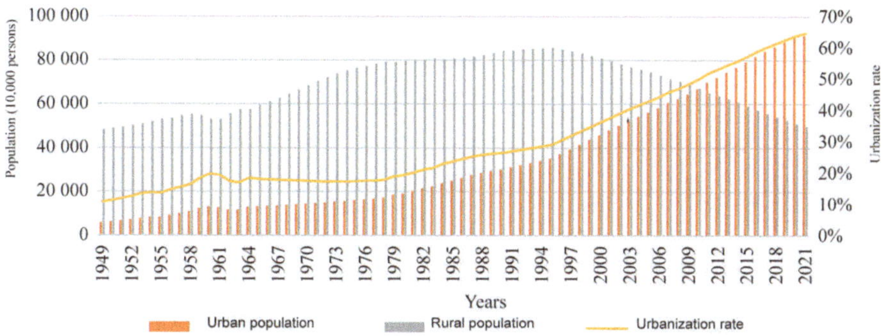

Fig. 1 Changes of urban population, rural population and urbanization rate in China over the years. *Source* National Bureau of Statistics, 2022

rapidly advancing model of urbanization, including the need for coordinated development between urban and rural areas, the substantial resource consumption and issues related to circular economy resulting from urbanization. As urbanization enters a more advanced stage, the increase in urban residents' income stimulates the continuous elevation of consumption capacity and levels. Additionally, industrialization and urbanization lead to a rapid increase in the demand for energy resources, placing greater strain on regional development in terms of resource consumption and environmental sustainability. There are two main aspects that highlight the existing irrationalities in China's urbanization development.

One issue is the coordination between people and land in urbanization. The core task of urbanization is to achieve population urbanization, and land is the crucial carrier of urbanization. The speed relationship between population urbanization and land urbanization reflects the coordination between the level of urbanization and resource consumption. According to the research report *"China: Toward Efficient, Inclusive and Sustainable Urbanization"* compiled by the Development Research Center of the State Council in 2014, from 1996 to 2012, the national construction land increased by an average of 4,827 square kilometers per year, with urban construction land accounting for an average of 2,380 square kilometers per year. From 2010 to 2012, the national construction land increased by an average of 6,353 square kilometers per year, with urban construction land accounting for an average of 3,433 square kilometers per year. From 2000 to 2011, the area of urban built-up areas increased by 76.4%, significantly higher than the 50.5% growth rate of urban population. This indicates that the large-scale expansion of urban built-up areas has not been accompanied by a corresponding scale of population urbanization. The speed of land urbanization is much faster than that of population urbanization, reflecting the extensive expansion of urban construction and excessive consumption of resources in the process of urbanization. The path dependence of land urbanization has led to the rapid expansion of urban space in the short term, but it has also brought about significant risks and conflicts of interest in urbanization operation.[2] These include challenges to

[2] Wang Liwei and Zhao Ming (2014).

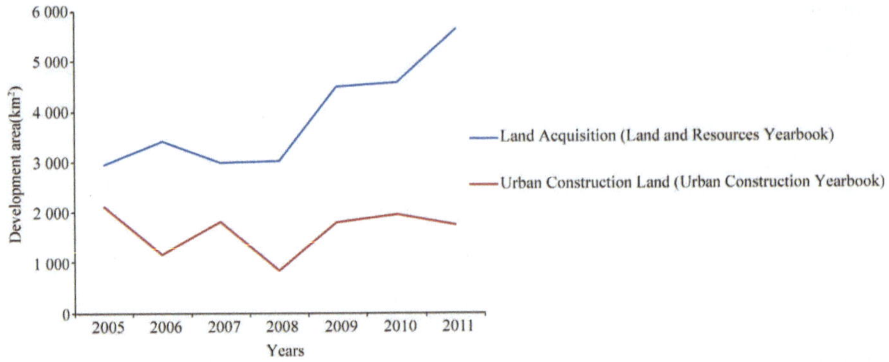

Fig. 2 Land requisition is faster than urban land development. *Source China* Promoting Efficient, Inclusive and Sustainable Urbanization[3]

the ecological environment carrying capacity and pressure on the protection of farm land under the national conditions of a large population and limited land (Fig. 2).

In addition to the extensive and inefficient use of construction land, the imbalance of urbanization development is also reflected in the unreasonable spatial distribution and scale structure of cities and towns, as well as the mismatch with the carrying capacity of resources and the environment. In some densely populated areas in eastern China, there is a tightening constraint on resources and the environment, while the urbanization potential of regions with stronger resource and environmental carrying capacity in central and western China remains untapped. The layout of urban agglomerations is not entirely reasonable, with insufficient coordination and cooperation among different cities within the agglomerations, resulting in sub-optimal cluster efficiency. Some major cities are facing population pressures in their central areas that exceed their overall carrying capacity. Meanwhile, small and medium-sized cities lack sufficient industry agglomeration and population, which hinders their full potential. Additionally, the large number of small towns with limited scale and weak service functions significantly increases the economic, social, and ecological costs. Moreover, there is a common presence of low-quality residential space construction in suburban and urban–rural mixed areas, which reflects the deviation between the rapid growth of urbanization rate and the quality of urbanization development.

The second issue is the coordination of urban–rural relations in the process of urbanization. The development path of urbanization in China determines that the stable development of rural areas is the bottom-line guarantee of urbanization and the key to promoting its healthy development. Internationally, the experience of developing countries, particularly in Latin America, shows that urban population growth is a double-edged sword. In its 2002 report, the United Nations pointed out, "Adequate supply of cheap labor (rural immigrants) guarantees the development of labor-intensive manufacturing, but a large and rapid increase in urban population

[3] Development Research Center of the State Council and World Bank (2014).

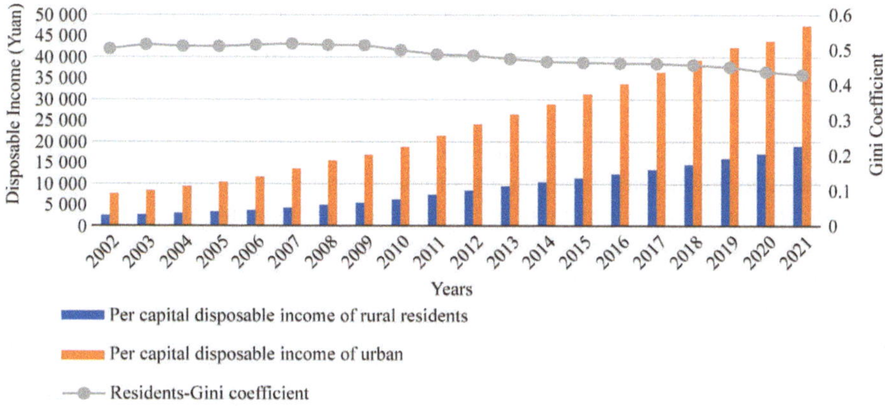

Fig. 3 Per capital disposable income and Gini coefficient in urban and rural areas of China over the years. *Source* National Bureau of Statistics, 2022

(especially rural immigrants with room for improvement in their skills) brings difficulties to the construction, management, and employment of large cities." In a joint report by the United Nations Development Programme and UN-Habitat in 2004, it further stated, "Traditional urbanization not only leads to increased poverty in rural areas, but also creates a large population of the urban poor. This social injustice, spreading rapidly from rural areas to cities, is one of the main causes of the 'Latin American trap'."

By analyzing the Gini coefficient of income for urban and rural residents[4] in China, we can make a preliminary judgment on the development gap between urban and rural areas. According to the per capita disposable income and Gini coefficient calculated by the National Bureau of Statistics, there has been a significant decrease in the Gini coefficient for urban and rural residents' income since 2010. In 2021, the Gini coefficient has dropped to 0.43, which is a decrease of about 0.1 compared to 2002. This indicates that the income gap between urban and rural areas is gradually narrowing, although it still remains at a relatively large level (Fig. 3).

1.3 Requirements for Developing New Urbanization

In response to the prominent contradictions and issues arising from the process of urbanization development, the country has been promoting a sustained and comprehensive top-level design since 2012. At the 18th National Congress of the Communist Party of China in 2012, it was proposed to "adhere to the path of new industrialization,

[4] Gini coefficient is an important analytical index used to comprehensively investigate the difference of income distribution among residents in the world. The value is between 0 and 1, and the closer the Gini coefficient is to 0, the more equal the income distribution tends to be.

informationization, urbanization, and agricultural modernization with Chinese characteristics." In 2013, the Third Plenary Session of the 18th CPC Central Committee adopted the *"Decision of the Central Committee of the Communist Party of China on Several Major Issues Concerning Comprehensively Deepening Reform,"* which put forward 60 reform decisions. In the same year, the Central Urbanization Work Conference was held, identifying five major prominent contradictions and issues accumulated during the rapid development of urbanization in China. In 2014, the *"National Plan on New Urbanization (2014–2020)"* was issued, explicitly proposing to follow the path of new urbanization with Chinese characteristics. In 2015, the Fourth Central Urban Work Conference pointed out the ten major problems in China's urban development. In the same year, the Central Committee of the Communist Party of China and the State Council issued the *"Integrated Reform Plan for Promoting Ecological Progress,"* emphasizing the need to prioritize ecological civilization construction and integrate it into all aspects and the entire process of economic, political, cultural, and social development. In 2016, the State Council issued the "*Opinions on Further Promoting the Construction of New Urbanization*," and in the same year, the *"Opinions of the CPC Central Committee and the State Council on Strengthening Urban Planning and Construction Management"* were issued. In the report of the 19th National Congress of the Communist Party of China in 2017, further clear requirements were put forward, from new urbanization to rural revitalization, and then to the integration of urban and rural development.

From this, it can be seen that since 2012, the country has been continuously promoting top-level policy design work to guide the healthy development of urbanization, while also constantly putting forward requirements for urban planning. Firstly, it is necessary to clarify the overall planning of territorial space. In May 2013, the Sixth Collective Study of the 18th CPC Central Committee's Political Bureau put forward the principle of "balancing population, resources, and environment and unifying economic, social, and ecological benefits" to comprehensively plan the development of territorial space, coordinate population distribution, economic layout, land use, and ecological environment protection, and scientifically layout production space, living space, and ecological space. Secondly, understanding the importance of planning. General Secretary Xi Jinping emphasized the importance of scientific planning during his inspection of Beijing's urban construction in February 2014, stating that scientific planning brings the greatest benefits, planning mistakes lead to the greatest waste, and planning haphazardly is the biggest taboo. Thirdly, it is necessary to promote the integration of multiple planning categories. In December 2013, the Central Urbanization Work Conference clearly stated the need to establish a spatial planning system, promote planning system reform, accelerate planning legislation, and form a unified, complementary, and coordinated planning system. This includes exploring the integration of economic and social development planning, urban and rural planning, and land use planning to create a planning blueprint for each county or city that can be consistently implemented. Fourthly, the role of planning as a guiding principle and bottom line constraint. In March 2019, the Second Session of the 13th National People's Congress emphasized the importance of adhering to bottom-line thinking and using territorial spatial planning as a basis. Urban, agricultural, and ecological spaces, as well as ecological protection red lines, the red line of permanent prime farmland protection, and the boundaries for urban development should not be crossed when adjusting economic structure, planning industrial development, and promoting urbanization. Fifthly, addressing the issue of urban expansion and extensive use of land resources. Both the Central Urbanization Work Conference in 2013 and the Central Urban Work Conference in 2015 emphasized that artificial urban development should not be pursued at the expense of economic development. It is necessary to strictly adhere to bottom-line principles, adjust the structure, and increase the level of land use efficiency in urban construction. Sixthly, addressing the issue of ecological and agricultural space being occupied by construction. The December 2015 Central Urban Work Conference highlighted the importance of mountains, water, forests, fields, and lakes as integral parts of urban life that should not be arbitrarily encroached upon or destroyed, and emphasized the need to integrate cities with nature. Seventhly, addressing the problem of neglecting the human aspect in urban work in the past. Greater attention should be given to the construction of urban living environments. Eighthly, addressing the issue of urban spatial structure. It is necessary to enhance the rationality of internal urban layout, allowing for the interweaving and organic combination of different functional areas such as residential, commercial, office, and cultural spaces. Ninthly, addressing the issue of prioritizing cities over rural areas. It is important to fully recognize the differences between urban and rural development and strengthen coordination. Without rural development, urbanization will lack a solid foundation. Tenthly, it is necessary to improve the scientific, forward-looking, and mandatory nature of planning.

Taking a closer look, the Central Urban Work Conference in 2015 proposed the concept of "1-Recognition, 5 Coordinations Principle". This means recognizing, respecting, and adapting to the laws of urban development. Urban development is a process where rural populations gather in cities and agricultural land is transformed into urban construction land in a corresponding manner. It is crucial to ensure that population and land resources are well-matched, and that the scale of urban development is in line with the carrying capacity of resources and the environment. Based on respecting the laws of urban development, we need to make overall plans from five perspectives: coordinating the spatial, scale, and industrial structures to enhance the overall effectiveness of urban work; coordinating the planning, construction, and management processes to improve the systematic nature of urban work; coordinating reform, science and technology, and culture to ensure sustainable urban development; coordinating production, living, and ecological layouts to enhance the livability of cities; and coordinating the government, society, and citizens to enhance the enthusiasm of all parties involved in promoting urban development and encourage multiple stakeholders to participate in urban governance.

2 Exploration of Transformation Development and Planning Reform

2.1 The Background of Transformation and Development

Currently, it is important to have a clear understanding of the unsustainability of the existing urban development model in order to address the numerous issues and risks that have accumulated in urban development. Transformational development has become an urgent necessity for social development, urban development, and planning. From the perspective of social development, China's economy has shifted from a phase of high-speed growth to one of high-quality development. We are currently in a critical period of transitioning development patterns, optimizing economic structures, and transforming growth drivers. As the country as a whole enters the ranks of middle- to high-income nations, it is crucial to surpass the middle-income trap and ensure a simultaneous improvement in the quantity and quality of urbanization. Looking at urban development, we must confront new trends such as population aging and rising consumption, and strive to create more sustainable and high-quality urban and rural living spaces through effective planning. From the standpoint of planning itself, its role must shift from solely serving economic development to providing a sound living environment. Planning should play a vital public policy role and facilitate the comprehensive management of urban and rural spaces across all domains, factors, and processes in order to truly achieve integrated development. Under the demands of development transformation, planning should prioritize people, balance efficiency and fairness, and play a comprehensive and coordinating role in the development of urban and rural spaces.

2.2 Requirements of Transformation and Development for Planning Reform

At the national policy level, the Third Plenary Session of the Eighteenth Central Committee of the Communist Party of China in 2013 proposed the establishment of a sound management system for the country's natural resource assets. In 2015, the Fifth Plenary Session of the Eighteenth Central Committee further emphasized the promotion of regional coordinated development, the shaping of orderly and free flow of factors, effective constraints on main functions, equal provision of basic public services, and the achievement of coordinated regional development that is sustainable in terms of resources and the environment. It also called for the advancement of the construction of main functional areas and the utilization and management of resources. These requirements are directly put forward by the central government in the context of planning reforms.

Looking at the development and evolution of urban and rural planning, it can be seen that the establishment of *the Urban Planning Regulations* in 1984 through administrative legislation created two types of planning, namely, overall urban planning and detailed urban planning. It also established the system of construction land permits and construction project permits within urban planning areas, delegating the power to manage local urban planning and construction to city governments. *The Urban Planning Law* implemented in 1990, as the first national law on the subject, established a system for the preparation and approval of urban master plans, detailed plans, and urban system plans. It also introduced the "One book, Two certificates" system, which includes the construction project site selection opinions, construction land planning permits, and construction project planning permits. The new Urban and Rural Planning Law enacted in 2008 established a comprehensive urban and rural planning system covering regions, cities, and villages, pushing forward the evolution from "urban planning" to "urban and rural planning". However, throughout this development process, the urban and rural planning system itself has shown certain limitations. These include the imperfect procedures for preparation, approval, and revision, a focus on the expansion of construction space, and a neglect of overall planning and management of all elements, particularly non-built-up areas and rural areas, leading to a long-standing technical gap in planning management.

Parallel to this is the institutional framework for land use. In 1986, *the Land Administration Law* was enacted, elevating the issues of land use and protection to the level of national legislation. In the same year, the National Land Administration Bureau was established. In 1998, revisions were made, emphasizing the need to implement the "strictest farmland protection system." The vertical management of land was strengthened, with the central and provincial governments assuming control over land administration rights, and lower-level land use plans being developed in accordance with higher-level land use plans. Furthermore, there was a focus on enhancing the management of land use plans and implementing total control over the amount of construction land. As a result, the Ministry of Geology and Mineral Resources, the National Land Administration Bureau, the State Oceanic Administration, and the National Bureau of Surveying and Mapping were merged to form the Ministry of Land and Resources.

Therefore, it can be observed that in the development of the past two core planning systems, comprehensive laws and regulations have been established, covering various levels from cities to regions, and from cities to rural areas. However, due to the different operational logics, administrative systems, and action mechanisms, the emphasis on bottom-up urban and rural planning and the emphasis on top-down land use planning have resulted in a cycle of "contradiction-coordination-conflict".[5] This situation is not only reflected in the specific content of the planning objects, but also in the process of resource management and planning management (Fig. 4).

To address the various contradictions in spatial planning, the Chinese government officially proposed the establishment of a spatial planning system and the implementation of multiple compliance starting from 2013. *The Decision of the Central*

[5] Zhao Min (2019).

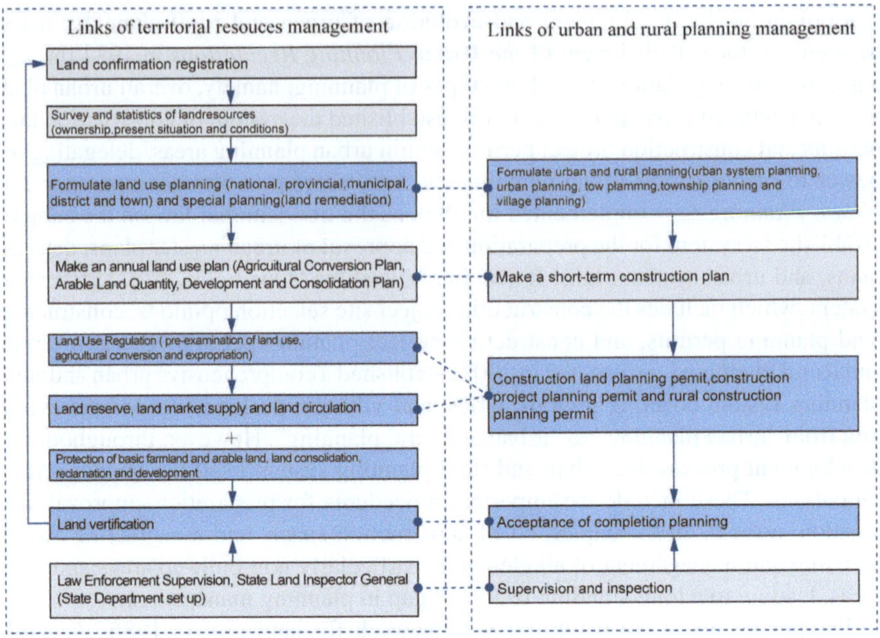

Fig. 4 Differences between territorial resources management and urban and rural planning management. *Source* The author draws according to *the Urban and Rural Planning Law* and *Land Administration Law*

Committee of the Communist Party of China on Several Major Issues Concerning Comprehensively Deepening Reform, adopted at the Third Plenary Session of the Eighteenth Central Committee of the Communist Party of China in 2013, clearly stated the need to establish a spatial planning system, define the boundaries for the development and control of production, living, and ecological spaces, and enforce land use regulations. *The National New Urbanization Plan (2014–2020)*, introduced in 2014, further emphasized the promotion of "multiple compliance" in economic and social development planning, urban and rural planning, and land use planning in eligible areas. Subsequently, various government departments including the Ministry of Housing and Urban–Rural Development, the National Development and Reform Commission, and the Ministry of Land and Resources put forward pilot programs for implementing multiple compliance and accumulated valuable experience. For instance, the pilot program for "multiple compliance" initiated by the Ministry of Housing and Urban–Rural Development in 2014 emphasized the importance of a comprehensive approach based on the principles of "development strategy + spatial layout + implementation strategy". It sought to achieve organic integration of various planning aspects by emphasizing "Five Determinations" (qualitative, quantitative, shaping, delimitation, and policy) and realizing "Five Unifications" (unified development goals, unified technical indicators, unified spatial coordinates, unified legend standards, and unified implementation platforms). However, during the exploration process, the issue of poor coordination between the work of different departments became apparent due to their relatively independent nature. Therefore, it is crucial to conduct comprehensive top-level design exploration at a higher level, building on the experience gained from pilot projects at the grassroots level in cities and counties. Only by coordinating the efforts of various central government departments and effectively managing the boundaries and connections between local departmental plans can this be successfully achieved.

3 Spatial Planning Reform Under Ecological Civilization Framework

3.1 Mandates in Ecological Civilization Framework Reform for Spatial Planning

In order to address the issues in land development and conservation in China, as well as to resolve the conflicts in spatial planning, the Communist Party of China Central Committee and the State Council issued the *Integrated Reform Plan for Promoting Ecological Progress* in 2015. This plan calls for the establishment of a clear property rights, multi-stakeholder participation, balanced incentives and constraints, and comprehensive system of ecological civilization institutions, consisting of eight core systems. Its goal is to advance the modernization of the national governance system and governance capacity in the field of ecological civilization. Among these core systems, the third one explicitly proposes the construction of a spatial planning system. Based on the reform of the ecological civilization system, the central government has put forward four general requirements for the establishment of a national spatial planning system. First, ecological civilization should be given a prominent position to optimize the spatial pattern of territorial land. Second, planning should be integrated into the governance system, with the spatial governance system consisting of spatial planning, land use control, and differentiated performance assessment. Third, through planning legislation and departmental integration, the modernization of the national governance system can be achieved. This is also a prominent manifestation of the modernization of the national governance capacity. In terms of the establishment of the planning system, it is necessary to integrate the various spatial plans currently prepared separately by different departments, and develop a unified spatial plan to achieve comprehensive planning coverage that is unified, interconnected, and hierarchically managed at the national, provincial, and municipal levels. Fourth, in terms of land space development and conservation, a sound land use control system should be established. The intensity indicators of development should be decomposed to the county-level administrative areas as binding indicators to control the total amount of construction land. Use control should be extended to all natural ecological spaces, and Ecological Redlines should be delineated and strictly adhered to, in order to prevent the Ecological Redlines from being compromised by inappropriate development and construction activities.

3.2 Exploration of Provincial "Spatial Planning" Pilot Project

Building on the pilot project of "integrating multiple regulations into one" at the city and county level in 2014, the central government has further advanced the reform pilot of "integrating multiple regulations into one" at the provincial level. Taking Hainan Province as an example, in order to formulate the *Integrated Plan of Hainan Province (Spatial Class 2015–2030)*, a comprehensive verification of the differences between hundreds of thousands of land use plans and urban–rural plans has been conducted through collaboration between the province and cities/counties, resulting in the establishment of a unified current data for the entire province. Through institutional integration and local legislation, a preliminary legal system for the reform of "integrating multiple regulations into one" has been established to ensure the implementation of the plan. Based on in-depth evaluations of resource and environmental carrying capacity and specialized research, the planning framework has been established from three perspectives: strategy, spatial layout, and implementation management and control. This framework adopts strategic planning as the guide, spatial planning as the core, and implementation management and control as the support, forming a comprehensive content system.[6] Specific contents include:

(1) Coordinate and harmonize various spatial planning based on the planning of main functional zones, and fully integrate them with the 13th Five-Year Plan for national economic and social development. Emphasize the coordination of spatial planning such as main functional area planning, ecological protection red line planning, urban system planning, overall land use planning, forest land protection and utilization planning, and marine functional zoning, and implement them into a unified blueprint.
(2) Establishing ecological space, agricultural space, development space, and the "Three Zones, Three Lines" of ecological protection red line, permanent prime farmland, and urban development boundaries, in order to optimize spatial layout. Emphasizing the protection of the ecological environment, efficient allocation of resources, improvement of infrastructure and public service facilities, and land use conservation and intensification as key aspects of the planning.
(3) To establish a unified information platform to provide technical support for spatial control. The Hainan Province has developed a comprehensive information management platform, integrating the *Integrated Plan of Hainan Province* into the business management information systems of various departments including development and reform, agriculture, land, housing and construction,

[6] Zhang Bing and Hu Yaowen (2017).

transportation, ocean affairs, water affairs, forestry, environmental protection, as well as municipalities and counties. This integration allows for the centralized management of spatial master planning information and data resources.

In 2017, the central government elevated the pilot level and expanded the implementation of the provincial "spatial planning" pilot program in a more comprehensive manner. According to the Notification on the *Pilot Program of Provincial Spatial Planning* in 2017 issued by the General Office of the CPC Central Committee and the General Office of the State Council, the main task of the pilot program is to "comprehensively identify and analyze the underlying conditions of land space based on the main functional zoning plan, establish urban, agricultural, and ecological spaces, as well as ecological protection red lines, permanent prime farmland, and urban development boundaries, and emphasize the control of development intensity and the implementation of key control measures, while coordinating various spatial planning initiatives and formulating unified provincial space plans, in order to accumulate experience and provide examples for the integration of multiple regulations, the establishment of a sound land and space development and protection system." The pilot program aims to achieve "four ones": a set of planning results, a set of technical specifications, an information platform, and a set of reform recommendations. The pilot program covers nine provinces (autonomous regions): Hainan, Ningxia, Jilin, Zhejiang, Fujian, Jiangxi, Henan, Guangxi, and Guizhou. Therefore, the core of the provincial-level reform is to comprehensively plan and integrate various spatial planning initiatives, including the main functional zoning plan, provincial urban system planning, overall land use planning, and marine functional zoning, into a cohesive blueprint, while establishing a unified information platform.

3.3 Reform of Territorial Space Development System

(1) Reform of Institutional Functions

The exploration and pilot construction in recent years have shown that, under the current conditions in China, merely exploring spatial planning techniques while neglecting the deeper issues of departmental planning and management coordination in spatial resource allocation makes it difficult to achieve the expected goals. In 2018, the Third Plenary Session of the 19th CPC Central Committee approved the "Deepening the Reform Plan of Party and State Institutions," which consolidated the responsibilities of spatial planning and management related to the protection and utilization of natural resources under the newly established Ministry of Natural Resources. It is hoped that this consolidation will effectively resolve the contradictions that have arisen in the past due to institutional mechanisms. The integrated Ministry of Natural Resources now assumes three major functions: planning control, resource investigation, and confirmation registration. This includes supervising the development, utilization, and protection of natural resources, establishing a spatial planning system and overseeing its implementation, fulfilling the responsibilities of all types of natural resource assets owned by the whole population, unifying survey and confirmation registration, establishing a system of compensated use of natural resources, and managing the surveying and mapping as well as geological exploration industries. Therefore, the institutional reform has solidified the transformation from land and resources management and urban–rural planning management to natural resources management. It emphasizes a top-level framework for comprehensive and systematic natural resources protection and utilization, embodying the full requirements of the nation's institutional reform (Fig. 5).

For the national land and space sector, two core institutions are established to be responsible for national land and spatial planning and use control. The first is the National Land and Spatial Planning Bureau, which formulates policies related to national land and spatial planning, undertakes the establishment of a spatial planning system, and supervises its implementation. It organizes the preparation of national land and spatial planning and related special plans and supervises their implementation. It is responsible for reviewing and approving local land and spatial planning submitted to the State Council for approval, and provides guidance and review for major national special plans related to land and space development and utilization. It also conducts suitability assessments for land and space development and establishes a monitoring, evaluation, and early warning system for the implementation of land and spatial planning. The second is the Land and Space Use Control Department,

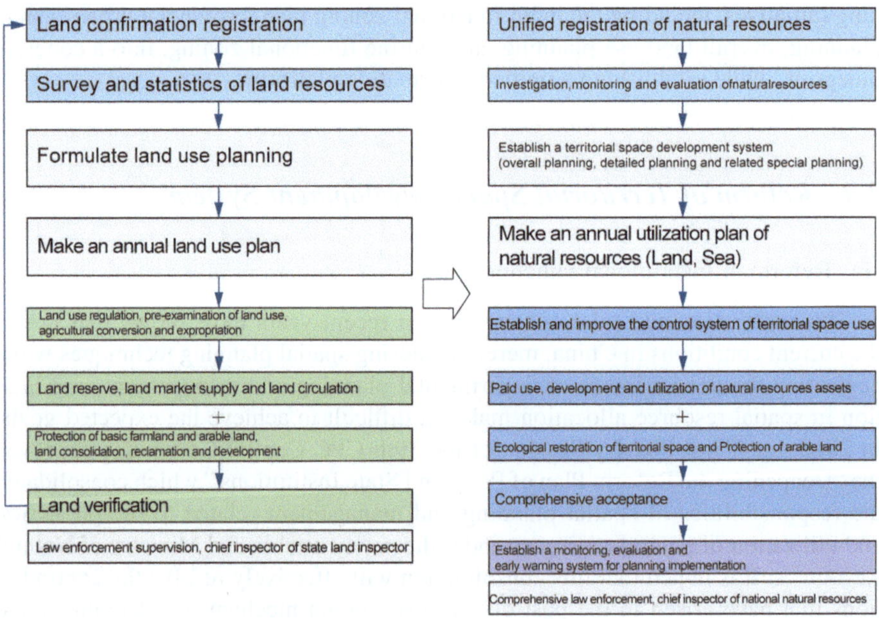

Fig. 5 Transformation of natural resource management system. *Source* The author draws according to the relevant contents of *the Provisions on Function Allocation, Internal Organization and Staffing of the Ministry of Natural Resources*

which formulates norms and technical standards for land and space use control. It proposes annual land and sea utilization plans and organizes their implementation. It organizes the formulation of policies for the conversion of land and space uses such as farm land, forest land, grassland, wetland, sea area, and islands, and provides guidance for land use pre-approval work for construction projects. It is responsible for reviewing and approving various types of land use conversions submitted to the State Council for approval. It also formulates and supervises the implementation of policies for land use control, such as urban and rural planning management (Fig. 6).

(2) Construction of Territorial Spatial Planning System

Based on institutional reform, the Central Committee of the Communist Party of China and the State Council issued the *"Opinions on Unifying the Planning System to Better Play the Guiding Role of National Development Planning"* in 2018. The guiding opinions first proposed to "rationalize the relationship between national development planning, national special planning, regional planning, and spatial planning," and to "establish a national planning system led by national development planning, based on spatial planning, supported by special planning and regional planning, and composed jointly by national, provincial, municipal, and county-level plans

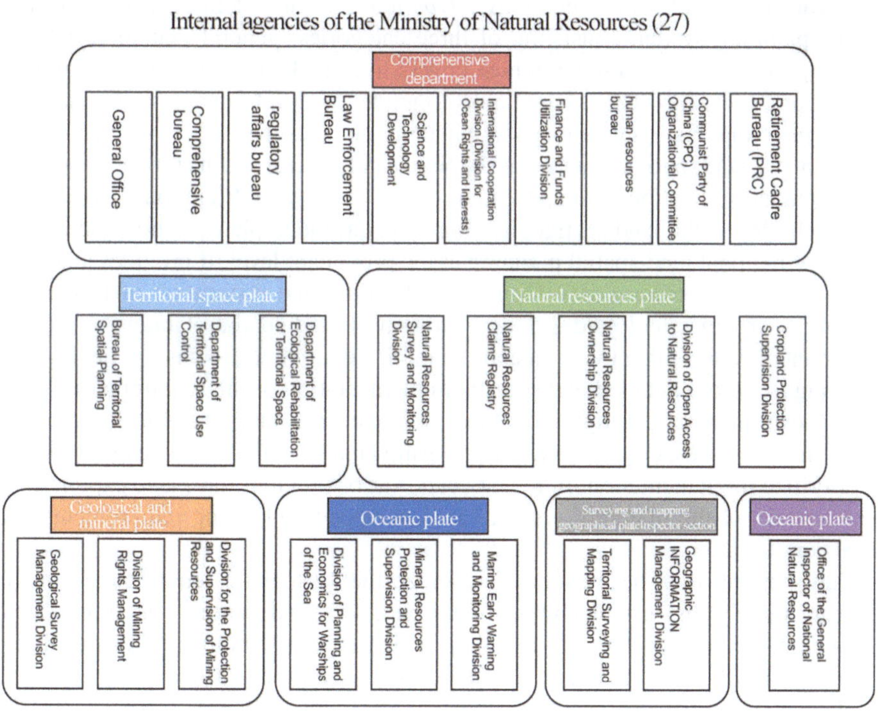

Fig. 6 Internal structure of the ministry of natural resources. *Source* The author draws according to the relevant contents of *the Provisions on Function Allocation, Internal Organization and Staffing of the Ministry of Natural Resources*

with accurate positioning, clear boundaries, complementary functions, and unified connection." In 2019, the *"Opinions of the Central Committee of the Communist Party of China and the State Council on Establishing the Territorial Spatial Planning System and Supervising its Implementation"* (hereinafter referred to as the *"Opinions"*) were issued, further clarifying the "establishment of the territorial spatial planning system and its supervision and implementation." Territorial spatial planning is regarded as the "guidelines for national spatial development and the spatial blueprint for sustainable development, and the legal basis for various development, protection, and construction activities." The main goals for the phased implementation of the territorial spatial planning system reform by 2020, 2025, and 2035 were also formulated. The release of the *"Opinions"* is of significant and far-reaching importance, marking the integration of spatial planning such as main functional area planning, land use planning, and urban–rural planning into the overall framework of the "Territorial Spatial Planning System," which is an important reform achievement and an innovative institutional construction. Since 2019, the Ministry of Natural Resources has issued the *"Notice on Fully Carrying out Territorial Spatial Planning Work,"* which further refines the requirements of the *"Opinions"* and continues to promote the research and formulation of planning procedures, technical guidelines, technical standards, and other aspects.

Regarding the planning hierarchy, *"Opinions"* clearly defines a national land spatial planning system consisting of three categories: general planning, detailed planning, and special planning, which covers five levels: national, provincial, city, county, and township. This system aims to guide the development of national land spatial planning. National land spatial planning involves the overall arrangement of the country's land space and serves as the policy and general outline for the protection, development, utilization, and restoration of the land space. It embodies a strategic approach. Provincial land spatial planning serves as the implementation and continuation of national land spatial planning at the provincial level. It provides a general outline for the protection, development, utilization, and restoration of land space within the province and guides the preparation of land spatial planning at the city and county levels, with a focus on coordination. City (county, township) land spatial planning is a more detailed and specific implementation of the requirements set by higher-level planning. It highlights local characteristics and adapts to local conditions. In some cases, city, county, and township land spatial planning can be merged to prioritize implementation. In recent years, various regions have also explored the optimization of the planning system based on specific planning requirements and local conditions.

Furthermore, it is necessary to recognize the systematic requirements of the reform, namely, the establishment of the national land spatial planning system is a comprehensive system construction, which is jointly built by the four major systems of the "five levels and three categories" preparation and approval system, the implementation supervision system, the laws and regulations policy system, and the technical standard system, forming a complete land spatial planning system (Fig. 7).

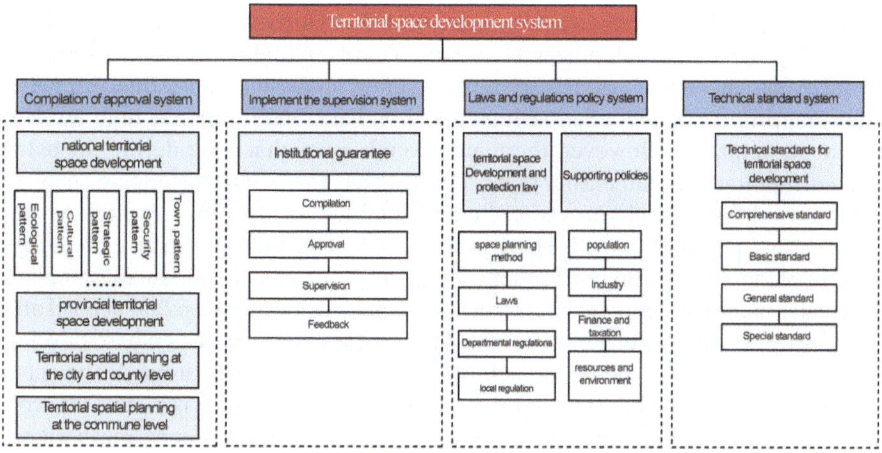

Fig. 7 Construction of territorial space development system. *Source* The author repainted it from Pan Haixia and Zhao Min. Construction process, basic connotation and main specificity of territorial space development system [J]. Urban and Rural Planning, 2019 (05): 4–10

4 Exploration on the Compilation of the Master Plan of Land and Space

4.1 Characteristics of Compiling National Land and Space Master Plan

Compared to national and provincial land spatial planning, the city and county land space master plan should prioritize implementation. It serves as an overall arrangement and comprehensive deployment for the development and protection of land space within the city and county, a blueprint for formulating space development policies, carrying out the protection, utilization, and restoration of land space resources, and implementing land spatial planning management. It also serves as the basis for developing relevant special plans and detailed plans. Based on international and domestic experience, the preparation of spatial planning at the city and county level inevitably involves research on regional issues and economic and social development.

In fact, the overall planning at the city and county level has undergone a long period of practical exploration in both the original urban master plan and the original land use master plan (referred to as "Two Regulations"). It is necessary to summarize their characteristics and shortcomings and fully integrate the technical features of the Two Regulations into the territorial spatial planning, while discarding the existing deficiencies. Regarding urban master planning, there are three key aspects. Firstly, it emphasizes strategic guidance, focusing on resource space guidance and control that is oriented towards development and construction. Secondly, it emphasizes comprehensiveness and integrity by taking into account multiple factors such as urban politics, economy, society, culture, and environment, in order to plan for the city's long-term development as a whole. Thirdly, it highlights the importance of publicity and inclusiveness. The formulation of urban master plans actually involves adjusting property rights relationships based on land, and it requires high levels of public participation. However, there are several issues that arise during the actual implementation of urban master plans. For instance, the urban master plans may not be approved for several years after being submitted, or they may expire and new plans are not promptly issued. There is also a lack of strong legal and rigid management of planning, resulting in frequent deviations from approved urban master plans, especially in the case of many prefecture-level cities where revisions are needed after only 5–6 years of implementation. In addition, there are instances where development space exists, but there is a lack of construction land indicators, or the actual construction land exceeds the land indicators. While importance is placed on urban spatial layout structure, the transmission of spatial use control is poor. Furthermore, while there is a focus on construction land planning, non-construction land planning is often overlooked.

In the past, the overall land use planning was characterized by emphasizing rigid control and management, focusing on the target control of land use indicators, and following a top-down approach to meet the demand. The control of land resources was implemented through a "combination of indicators and boundaries" approach, which was simple, effective, and legally binding. However, the issue with this approach is that it did not thoroughly consider the real situation of urban development, lacked a strong strategic guiding role in urban development, and failed to provide specific spatial arrangements for the functions within the city.

Therefore, based on the reform requirements of spatial planning, the overall planning of national land space should shift from individualistic approaches to comprehensive and all-encompassing implementation of territorial space governance. In the face of complex spatial governance tasks, the overall national spatial planning should not only focus on delineating "three zones and three lines," but also incorporate the practical, effective, and manageable aspects of urban and rural planning as well as land use planning into the national spatial planning. This will uphold and further enhance the original roles of "strategic guidance" and "bottom line control",[7] and achieve the integration of planning, construction, and management, thereby guiding the high-quality and sustainable development of urban space.

4.2 Discussion on the Technical Route of the Master Plan of Land and Space

Guided by the concept of "ecological civilization", the territorial spatial planning is supported by the goal-oriented approach, focusing on both constraint-oriented and problem-oriented dimensions. The implementation of constraint orientation and

[7] Dong Ke and Zhang Jing (2018).

problem orientation is realized through specialized research. Drawing on the understanding of the characteristics of the overall master plan of land and space in cities and counties, as well as practical exploration, a technical roadmap for the preparation of the master plan of land and space in six stages is summarized (Fig. 8).

(1) Guided by three orientations. Goal-oriented, constraint-oriented, and problem-oriented are the three main orientations of territorial spatial planning, which are consistently applied throughout the planning process. Goal orientation emphasizes high-quality development, rather than simply protecting without development or developing without protection. It involves clear strategic positioning, development goals, and indicator systems. Constraint orientation emphasizes the foundation of ecological civilization, strengthening bottom-line security, and assessing the carrying capacity of environmental resources, suitability of spatial development, and various bottom lines of territorial space. Problem orientation aims to grasp the core contradiction between spatial development and economic and social development, including issues such as economic structure, quality of industrial space, regional balance, and ecological environment, with a focus on guiding the overall layout of productivity through spatial planning. Spatial planning under these three orientations serves as the spatial support for economic and social development.

(2) Establish a spatial pattern. Under the three major guiding principles, clearly define the overall spatial pattern of the entire region, establish the relationship between the three-zone system of ecology, agriculture, and urban areas, coordinate the delineation of ecological protection red lines, permanent prime farmland protection lines, and urban development boundaries, and establish an overall spatial development and protection pattern for the entire region based on goals and constraints.

(3) Clarify the layout. Building on the spatial pattern, further propose a protective layout and a developmental layout, which are extensions of constraint-oriented and problem-oriented approaches. The protective layout mainly includes layouts for the protection of natural resources, preservation of cultural heritage, environmental functional zoning, ecological protection zoning, and comprehensive land improvement plans. The developmental layout clearly defines urban and rural development, setting goals for urban development, functional layout, urbanization rate, industrial development and layout, as well as proposing guidelines for rural development and layout, development of characteristic villages, and rural aesthetics.

(4) Refine the functions. Functional layout is the concrete manifestation of the protective layout and developmental layout, providing specific spatial arrangements for all elements in the city area. The functional layout serves as the comprehensive spatial plan, with a focus on functional sector layout, including urban functional layout, agricultural land functional layout, village land layout, and ecological land layout. It further implements the planning requirements of the protective layout and developmental layout.

Fig. 8 Technical roadmap of master land and spatial planning of cities and counties. *Source* Self-drawn by the Author

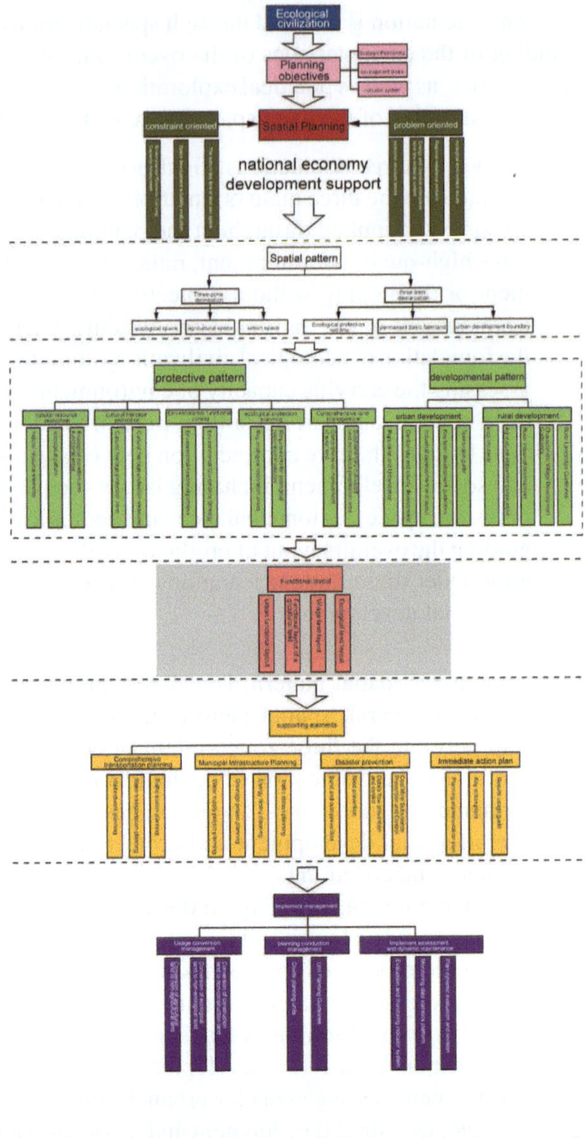

(5) Ensure support. Supportive element layout is designed to ensure the smooth implementation of the functional layout by providing material elements and action plans. This includes comprehensive transportation planning elements, municipal infrastructure elements, disaster prevention measures with local characteristics, as well as short-term action plans.
(6) Implement management. The contents of the territorial spatial planning are ultimately implemented through effective management, including the management of land use conversion, planning transmission, implementation evaluation, and dynamic maintenance. Land use conversion management includes the

management of agricultural land conversion to non-agricultural land, ecological land conversion to non-ecological land, and construction land conversion to non-construction land. Planning transmission management mainly involves the division of planning units and the guidance of unit planning. Implementation evaluation and dynamic maintenance include the evaluation and monitoring of indicators, the establishment of a monitoring data statistical platform, and the dynamic evaluation and revision of planning.

5 Several Opinions

5.1 Prospect and Reform of Territorial Space Development

Space serves as the carrier for all activities, including administration, market, and society. Space governance is a series of institutional arrangements that directly or indirectly impact the structure and process of government governance, market governance, and social governance through controlling and guiding the elements of territorial space. It fully embodies the strategic intent and value orientation of the state in optimizing the development pattern of territorial space and promoting sustainable economic and social development. Advancing spatial planning reform is necessary to better serve the future transformation and development of urbanization in our country, as well as to achieve a higher level and more modernized approach to space governance. In light of the future direction of territorial spatial planning reform, the following three considerations are proposed.

Firstly, how to construct a territorial spatial planning system with Chinese characteristics, and fully leverage the role of territorial spatial planning in economic and social development and urbanization, requires sound top-level design and the establishment of a scientifically-based technical platform. It is also necessary to reasonably delineate the jurisdictional boundaries of various relevant management areas and their governing bodies.

Secondly, in terms of territorial spatial planning, it is crucial to consider the background environment of natural resources, and thus it is necessary to strike a balance between the development and protection of natural resources and space resources. Additionally, cities and towns are artificial environments, and the allocation of space resources should shift from a mere quantitative dimension to a multidimensional dimension that combines both quantity and quality, reflecting its diverse features. Territorial spatial planning should not only focus on the rational allocation of space resources with natural resource attributes, but also adhere to the principles of fairness and efficiency in order to regulate and adjust the spatial order and distribution of benefits for economic and social development.

Thirdly, the future management of territorial spatial planning involves not only "land-use regulation" but also "construction project planning and management." In light of the "planning, construction, and management coordination" mentioned as one of the "five coordination" in the 2015 Central Urban Work Conference, further consideration is needed to determine the path and implementation platform for coordinating the three major aspects of planning, construction, and management, thereby enhancing the systematic nature of urban work. From current practices, the relationship between territorial spatial planning and urban construction management has not been fully clarified. It is necessary to effectively allocate responsibilities, streamline institutions and mechanisms in order to achieve the goal of "coordinating planning, construction, and management to enhance the systematic nature of urban work.

5.2 Interdisciplinary Supporting the Construction of Territorial Space Development System

Territorial spatial planning is a completely new plan that is formed based on the reform of the ecological civilization system, rather than a simple addition of urban and rural planning and land use planning. This calls for higher education institutions to actively respond to the reform of the territorial spatial planning system, harness the collaborative efforts of multiple disciplines, and contribute to the construction of the territorial spatial planning system.

Prior to 2011, "Urban and Rural Planning" was a sub-discipline under the first-level discipline of architecture. In 2011, the first-level discipline of architecture was divided into three disciplines: architecture, urban and rural planning, and landscape architecture. The discipline corresponding to "land use planning" is "Land Administration," which is a sub-discipline under the first-level discipline of "public management." In other words, these two planning fields belong to different first-level disciplines: urban and rural planning and public management.

Disciplines are relatively independent knowledge systems. First and foremost, "independence" implies the concept of irreplaceability, meaning that a specific discipline cannot be substituted by others. Secondly, a "knowledge system" is not equivalent to a "operational system" but rather an organization based on certain theories. Therefore, it can be said that the division of disciplines follows the logic of the knowledge system itself. In this sense, urban and rural planning and land administration cannot simply be amalgamated into a single discipline. This is also why the "construction of disciplines related to land spatial planning" is specifically mentioned in the "Several Opinions" rather than simply referring to the "construction of the discipline of land spatial planning". Therefore, under the objective of establishing a national spatial planning system, from the perspective of discipline construction, careful consideration needs to be given to how to construct related disciplines to support the establishment of the national spatial planning system and the operational system of territorial spatial planning, including the four major systems of the approval system, the implementation and supervision system, the regulatory policy system, and the technical standard system for territorial spatial planning.

In fact, General Secretary Xi has long pointed out the direction. On May 2, 2018, during a symposium at Peking University, General Secretary Xi stated, "We must make great efforts to establish interdisciplinary groups and strong scientific and technological research teams, strengthen collaborative innovation among disciplines, and support original, systematic, and leading research.“ During the study session of the Central Political Bureau on October 16, 2020, General Secretary Xi also emphasized the need to “strengthen interdisciplinary integration and innovation in multiple technological fields.“ Territorial spatial planning, which emerges from multiple compliance, possesses the characteristics of a complex and vast system, surpassing the scope of a single discipline. Practice has also demonstrated that neither urban and rural planning nor land administration alone possess the necessary knowledge, methods, and tools to address the challenges of territorial spatial planning. Therefore, a collaborative effort across multiple disciplines is required to tackle these issues. Interdisciplinary collaboration has the potential to break down existing disciplinary boundaries, restore the integrity of science, and integrate fragmented knowledge systems.

Therefore, the discipline system support required for territorial spatial planning should be comprehensive and systematic. The relevant disciplines of territorial spatial planning include urban and rural planning, land administration science, etc. Its knowledge system should possess openness and comprehensiveness, and through the cross-fertilization of multiple disciplines, continuously enhance the ability of planning disciplines to respond to economic and social changes. In the practice of territorial spatial planning, the integration of professional knowledge can be promoted through mutual support and mutual learning, in order to address new challenges in the modernization of national spatial governance and the construction of an ecological civilization.

References

Development Research Center of the State Council and World Bank. (2014). *China:Promote efficient, inclusive and sustainable urbanization* Beijing: China Development Press.

Dong Ke, Zhang Jing. (2018). *The reform goal and path of overall urban planning* China City Planning Review, (01): 50–57.

Peng Zhenwei. (2021). *Significance of Township Land Spatial Planning in Land Spatial Planning system* Development of Small Cities & Towns, 39 (11): 1.

Wang Liwei, Zhao Ming. (2014). *Systematic Logic of Urbanization Evolution in China; Based on the perspective of the relationship between man and land.* Urban Planning Review, 38(04): 17–22 +33.

Zhao Min. (2019). *Discussion on the logic and operation strategy of the construction of territorial space development system* China City Planning Review, (04): 8–15.

Zhang Bing, Hu Yaowen. (2017). *Exploring scientific spatial planning: Thoughts based on the overall planning of Hainan Province and the practice of "integrating multiple regulations into one"* Planner, (02): 19–23.

Modern Theoretical Framework of Spatial Planning

Wu Zhiqiang

1 What is Spatial Planning

1.1 Planning Distinguishes Human Activities from Animal Instincts

Humans stand out in the animal kingdom not because of their powerful instincts but because of their intellectual capabilities. Homo sapiens surpassed animal instincts by developing the ability to plan for shared visions, create complex tools under common goals, discover technological patterns, engage in language communication, and perform large-scale collective actions. This encompasses 3 key abilities:

(1) Visoning- The ability to envision a shared vision. Through negotiation and imagination, humans can establish a collectively recognized vision that does not yet exist in reality.
(2) Approaching-The ability to set paths to achieve the goal. This involves systematically dividing large-scale individual efforts, organizing groups, and advancing plans in stages.
(3) Coordinating-The ability to communicate, coordinate, and adjust. During the implementation of the Vision, humans can communicate about potential scenarios that have not yet occurred, continuously optimizing the goals, paths, and divisions of labor until the vision is realized, or even achieving results beyond the original vision.

These three abilities are the essential capabilities that distinguish humans from other animals and also constitute the three core elements of planning: visioning, path implementation, and communication coordination. Therefore, understanding the

W. Zhiqiang (✉)
College of Architecture and Urban Planning, Tongji University, Shanghai, China
e-mail: wus@tongji.edu.cn

Z. Peng (ed.), *Theories and Frontiers in Territorial Spatial Planning*, Urban Sustainability, https://doi.org/10.1007/978-981-96-9825-7_2

fundamental differences between human and animal abilities helps us comprehend why planning is an inevitable outcome of human activity. Conversely, any abilities or instincts that contradict these three core elements of planning are rooted in animalistic tendencies and intuition.

Therefore, before modern civilization, the competition between different human societies could almost entirely be understood as a competition for survival land and the resources for living and production found on it. Until the emergence of Marxism, Western thought generally recognized that "land is the mother of wealth, and labor is the father of wealth." Marxism expanded the elements upon which capitalism generates wealth, asserting that in addition to land and labor, capital and other factors also play a role.

Today, as wealth continues to be created globally, land has expanded into space, labor has evolved into intellectual labor, and capital has extended into the broader financial sector. Information, technology, and organizational management models have also become key elements in wealth creation within a specific space.

1.2 Cities Are Symbols of Civilized Society

Around 8,000 years ago, humanity entered a civilized society, and cities emerged alongside it. In other words, cities are symbols of civilized society and carriers of human civilization. Only with cities can we truly say that human civilization exists.

This idea can be corroborated by the common origin of the words "civil" and "city," with the Latin word civic meaning both "city" and "citizen." In this sense, the English word "civilization" can be translated as the "process of becoming civilized," or more directly as the "process of urbanization and becoming civic-minded." Urban space is the carrier of civilization. The Chinese understanding of "civilization" begins with the enlightenment brought by written language, rather than with the concept of urbanization and civic-mindedness. Therefore, when the Chinese explore their own civilization's history, we often consider the earliest origins of written language as the starting point of our civilization.

Given the different understandings of civilization mentioned above, UNESCO (United Nations Educational, Scientific and Cultural Organization) generally recognizes three common characteristics of a civilization: the presence of urban development, the existence of written records, and the establishment of governance structures. To respect the differences among various civilizations and to enhance interaction, cooperation, and integration among them within the UN framework, the United Nations has also established the United Nations Alliance of Civilizations (UNAOC).

Chinese civilization is a diverse and unified civilization. Since the Liangzhu culture, there have been remnants of urban construction. Oracle bone script serves as a crucial foundation of Chinese civilization, and the artifacts unearthed from the Sanxingdui site testify to the existence of an early social ruling class and its governing system, reflecting the governance structure within ancient Chinese civilization.

1.3 Space vs. Land

The complexity of "Space" compared to "Land" can be understood from the following three aspects:

(1) Spatial Dimension: Space encompasses depth and height, including underground, ground level, above ground, and the air.
(2) Production Dimension: Beyond the primary sector, Space also includes the more significant secondary and tertiary sectors, as well as even higher dimensions of production organization and relationships.
(3) Constituent Elements Dimension: Space is a multi-faceted composite that includes not only natural and man-made physical spaces but also social spaces and the emerging digital intelligent spaces.

Therefore, we now generally refer to "space" as composite, and it has entered three dimensions: physical space, social space, and digital space. These three dimensions operate and advance in an integrated manner within a given time period, indicating that human civilization is entering a higher dimension. Consequently, spatial planning has become more multidimensional, more systematic, and more complex, demanding a higher level of civilization to effectively plan and govern.

Typically, when planning for agriculture, animal husbandry, and mining, we use the term "land planning." When planning for urban development, social, and economic growth, we prefer to use "spatial planning." "Spatial planning" should encompass "land planning," with "land planning" serving as a fundamental and important support for "spatial planning."

Spatial planning is a product of civilization, and different stages of civilization correspond to different forms of spatial planning. With the advent of industrial civilization, as urban spaces became more three-dimensional and the elements of urban wealth began to flow rapidly, the planning of large cities emerged as a profession, marking the origin of modern spatial planning. Modern spatial planning gradually evolved from the spatial planning of large urban areas to the planning of medium and small cities, and eventually extended to the planning of agricultural regions. This evolution has made spatial planning encompass the comprehensive planning of human living spaces in both urban and rural areas.

In the 1970s, as environmental issues became increasingly prominent in Western countries, environmental impact planning was gradually integrated into spatial planning. In the 1980s, with the intensification of social problems, social development planning was also progressively incorporated into spatial planning. By the 1990s, spatial planning had become more systematic and comprehensive, with watershed planning, regional planning, and coastal planning all being included in the spatial planning framework. The globalization of the 1990s further spurred spatial planning to play a role in organizing global production, gradually expanding its scope to encompass the relationships, exchange of information, and material flows between transnational and intercontinental urban clusters.

1.4 The Origin of Modern National Spatial Planning

The spatial planning of modern nations first emerged in modern large cities. The spatial issues triggered by the Industrial Revolution initially manifested in these large cities. The direct symptoms of these urban problems were closely related to the massive increase in population concentration and spatial density.

The problems in large cities caused by mass industrial production primarily manifested in several ways: frequent urban epidemics, frequent city fires, spatial segregation of social classes, increased gang-related crime, and concentrated industrial pollution.

Tracing the causes of these problems directly influenced how space was managed: limiting urban sprawl by setting boundaries for expansion; controlling urban density by regulating city density and building height; creating wide boulevards to establish firebreak corridors in central areas; constructing sewage pipelines to develop modern urban sewage systems; and enacting neighborhood and street security laws to prevent organized crime and mass incidents in large cities.

As a result, five fundamental goals of large city spatial planning were established:

(1) Safety: Prevent the spread of infectious diseases and the spread of fires.
(2) Sanitation: Ensure unified sewage disposal through underground pipelines and carry out large-scale municipal construction, including stormwater systems.
(3) Aesthetics: Develop urban design focused on the facades along streets and the design of nodal plazas.
(4) Management: Implement controls on building height, density, and form, as well as measures to combat organized crime, using neighborhoods as the basic unit of management.
(5) Economy: Consider how to control construction costs during the implementation of large city spatial planning.

From the concentration and expansion of urban populations to the crowded living conditions of the proletariat in cities like London and Paris; from the greedy expansion of industrial land by capitalists to the demands of the British royal family and German aristocracy for ensuring the quality of urban living spaces; from the looting and riots by the urban proletariat to the establishment and strict control by urban police forces; from decades of cholera outbreaks in towns to the stench in city centers caused by unmanaged waste disposal; from dreams of a return to the Victorian era to the urban struggles organized under proletarian revolutionary theory… In Europe's first major cities, modern industrial civilization not only perplexed the urban proletariat and citizens but also troubled capitalists and the royal families. Problems had already erupted, with symptoms visible everywhere. Should these issues be addressed through technical solutions or through management? Should they be approached from the perspective of an ideal society, or should people seek to escape from the large cities?

Spatial planning was born in this context, emerging in large cities. It was developed under the guidance of social ideals, in response to the concentration and eruption of urban spatial problems that urgently needed addressing, and through the collective call and push from different social classes within the city. Broadly speaking, even before the Industrial Revolution, in feudal societies and even in tribal spaces, there was some form of spatial planning. However, the focus of this Chapter is specifically on the modern spatial planning system.

The spatial planning in the developed industrial countries of Europe is fundamentally driven by national modernization and began to emerge when the urbanization rate in these countries reached around 60%. This led to the development of comprehensive national spatial plans that cover the entire country. Notable examples include the spatial planning systems of Germany, the United Kingdom, France, Luxembourg, and Belgium. These nationwide spatial plans are generally referred to as National Planning. They not only involve spatial layout and arrangements at the national level but also include spatial planning at various levels such as provinces, states, cities, towns, and villages.

From a multidisciplinary perspective, national spatial planning encompasses a wide range of areas, including national natural resource conservation plans, regional and river basin flood safety and disaster prevention plans, national transportation network planning, national economic and industrial layout planning, the regional layout of large-scale infrastructure (such as airports, reservoirs, highways, shipping routes, power stations, etc.), as well as the distribution of population, social development, and the regional layout of educational and medical facilities. Essentially, these plans cover the overall strategy and spatial layout of a country's ecology, production, and living spaces, making them classic examples of national spatial planning worldwide.

At the national level, these spatial plans typically begin with abstract national strategies and frameworks, which are followed by more detailed spatial implementation plans at lower levels. At key points, there are detailed spatial designs, supported by a comprehensive set of spatial governance systems, including the approval processes for the protection, development, and use of space.

After 1990, with the advancement of European integration, there emerged a significant number of transnational plans that crossed national borders, known as Transnational Planning. This marked a new stage in national spatial planning.

1.5 The Subject and Object of National Spatial Planning

Territorial Spatial Planning Requires the Unification of Both Subject and Object.

The first part is the unification of the objects. For a long time, the three major spaces in China's national spatial planning (urban space, agricultural space, and ecological space) have not been unified, leading to contradictions and conflicts between various plans. In 2018, the Ministry of Natural Resources of the People's Republic of China was established, tasked with creating and overseeing the implementation of a spatial planning system. This move responds to the objective need for the unification of multiple types of planning within China's spatial planning, playing a guiding and coordinating role in various specialized plans. The "unification of multiple plans" in national spatial planning is not simply the merging of different spatial plans into a unified text with different chapters. Instead, it requires the pursuit of a "philosophical unification" of these plans, beyond just the formal integration. This involves further integrating the theories, standards, and methods of the individual plans for the three major spaces.

The second part is the integration of the subjects. The fundamental principles and doctrines of planning for these three spaces have not yet been unified. Ecological space emphasizes the absolute importance of ecology, often without considering social development, historical culture, and the needs of the people. Agricultural space focuses more on farmers' income, rural characteristics, and local economies. Urban space development emphasizes comprehensive coordination, seeking harmony between history and the future, and between humans and nature, representing a complex relationship. Therefore, although we discuss these three spaces, the unification of core values is the ideological prerequisite for the future unification of the objects. All types of spatial planning need to adhere to the value system of national spatial planning.

2 The City Region and City Generations

2.1 *Regional Development*

The development of a city is not isolated. Particularly when we view a city as a symbol of civilization, we must consider its role within a larger regional context. The development of a civilization depends not only on central cities but also on the broader civilizational space that forms within a region. Civilization relies on cities for creation and innovation, and it is within cities that inter-city exchange and interaction of civilizational genes occur. This is precisely what distinguishes cities from the relatively homogeneous cultural genes of rural areas. In other words, cities bring together people from diverse cultural backgrounds in a more densely populated space. Therefore, cities are more likely to foster innovation and promote technological advancement. We can observe that cities generate new ideas, new cultures, new technologies, new economies, and new systems through continuous civilizational exchange.

The city is not only a core-driver space for the continuous creation of civilization but also a center for the preservation and protection of civilizational achievements. Through governance systems and wealth security, cities safeguard the spiritual and material creations of civilization, including museums, theaters, plazas, sports venues, and other places for spiritual activities and knowledge transmission, such as Confucian temples, academies, and universities.

In a civilization, a city, in addition to its roles in innovation, interaction, and preservation, also serves as a radiating hub. The spread of civilization often occurs between cities and between cities and rural areas, continuously facilitating point-to-point and point-to-area cultural diffusion.

The identification with a civilization is defined within a specific region. Therefore, as the central point of a civilization, a city inevitably forms an integrated whole with its surrounding region. As a feature of civilization and a reflection of the natural elements of its region, the natural conditions of the area where a civilization is located (such as oceans, plateaus, hills, river networks, latitude, and longitude) have long and profoundly influenced the civilization. Thus, if we consider space as the carrier of civilization, spatial planning in a region is essentially the planning of the spatial conditions for the survival and development of that civilization. It involves both actively adapting to the natural environment of the region and engaging in intelligent interaction with that environment.

2.2 *Intergenerational Transformation of Cities*

Cities can be divided into different generations based on the stages of civilization development. Generally, cities that existed before modern industrial civilization are collectively referred to as pre-industrial cities. These pre-industrial cities can be categorized based on their specific functions within civilization: cities as trade centers, cities as administrative centers, cities as spiritual centers, and cities as defense centers.

After the advent of modern industrial civilization, cities began to engage in large-scale production, and we refer to these as industrial cities. In Western countries, the 1970s are often marked as the transition from industrial civilization to post-modern civilization. Similarly, Chinese cities truly entered the phase of large-scale industrial development after the reform and opening-up period. Leveraging a large number of pre-industrial cities across the country for civilizational upgrading, China transitioned into modern civilization and industrial civilization. This process also led to the emergence of a number of new industrial cities. These new areas, which appeared on the outskirts of some large cities, represent a unique form of industrial civilization and are often referred to as new districts, development zones, or high-tech zones.

From a global perspective, industrial civilization leads to the emergence of technological civilization, ecological civilization, and intelligent civilization. These are not only higher forms of civilization triggered by industrial civilization but also necessary for maintaining and sustaining its development. In terms of the overall civilizational progress of the nation, Chinese cities are currently at different stages of civilizatioal development. Some cities remain in the pre-industrial stage, most are in the early stages of industrial civilization, and a few cities, after 40 years of industrial development, are now transitioning into a new generation of civilizations, including technological, ecological, and intelligent civilizations.

Based on the different stages of urban civilization development, cities can be categorized into different generations. In summary, cities can be classified into the following generational categories: pre-industrial cities, industrial cities, and post-industrial cities characterized by intelligence, innovation, and ecological civilization.

3 The Theory of City-Being

A city is a living entity (City-being), and its development can be divided into eight life stages: gestation, childhood, adolescence, young adulthood, maturity, old age, decline, and regeneration.

In the formation of my theory of City-being, I was influenced by Professor R. Thompson's "Export Model," which divides city development into five stages, as well as Professor Sorberg's three-stage division, both of which I encountered during my doctoral studies in Germany in the late 1980s.

3.1 Stage One: Gestation Period

Cities in the gestation period primarily refer to traditional feudal cities that existed before the birth of modern cities. These cities served four main functions: governance, spiritual center, trade, and military. They did not engage in the export of material goods, focusing instead on trade and interaction. Cities in this stage concentrated the consumption of surrounding areas and were not necessarily small in scale (see Figs. 1, 2 and 3).

Cities in the gestation period are prone to urban issues:

(1) Lack of modern awareness and innovative spirit.
(2) Pursuit of luxury and resistance to change.
(3) Pride in historical achievements, resisting the need for innovation.

Lack of industries driven by modern innovation.

Fig. 1 Beijing before 1949. *Source* https://www.163.com/dy/article/GJCDCGSB0552A5FC.html

Fig. 2 Ancient Chang'an. *Source* https://m.sohu.com/a/358623037100238527/?pvid=0001153wa

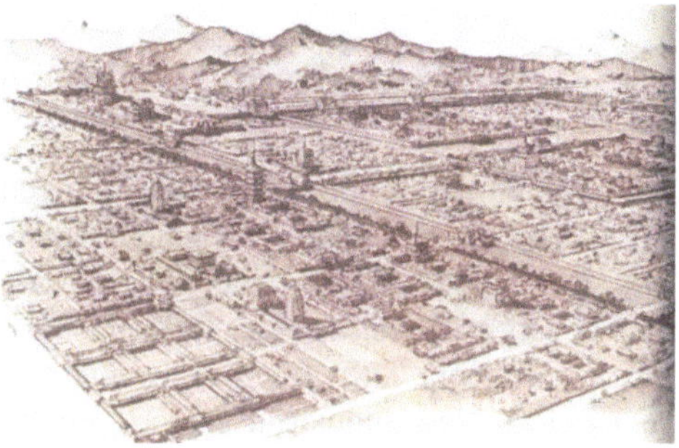

Fig. 3 Virtual restoration of Luoyang City from the Northern Wei Dynasty. *Source* https://www.sohu.com/a/153267691 99904236

3.2 *Stage Two: Childhood*

In the childhood stage, the economic lifeline of a city is often determined by a single factory or enterprise. For example, cities like Panzhihua in Sichuan and Xinxiang in Henan, when heavy industry served as the core industry, were typical of this stage.

Cities in the infancy stage are prone to the following urban issues:

(1) Pride in a single product that represents the city's identity.
(2) Core technology, capital, and markets are controlled externally.

If the city's main enterprise collapses, economic and employment problems emerge, and the city risks reverting to a consumption-based city, similar to an agricultural society.

3.3 Stage Three: Adolescence

In the adolescence stage, a city's economy is composed of multiple enterprises, with historical continuity among them, and local industries dominate the regional market. The city's economy evolves from being reliant on a single enterprise to being supported by various businesses with intergenerational connections.

Cities in the adolescence stage are prone to the following urban issues:

(1) Modern industries provide employment opportunities, but remnants of the agricultural society are still mixed within the industrial zones.
(2) The city lacks a clear development strategy.

The urban landscape is mixed, and the environment shows pollution that is concentrated in specific spots.

3.4 Stage Four: Young Adulthood

In the young adulthood stage, a city's economy has matured, with key industries possessing research, development, and innovation capabilities, enabling independent and sustained growth. Product generations are continuously updated.

The city becomes a center for product research and innovation based on its mainstream products, attracting young talent, capital, and emerging industries through innovation. New industrial sectors, targeting the local market, bring larger-scale industrial development and create more jobs, thereby expanding the local market. As a result, the city's foundational industries undergo continuous technological upgrades (see Fig. 4).

Cities in the young adulthood stage are prone to the following urban issues:

Fig. 4 Villages within the City of Shenzhen. *Source* https://www.sohu.com/picture/202297247?request_id=1639067097175202297247

(1) The city lacks a refined urban management system that extends to the community level.
(2) Urban safety, particularly community safety, faces significant challenges.

3.5 Stage Five: Maturity

Cities in their maturity phase enter a state of continuous spatial expansion, marked by three prominent characteristics: the market of emerging industries expands outward to peripheral areas; the city diversifies its product output; and the city's capital and industries generate new gravitational pull.

However, mature cities are prone to the following urban problems:

1. The city becomes highly concentrated as it continually attracts new capital.
2. The increasing diversity of elements leads to mixed communities and cultural conflicts.
3. The distance between new industrial areas and the old city becomes too great, resulting in pendulum-like traffic patterns.
4. As the city grows larger, higher, and denser, it loses long-term planning for its self-management, self-organization, and future vitality.

3.6 Stage Six: Old Age

In the old age phase, a city's specialized production and services have reached a monopolistic position nationally or internationally. However, its original functions are gradually being taken over by other cities, while new functions have yet to be introduced. World-class cities in the 1980s, such as New York, London, and Shanghai, are examples of cities at this stage.

Cities in their old age are prone to the following urban issues:

(1) The city often takes pride in its historical glory, while its underground infrastructure is steadily deteriorating, sometimes even facing localized infrastructure failure.
(2) The gap between the city's concentrated luxury and its patchy areas of decline grows increasingly wider.

3.7 Stage Seven: Decline

In the decline phase, a city's original functions are entirely replaced by other cities, and it is unable to stimulate innovation through the introduction of new functions. The city's living and production spaces lack vitality, making it unattractive for young people to gather, live, work, or innovate, leading to a rapid population decline.

Cities in decline are prone to the following urban issues:

(1) The production and living costs for entrepreneurs rise sharply.
(2) The city's ecological environment experiences widespread deterioration across different communities.
(3) The city's ability to withstand economic, natural, and social shocks becomes increasingly fragile.

3.8 Stage Eight: Regeneration

Once new urban functions are introduced, especially when pioneering technologies drive the development of emerging and creative industries, the city experiences a new lease of life, building on its existing foundations. Examples include the redevelopment of the Brooklyn Navy Yard in New York, the development of the Pudong New Area in Shanghai, the West Lake and Qiantang River area development in Xiaoshan, Hangzhou, and the development of Tongzhou as Beijing's sub-center and Xiong'an New Area in Hebei (Fig. 5).

Issues to be addressed in the regeneration of a city:

(1) The city's old spaces may have inefficient use.
(2) The new urban functions may not receive the proper locations they deserve.
(3) The original "genetic code" of the city must be preserved, while the growth of the new life must also be nurtured. The city government, existing enterprises, and citizens will need to strike a balance between the old and the new, as any imbalance could lead to urban instability.

4 Urban Internal System Ecology

4.1 Urban Illnesses

Spatial planning refers to the overall arrangement of the relationships between internal elements of a space and their connection to external elements of the space, ensuring an orderly functioning over time.

As civilization progresses, cities, as complete systems, see a gradual increase in the number of elements, and the relationships between these elements grow increasingly complex. However, the fundamental principle remains unchanged: no single element within the system should dominate, as this would hinder the development of other elements and ultimately damage the city's growth as a carrier of civilization.

The formation of almost all urban problems is due to the excessive exaggeration and expansion of certain elements within the city, while other elements do not receive sufficient development. The specific manifestations include:

(1) Emphasizing only the administrative governance function of the city while neglecting its economic, production, and trade functions leads to a prolonged state of parasitism, lacking economic vitality and momentum.

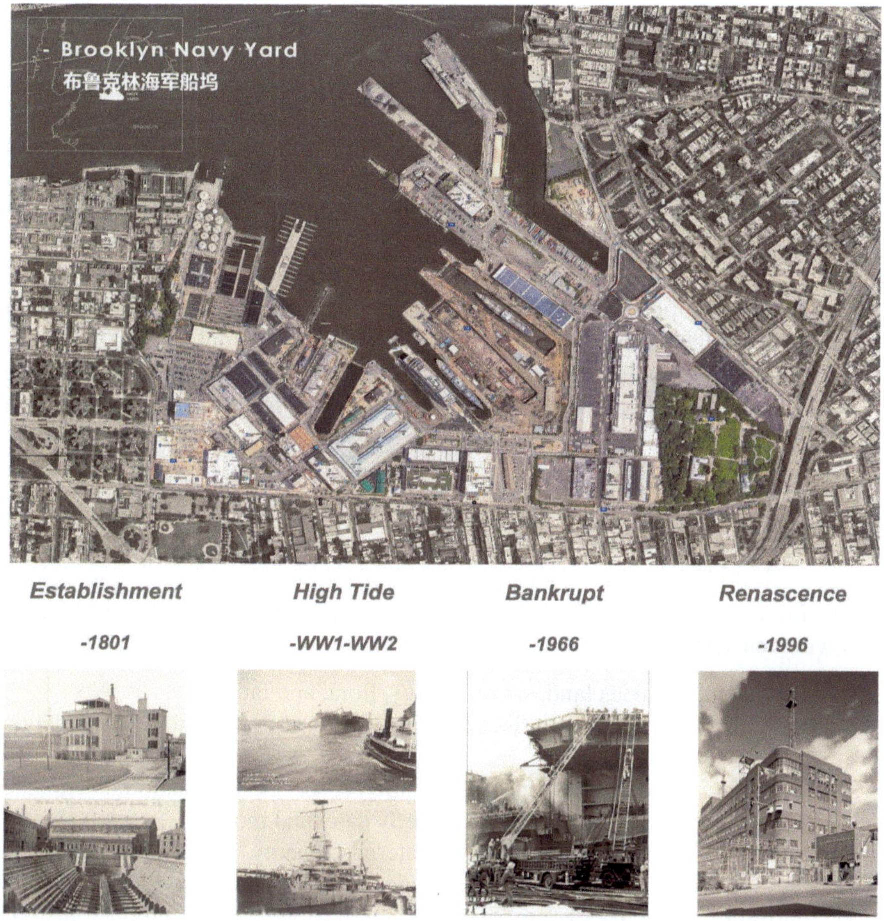

Fig. 5 Evolution process of the Brooklyn Navy Yard. *Source* https://brooklynnavyyard.org/about/history

(2) Emphasizing only the production function of the city while neglecting the supporting functions of education, transportation, and social interaction results in a lack of talent, skilled labor, and technological exchange in the city's production. This causes the economy to experience only short-term booms, and after the production surge of one generation of products, the city's economy quickly weakens, leading to both over expansion and contraction of the urban economy.

(3) Overemphasizing the preservation of historical culture while neglecting the need for professionals specialized in cultural preservation, overlooking the interaction within the cultural industry, and failing to ensure long-term economic support leads the city into a pathological state that denies innovation and hinders the development of civilization.

(4) By focusing solely on providing a high standard of living and generous welfare benefits, while neglecting and undermining the motivation of urban innovators and entrepreneurs, some cities—particularly in developed regions—face unsustainable outcomes. The high cost of innovation stifles the enthusiasm of young entrepreneurs, causing them to relocate to cities with better innovation ecosystems. This ultimately creates a facade of prosperity underpinned by high welfare but masks the city's inability to sustain long-term growth driven by innovation and entrepreneurship.

(5) By emphasizing only the spatial aesthetics above ground while neglecting the renewal and upgrade of the city's underground infrastructure, the ability of the city to handle material and energy flows becomes outdated. This results in weakened flood prevention and drainage capabilities, reduced efficiency in water and gas supply, and even the emergence of potential safety hazards and disasters.

4.2 The Ten Urban Systems

Analyzing a city based on its functions, the key elements can be divided into the following ten systems:

(1) Ecosystem: Includes land, water, energy, flora, and fauna.

(2) Residential System: Comprises housing, front yards, and similar residential spaces.

(3) Livelihood System: Encompasses commerce, services, education, healthcare, entertainment, social interactions, elderly care, and more.

(4) Production System: Includes raw materials, energy, production collection, transportation, processing, sales, storage, logistics, and supply.

(5) Innovation System: Covers scientific innovation, design, product display, interaction, and transactions.

(6) Cultural System: Includes performance venues, libraries, museums, cultural centers, youth centers, art studios, conference centers, and exhibition halls.

(7) Safety System: Encompasses security systems, military presence, road safety systems, and fire protection systems.

(8) Municipal System: Involves bridges, rail transit, gas and heating infrastructure, and more.

(9) Management System: Covers community and street management.

(10) Digital and Smart System: Includes digital centers, sensing, computing, decision-making, connectivity, and learning.

5 The Spatial Relationships Between the Various Elements of a City

The elements of the ten urban systems each have specific spatial requirements within the urban space. Every function of a city needs space to be accommodated, and the relationships between different functions can be categorized into four types: integrative relationships, complementary relationships, avoidance relationships, and conflicting relationships. The elements within a city are interconnected, and there are no situations where spatial relationships do not exist—only direct or indirect relationships.

5.1 Direct and Indirect Interaction Relationships Between Urban Elements

Generally, direct interaction relationships are observable, such as the integrative relationship between industrial production zones and research centers, the complementary relationship between residential areas and schools or kindergartens, the avoidance relationship between pollution-heavy industrial zones and residential areas, and the conflicting relationship between hazardous materials storage zones and city centers.

Indirect interaction relationships, however, require systematic urban studies to uncover the multi-layered effects between two functions through the connections of multiple elements. How to manage and balance these numerous element relationships simultaneously remains an eternal challenge in humanity's exploration of urban systems.

5.2 Growth in the Number of Spatial Elements and Demand

As the number of urban elements increases, the spatial relationships between these elements give rise to numerous dynamic models. If the connections between various elements are organized and expressed, the spatial relationships within the city's operational system that influence one another become infinitely complex. The complexity of these relationships turns urban dynamic modeling into a challenging problem within the field of complex science.

Urban spatial planning is a process in which humanity continuously refines the relationships between different elements within a city. As this process unfolds, the elements involved in urban development increase in number alongside the progress of civilization. The more advanced a city's civilization, the more complex its elements become, requiring more specialized knowledge and stronger governance capabilities to support the city's growth.

6 Maintaining the Life of the City-Being

6.1 Fundamental Issues in City Revitalization

A city is a living City-being. We emphasize "city revitalization" rather than just "urban renewal." "City revitalization" not only refers to the renewal of physical spaces but also includes the shared prosperity of the people living and working in it.

City revitalization and redevelopment are crucial both at the individual level, impacting whether people's lives are happy and fulfilling, and at the city level, affecting the sustainability and overall health of the city's life. Horizontally, it relates to the city's economy, employment, ecology, and disaster prevention.

City revitalization involves several fundamental questions: What is the theory behind city revitalization and redevelopment? Why do cities need revitalization? How can city revitalization be measured? What are the key elements of city revitalization? How should cities undergo revitalization? Who are the stakeholders initiating the demand for city revitalization? What principles should guide city revitalization?

6.2 Principles of Synergy in City Revitalization

The issue of city revitalization has reached a critical point where the academic needs to make breakthroughs and innovations in theory, thought, methodology, and technology. Scholars, over the world, have invested significant thought on city revitalization into the construction of the urban planning discipline and have developed certain theoretical outcomes. How can we address the different stages of urban development in China, propose relevant thoughts and knowledge, and integrate currently fragmented fields (including physical, spiritual, cultural, technological, and ecological aspects) to create a connected network, building a comprehensive theoretical and knowledge system that supports practice and policy-making?

There are eight crucial principles of synergy:

(1) Synergy between scientific diagnosis and public perception.
(2) Synergy between the construction of new urban areas and funding for the revitalization of old cities.
(3) Synergy between the improvement of living service facilities in old city revitalization and the enhancement of new urban areas.
(4) Synergy between local funding for old city revitalization and national financial support.
(5) Synergy between the city's government and the administrative functions responsible for old city revitalization.
(6) Synergy between local traditions and international design experience.
(7) Synergy between international organizations and local cultural exchanges.
(8) Synergy between public employment and the revitalization of urban living.

In the life cycle of a city, the existence of physical structures can outlast the lifespan of an individual. The varying needs of people at different stages of life, as well as the needs of different generations, drive changes in the demand for physical infrastructures. This shift in demand makes it impossible for cities to remain unchanged; the transformation of cities is inevitable.

The absolute changes in the life cycle of a city are not limitless; they have boundaries and follow certain patterns. Changes that go beyond these boundaries would cause the city to lose its genetic core and no longer be the same city. Therefore, the concepts of absolute change and absolute boundaries represent two extreme states in urban planning and design. Exceeding these boundaries disrupts the city's genetic identity, and without this core, the city transforms into some other city entirely different. Conversely, opposing iteration and development means using yesterday's infrastructure to meet today's needs, which results in a city being unable to best support its vitality, turning its spaces into obstacles to the city's growth.

The absolute change factors in cities, communities, and their aging buildings include: Demand-based absolute changes, such as the needs of different life stages and generations of people using the space. Technological changes, such as upgrades in lighting, fire prevention, air conditioning, communication tools, projectors, and screens in old buildings, which do not alter the core genetic identity of the space. On the contrary, these changes can serve to protect and enhance the core identity. Changes in material composition of cities, communities, and individual buildings, including new energy-efficient windows and doors, pollution-reducing flooring and wall materials, improved plumbing systems, and modern restroom materials. These three aspects are elements that continuously improve and evolve over time.

The immutable genetic identity of space includes the following elements: Architectural design style; Spatial organization patterns and structural texture; Symbols of social memory that are deeply recognized by society, such as the characteristic image of a space or building, including its color, roof, wall materials, and construction methods. These elements represent the genetic identity of cities and communities, preserved and solidified in collective memory.

Therefore, the genetic identity of a space can be divided into two main types:

Internal genes: These are inherent characteristics unique to the space itself, which are communicated outwardly and recognized by society.

External genes: These are characteristics imposed from the outside, recognized and accepted by society, and must remain unchanged during revitalization.

The internal and external genetic identities are interactive and influence each other in shaping the core essence of the space.

Within the internal genes of the genetic identity, there is a crucial element—the gene of technological progress and era-defining markers. Throughout the historical evolution of human creation in urban districts and architecture, people embed their ideologies into spaces, including the latest scientific and technological innovations, the newest innovative trends in culture and art, the innovative philosophical reflections on the relationship between production, living spaces, and nature, as well as the latest innovative construction techniques. These advancements are solidified and encapsulated within the space, forming a lasting part of its identity.

These four innovations—technological, artistic, ideological, and construction advancements—can be considered milestones of human civilization's progress in space. They are the most critical elements of the genetic identity that need to be recognized and explored in the process of urban revitalization and the renewal of old buildings. Without analyzing this aspect, any urban regeneration is superficial, merely a patchwork that lacks a vision for the progress of civilization, and may even result in the loss of the core genetic identity during the process.

The quality of urban revitalization can be judged from this perspective: whether it merely brings superficial brightness or fundamentally preserves life and value. Therefore, when evaluating different revitalization proposals, the depth of the designer's understanding and analysis of these four key innovations can reveal the depth of their work and thought.

How can the contradiction between the iterative lag of physical spaces and the iterative demands of users be reconciled? Here, the "users" are not individuals but an entire society, one in which professionals emerge—those who capture, understand, and manage future needs and technologies. This leads to the creation of urban planners in the field of cities and architects in the field of buildings. The standard for measuring the capabilities of planners and architects lies in their ability to foresee future human needs and future technologies. The more accurately one can anticipate the future, the more solid their expertise, allowing them to excel in their field. Their works are endowed with greater timeless vitality.

7 New Infrastructure for New Spaces

7.1 Definition and Characteristics of New Infrastructure

For thousands of years, human beings have created cities, revering space itself. With the development of urban technology, new infrastructure has enabled a millennial leap in human urban construction. New infrastructure (New Infra) transforms the way cities interact with space, allowing urban spaces to begin "revering" humans—sensing human needs, emotions, and responding instantly. New Infra shapes new urban spaces.

The essence of infrastructure is to maintain the exchange and processing of energy, materials, and information both within and outside the city's life system. New infrastructure, on the other hand, refers to infrastructure driven by the latest generation of urban technologies. This new generation of urban technologies, in the current era, specifically refers to the "Big Data, Artificial Intelligence, Mobile internet technology, Cloud Computing, and Blockchain" technologies of the seventh wave of technological advancements (see Table 1). However, considering the current development status and level of technological application in China's smart cities, it should also include biotechnology and sustainable development technologies. The construction of new infrastructure can be narrowly understood as the process of building infrastructure. However, in a broader sense, it should encompass the entire lifecycle of the infrastructure, including its operation, management, and maintenance after it is built.

The revolutionary changes in new infrastructure involve three stages: before, during, and after.

Before: The latest concepts and ideas.

During: Technological innovation.

After: The integration of concepts, technology, capital, construction, operation, usage, and management, creating a seamless connection between all these elements.

7.2 Key Points in Spatial Planning for New Infrastructure

1. **Characteristics of New Infrastructure in New Spaces**

The characteristics of new infrastructure in new spaces are:

(1) Based on the essence of infrastructure. New infrastructure represents the intelligent systems within urban spaces, forming part of the broader smart urbanization process, similar to the smartification of city streets and innovation zones. New infrastructure is the intelligent transformation of the physical and informational transportation systems within cities. It serves as a support system for the survival and sustainable development of urban and rural areas, underpinning the operation of society, the economy, and the environment.

(2) The sensing system is divided into two categories.Internal sensing addresses the coordination between urban operational systems, serving as the fundamental premise for ensuring the health and survival of the city within an intelligent system.External sensing addresses the coordination between the city and its surrounding environmental communities, providing the essential guarantee for the city's long-term and sustainable prosperity.

(3) Urban applications can be summarized into three key aspects: Enhancing traditional infrastructure with digitalization and intelligence, upgrading its functionality. Translating next-generation technologies into practical facilities, making them applicable in cities and improving people's quality of life. Planning necessary urban spaces for new infrastructure, ensuring these facilities can operate effectively and meet their functional requirements.

2. **Reshaping the Global Competitiveness Landscape of New Infrastructure**

Currently, countries around the world are focusing on the research and application of next-generation infrastructure. The United States is vigorously promoting the integrated application of technologies such as the Industrial Internet, 5G, and Artificial Intelligence in urban spaces. The European Union is advancing the development of digital infrastructure as a core strategy. Singapore's "Smart Nation 2025" plan proposes the construction of efficient and secure communication infrastructure.

Table 1 Technological waves and urban space iteration

Technological Wave	Representative technologies emerging within cities	Typical applications in cities	Time of emergence	Examples of corresponding types of urban spaces
First Wave	Mechanization of labor, blast furnace smelting technology	Spinning machines, water-powered looms, blast furnaces for iron smelting	1785—1845	Industry: Factories, freight stations, warehouse districts, commercial areas Sanitation: Urban water supply and drainage systems, sewage treatment, waste collection, and disposal stations
Second wave	Steam power technology, steel manufacturing technology, chemical materials technology	Steam engines, trains and railways, vulcanized rubber	1845—1900	Energy: Thermal power plants, hydroelectric power stations Transportation: Train stations, subways, railways; ports, ferries, bridges Architecture: Steel structure buildings, high-rise buildings
Third wave	Electric power technology, internal combustion engine technology, radio technology	Elevators, traffic lights, automobiles, landline telephones, telegraphs, radio stations	1900—1945	Energy: Various types of power plants, transmission corridors Transportation: Intercity highways, tunnels, elevated light rail, parking lots, passenger terminals Communication: Telephone exchange centers, signal relay stations, broadcast signal towers

(continued)

Table 1 (continued)

Technological Wave	Representative technologies emerging within cities	Typical applications in cities	Time of emergence	Examples of corresponding types of urban spaces
Forth wave	Electronic technology, petrochemical production technology, aerospace technology	Cable television network, airplanes	1945—1900	Industry: Oil refining and processing plants, electronic industrial zones Transportation: Airports (runways, transportation zones, maintenance areas, business areas, etc.) Research: Spacecraft development and launch sites Communication: Communication base stations, signal towers
Fifth wave	Biotechnology, information technology, gene cloning, genetically modified crops, data technology	Email, urban–rural internet connectivity	1990—	Research: Specimen laboratories, gene banks, biotechnology centers Communication: Communication satellites, communication networks, fiber optic cable channels

(continued)

Table 1 (continued)

Technological Wave	Representative technologies emerging within cities	Typical applications in cities	Time of emergence	Examples of corresponding types of urban spaces
Sixth wave	Renewable energy technology, sustainable technology, carbon neutrality technology	Building rooftop solar power generation, wind power plants, geothermal heat pumps, sponge cities, urban underground reservoirs	2000—	Energy: New energy power plants, tidal power stations; rooftop power generation, distributed energy Transportation: Charging stations Ecology: Integrated rainwater utilization utility corridors
Seventh wave	Big data technology, artificial intelligence technology, mobile internet technology, cloud computing technology, blockchain technology	Smartphones, self-driving cars	2020—	Smart Cities: Urban sensing devices, city data centers Communication: 5G base stations

Source Self-Painted by the author

Nations and regions are aligning with the development of smart cities, formulating urban strategies and action plans to drive the growth of new infrastructure.

Since 2015, various sectors of Chinese society began discussions on next-generation urban infrastructure. In December 2018, China's Central Economic Work Conference officially proposed the concept of "new infrastructure construction." Over the following two years, the Chinese government issued several policies aimed at leveraging new infrastructure to drive urban development.

In 2020, various regions across China introduced related measures: Beijing, Shanghai, and 12 other provinces (autonomous regions and municipalities) issued specific action plans, Shandong Province released guidance opinions,[1] and Anhui Province published a catalog of new infrastructure products.[2] Meanwhile, Hunan, Jiangxi, Hebei, and Henan provinces focused on "digital economy development plans",[3] Qinghai and Guizhou provinces emphasized "5G industry development

[1] Guidance on Digital Infrastructure Construction in Shandong Province.

[2] Catalogue of Technical Products and Services for New Infrastructure Construction in Anhui Province.

[3] Hunan Province Digital Economy Development Plan (2020–2025), Jiangxi Province Three-Year Action Plan for Digital Economy Development (2020–2022), Hebei Province Digital Economy

opinions",[4] Liaoning Province promoted "Industrial Internet action," and Hubei and Tianjin focused on "post-pandemic revitalization measures".[5] Hainan Province guided the construction and development of new infrastructure through the "Smart Hainan Master Plan".[6]

3. **Urban Space as the Carrier of New Infrastructure**

New infrastructure is driving the continuous emergence of new types of spatial nodes. Currently, most discussions around new infrastructure focus on investment and financing analysis at the business level (to B), policy formulation at the government level (to G), and product development at the consumer level (to C). However, technology can only fully realize its functions when integrated into physical spaces. Therefore, regions must use space (to Space) to integrate business, governments, consumers, guiding the construction of new infrastructure. Under the influence of each technological wave, new types of infrastructure and emerging urban spaces continuously arise and evolve.

4. **New Infrastructure Drives Urban Smart Upgrades**

Smart cities are the most extensive application scenario for new infrastructure. As the smart demands of governments, enterprises, and citizens continue to grow, and with the rapid development of 5G, the Internet of Things, big data, and artificial intelligence, the original concepts, pathways, and operating models for smart city development are increasingly revealing various issues that limit the progress of urban smartification. Accelerating the construction of new infrastructure driven by next-generation information technologies is essential for integrating these advancements into all systems of the economy and society. By leveraging new technologies to navigate the inherent order and interests within cities, it creates an opportunity to reshape the smart governance system and advance smart city development. With new infrastructure as a solid foundation, the development of next-generation smart cities can be accelerated.

The next-generation urban technologies driving new infrastructure are developing rapidly, and new infrastructure must be closely aligned with technological iterations. Planning and construction should be moderately forward-looking; otherwise, the construction cycle of the infrastructure will lag far behind the technology iteration cycle, resulting in a persistent mismatch between infrastructure and technological advancements.

Development Plan (2020–2025), Henan Province Digital Economy Development Work Plan for 2020.

[4] Implementation Opinions on Accelerating the Development of 5G Industry in Qinghai Province, Notice on Accelerating the Construction and Development of 5G in Guizhou Province.

[5] Several Measures of Tianjin Municipality for Effectively Responding to the Impact of COVID-19 to Promote Investment, Expand Consumption, and Stabilize Operations, Hubei Province Post-Pandemic Revitalization Plan for Addressing Shortcomings and Strengthening Functions: Ten Major Projects Three-Year Action Plan (2020–2022).

[6] Wu Zhiqiang et al. (2021).

In the context of the global wave of digitalization, informatization, and intelligent transformation, the construction of new infrastructure requires active international cooperation. By leveraging the "Belt and Road Initiative," efforts should efficiently integrate the strategies of "going out" and "bringing in." At the same time, it is essential to fully integrate five key entities—technology research and development, product manufacturing, infrastructure operation, government decision-making, and user consumption—to comprehensively build an innovative ecosystem for the development of new infrastructure.

8 Decision-Making Process in Spatial Planning

8.1 The Loop Process of Spatial Planning

Urban planning involves a complete loop process, including sensing, field surveys, diagnosing and analyzing urban issues, developing urban strategies, selecting and comparing different pathways to achieve urban goals, determining the driving forces of urban development, allocating measures and policies, making decisions about the timing and spatial aspects of urban planning, defining the implementation mechanisms and stakeholders after the plan is completed, evaluating performance during the city's operation, and addressing future planning optimizations. This loop eventually leads back to the next round of planning.

This process also includes democratic decision-making, scientific decision-making, and legal governance, as well as pre-assessment and pre-evaluation of decision outcomes. Traditionally, these complex systems were navigated through intuitive judgment by individual experts or planning teams based on their accumulated knowledge. However, with the advent of digital technologies like "Big Data, Artificial Intelligence, Mobile internet technology, Cloud Computing, and Blockchain," urban planning is increasingly moving toward a more scientific and technological approach to address complex, quantitative learning processes.

8.2 Artificial Intelligence-Assisted Planning Decisions

Artificial intelligence serves as a powerful technological support for urban governance. AI can be used to address challenges in urban governance, such as sensing complex systems, data collection, pattern discovery, development forecasting, and decision-making predictions. In this regard, China's urban planning community has been at the forefront of continuous exploration, leading globally. However, from the perspective of AI's historical development, its application in urban governance is still in the early stages. We aim to build a comprehensive AI-supported system for digital complex science in urban planning, leveraging the digital world to address the numerous challenges arising in the physical and social worlds. This system would help resolve issues in the planning decision-making process and tackle the complexity of spatial ecology, integrating four key ecosystems: natural ecology, social ecology, innovation ecology, and digital ecology.

In the future, as AI continues to be widely applied, and with the rapid development and enhancement of AI itself, as well as its increasing incorporation into urban pattern learning and planning decision-making, urban planning will become more robust.[7] AI will undoubtedly empower urban planning and design, contributing wisdom and insights to the planning decision-making process.

9 Competition and Diagnosis of Spatial Efficiency

9.1 The Conceptual Foundation of Urban Diagnosis

1. **Rational Thinking in Urban Planning**

Peter Hall divided the development of urban planning theory into seven stages. Modern urban planning' s study of urban issues began with the phase of "observing cities from a pathological perspective" (1890–1901), followed by phases focused on aesthetics (1901–1905), function (1916–1939), imagination (1923–1936), renewal (1937–1964), and pure theory (1975–1989). Eventually, the field returned to the "pathological observation of cities" (1980–1989). This demonstrates that the concept of "urban diagnosis" has always subtly influenced the study of urban issues, shaping research throughout various stages.

Artificial intelligence technology offers the potential for a historic technological transformation in the instrumental rationality of urban and rural planning: through the super-sensing of big data on urban and rural development dynamics; intelligent discovery of the laws of urban and rural development conditions; rational diagnosis of problems in the development process; and shared public participation in decision-making. This enables the rapid achievement of urban diagnosis outcomes that "reveal public demands, detect movement of the flows of urban elements, capture dynamics, understand rationality, uncover relationships, and preserve cultural context," thereby supporting urban and rural planning to progress in a more rational and reasonable direction.[8]

2. **The Concept of City-Being**

From the perspective of life characteristics, our understanding of cities has evolved from viewing them as mechanical systems to organic systems, and ultimately as living life-being. Since the late twentieth century, advancements in the field of microbiology have provided new explanations for macro life phenomena. This has directly led to

---

[7] Wu Zhiqiang (2018).

[8] Wu Zhiqiang and Li Xin (2016).

a revolution in philosophical cognition and gradually influenced the understanding across various scientific fields. In urban studies, people have begun to recognize that cities and lives share many similarities, such as complexity, self-organization, and autonomy. These comparisons have laid the foundation for connecting the two concepts and have provided a new cognitive basis for developing urban theory.

A city is a living, highly intelligent, and open complex mega-system. This is reflected in the following ways:

(1) The city has subsystems similar to those of lives. The elements that make up a city—land, people, transportation, buildings, energy, information, resources, and more—can be compared to the components of a life in medicine. The proper functioning of these subsystems supports the healthy development of the entire city.
(2) A city undergoes growth, decline, and self-renewal processes over time and space, and it possesses the ability for self-regulation. The development, mutation and renewal, decline, and evolution of a city can all be understood from the perspective of life characteristics.
(3) Future cities exhibit the characteristics of a "living being." This does not refer to biological attributes in a physiological sense, but rather signifies the presence of inherent life principles, as well as values, spirit, and cultural consciousness.

3. **The Concept of Revealing Laws Through Data**

The fundamental assumption in urban diagnosis is that a city has a normal/healthy state. Diagnosis involves identifying deviations from this normal/healthy state through data and individual cases, with the goal of finding pathways to restore the city to its optimal condition.

The Big Data, Artificial Intelligence, Mobile Internet, Cloud Computing, and Blockchain technologies provide enhanced tools and methods for urban planning and diagnosis. These technologies free urban planning from an idealistic or problem-driven working mode, allowing it to summarize the laws of urban life development and identify specific values that deviate from the normal/healthy state. By utilizing these laws, it becomes possible to accurately predict and simulate the impact of future disasters on cities.

Intelligent urban planning and governance, guided by the laws of City-being and supported by intelligent technologies, lead cities toward a healthier and more sustainable future. With the help of big data, artificial intelligence, and other technologies, multi-scale spatial data—ranging from global to city to community levels—can be sensed, collected, and presented, ensuring open and transparent information. Comprehensive data sensing, monitoring, and analysis support in-depth studies of the spatial transmission characteristics, mechanisms, and evolutionary laws of urban elements, allowing for the prediction of their temporal and spatial evolution trends. Additionally, through core AI algorithms like machine learning, urban planning can enhance dynamic monitoring of urban operations and simulate future scenarios for more accurate forecasting.

4. **The Concept of "Forms and Flows Interaction"**

In urban planning, urban space and land use, i.e. "forms"—are the primary objects of planning control and intervention. Most countries around the world have their own standards for urban land use classification and planning control principles. In contrast, planning tools often cannot directly influence "flows"; instead, they must intervene in "forms" to indirectly impact the various flows within the city.

Based on the understanding and assessment of the "form-flow" relationship in urban planning, two different approaches or methods can be identified:

The first approach treats "forms" as the ultimate focus of urban planning. It uses morphology, political imagery, spatial sequences, and even graphic patterns as the basis for studying and refining urban forms, while considering the "flows" within the city as secondary elements, merely byproducts of the urban form.

The second approach, while also seeing the "form" of the city as the primary object of management and intervention, considers the smooth operation of "flows" within the city to be a deeper goal of planning. This perspective views the city as a living being and distinguishes the flows within the city from unconscious flows in nature, such as water and air currents. It emphasizes that the flows in a city possess a unique vitality, movement intention, and value standards, with their origin rooted in the needs of the people within the city.

These two different approaches respectively exhibit the characteristics of "form determining flow" and "form-flow interaction." The former has persisted for thousands of years and continues to influence urban planning today, while the latter reflects a greater focus on people and is closer to the essence of urban functioning. The "form-flow interaction" approach has only emerged in the past century, as it is both a product of social democratic progress and a concept that has gradually matured alongside advancements in cognitive science and technology.[9]

9.2 Constructing the System of Urban Diagnosis

As a carrier of civilization, a city's development focuses not only on its material aspects but, more importantly, on its social, cultural, and ecological dimensions. Therefore, in the urban development process, it is essential to balance and manage the relationship between the man-made urban space as a carrier of civilization and the surrounding natural environment. As a complex living entity, a city has various systems with elements that need to be matched. Diagnosing the connections between different domains requires the introduction of complex science to identify correlations and breakthroughs between these systems.

As a continuously evolving life-being, a city requires planners to adopt a dynamic perspective in the diagnosis process. Without incorporating temporal dynamics, diagnosing individual elements in isolation cannot distinguish between deviations from normal that are self-correcting and those that are not.

[9] Based on the author's speech, "Urban Planning Methods of Form Determined by Flow," delivered at the 2015 World Congress on Urban Development and Planning.

1. **Theoretical Framework of Urban Diagnosis**

As a living being, a city possesses the following four major governance capacities:

(1) City Survival Power (SP): This refers to the city's ability to defend against disasters and ensure a baseline level of safety. It includes governance capacities related to disaster early warning, emergency response, and post-disaster recovery.
(2) City Operation Power (OP): This refers to the city's ability to provide its residents with seamless mobility and efficient economic and social interactions.
(3) City Innovation Power (IP): This refers to the city's ability to inspire innovation in knowledge, technology, industry, culture, and institutional structures. It enables citizens to experience social respect, personal liberation, and collective happiness through multidimensional innovation.
(4) City Collaboration Power (CP): This refers to the city's ability to coordinate governance with surrounding communities. Through this collaboration, the city ensures individual safety, order, efficiency, and innovative happiness.

Together, these elements form the CIOS theoretical framework for urban diagnosis.[10]

2. **Technical System of Urban Diagnosis**

As a living being, a city comprises various life systems. The diagnosis system can be divided into 72 different diagnostic sub-items, each falling within 9 key domains: region, ecology, industry, power, society, culture, mobility, space, and safety & health. These domains form the basis of the technical system for urban diagnosis, allowing for a comprehensive analysis of the city's health and functionality.

(1) **Diagnostic Dimension 1: Regional Diagnosis**

Regional diagnosis employs machine learning, visual analytics, spatial econometric models, and other technical tools to build innovation networks, information networks, industrial economic networks, and transportation networks. These tools are used to analyze and diagnose a city's network structure and connectivity within a regional community.

For example, in the Yangtze River Delta region, a comprehensive analysis can be conducted on cities' collaborative networks, focusing on innovation, economy, high-speed rail, and information. By grouping and classifying the networks, the structure and patterns of urban communities can be examined, and the hierarchical status of each city within the urban network can be diagnosed. This analysis provides strategic guidance for the future development of cities (see Fig. 6).

(2) **Diagnostic Dimension 2: Ecological Diagnosis**

[10] Wu Zhiqiang et al. (2020).

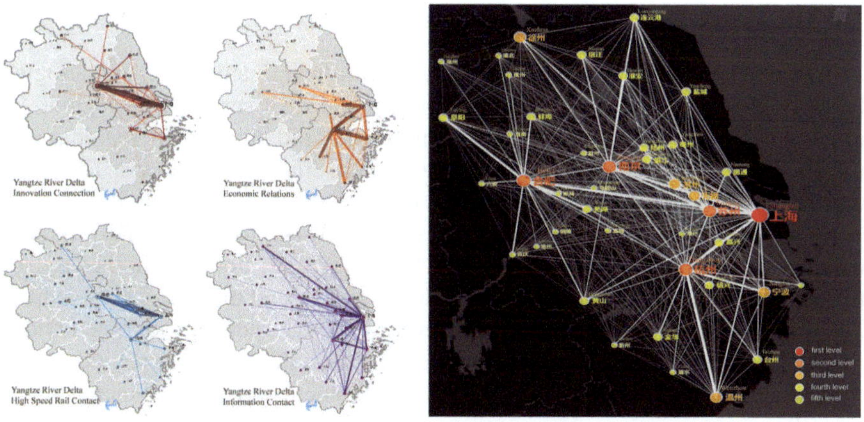

Fig. 6 Network diagnosis of the Yangtze River delta region. *Source* by the author

In the rapid process of urbanization, excessive land development and the large-scale encroachment of ecological patches have led to problems such as reduced biodiversity, fragmentation of ecological patches, and environmental instability. Ecological diagnosis encompasses various aspects, including urban energy and environmental diagnosis, urban water environment diagnosis, urban carbon neutrality diagnosis, urban atmospheric environment diagnosis, urban biodiversity diagnosis, and urban ecological pattern and suitability evaluation. Through the use of big data and artificial intelligence, ecological diagnosis reveals the characteristics and patterns of key urban ecological elements such as energy, water, materials, air, land, and biodiversity. These insights provide constructive recommendations for scientific ecological governance (Fig. 7).

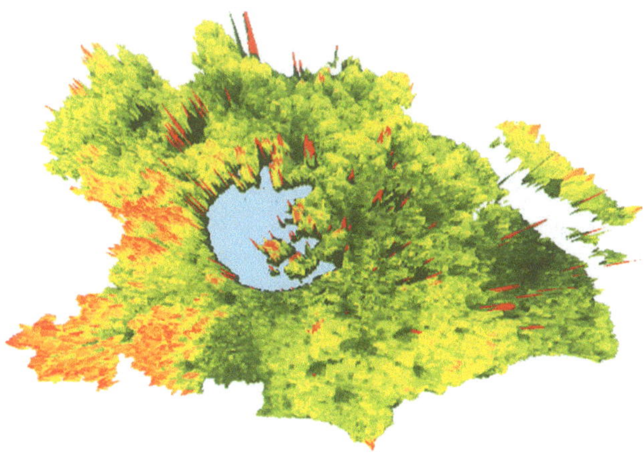

Fig. 7 Ecological diagnosis of carbon sequestration in the Taihu Lake region. *Source* by the author

(3) **Diagnostic Dimension 3: Industrial Diagnosis**

Industrial diagnosis leverages big data and artificial intelligence technologies to build an industrial knowledge graph for cities and their surrounding regions. This process identifies a city's advantageous industries and strengthens regional industries by precisely constructing and analyzing industrial network maps. The focus is on clarifying key industrial nodes around national strategic emerging industries and optimizing the industrial structure of cities and their regions. Additionally, industrial diagnosis extracts enterprise information from various sectors within the city, identifying their exact locations. Using spatial clustering and dimensionality reduction algorithms, it identifies industrial clusters for different sectors. Combined with the industrial chains derived from the knowledge graph, this analysis provides targeted recommendations for the city's industrial development and reinforcement strategies (Fig. 8).

(4) **Diagnostic Dimension 4: Dynamic Diagnosis**

The mechanism of urban development dynamics is the fundamental driving force behind urban growth, the essential factor for transforming urban development models, and the key support for advancing the urbanization process. It is also a critical factor in the transition from labor urbanization to intellectual urbanization. Research into urban development dynamics helps to understand and grasp the core drivers of urban development, ensuring the sustainability of cities.

Based on dynamic data collection and artificial intelligence technologies, the city's development patterns can be studied to construct a city dynamic development model. This model comprehensively examines the mechanisms of urban dynamics, diagnosing aspects such as city intelligence, investment, innovation capacity, dynamic engines, dynamic mechanisms, dynamic efficiency, sustainability, and the relationship between innovation and key elements of innovation. By identifying the problems

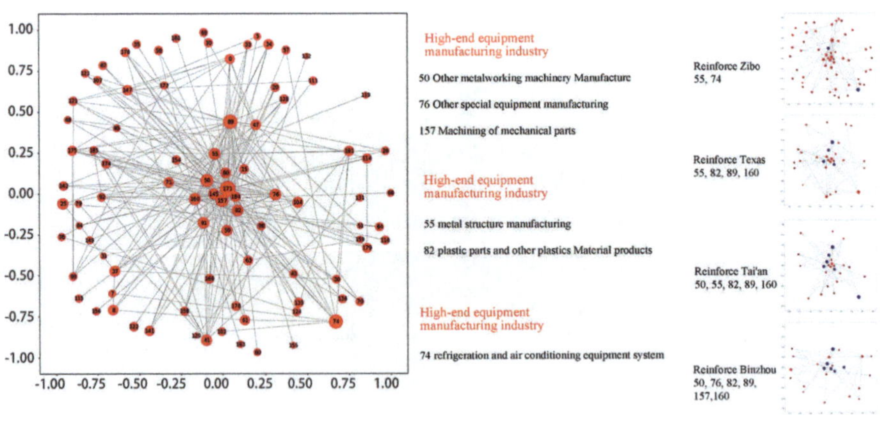

Fig. 8 Industrial knowledge graph of Jinan and strengthening industries in the Taishan Urban cluster. *Source* by the author

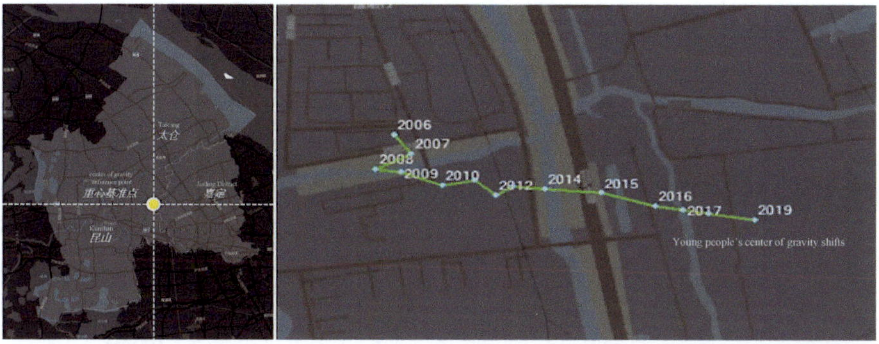

Fig. 9 Evolution of youth population shift in the Jiading-Kunshan-Taicang region. *Source* by the author

and key factors of urban development, this approach provides scientific evidence to support the city's future sustainable and efficient development.

(5) **Diagnostic Dimension 5: Social Diagnosis**

The socio-economic development of a city largely depends on its innovation capacity, which fundamentally relies on its population, including population size, geographic concentration, and age structure. Social diagnosis, based on big data collection and processing technologies, utilizes social media data, user-generated data, mobile terminal data, smart transportation data, and various remote sensing data to analyze population evolution laws, distribution characteristics, and age structure in cities.

Aimed at enhancing innovation capacity, social diagnosis provides scientific forecasts for future population size, spatial distribution, and age structure. By leveraging the law of "population driven by innovation capacity", it also offers a deep analysis of talent development characteristics within the city, diagnosing the population's innovation capacity to provide insights for high-quality, technology-led urban development (Fig. 9).

(6) **Diagnostic Dimension 6: Cultural Diagnosis**

In a broad sense, urban culture encompasses a cultural guidance system, which refers to the cultural leadership that influences a region, the nation, or even the world, as well as advanced spiritual cultural products and activities. It also includes a social knowledge system, which refers to scientific and cultural education bases that produce and disseminate knowledge, along with cultural, entertainment, and sports systems that foster creativity and physical recovery.[11] Cultural diagnosis involves various aspects such as the diagnosis of a city's historical and cultural value, historical cultural nodes, the development of cultural and creative industries, and the vitality of residents in urban neighborhoods (Fig. 10).

[11] Wu Liangyong (1996).

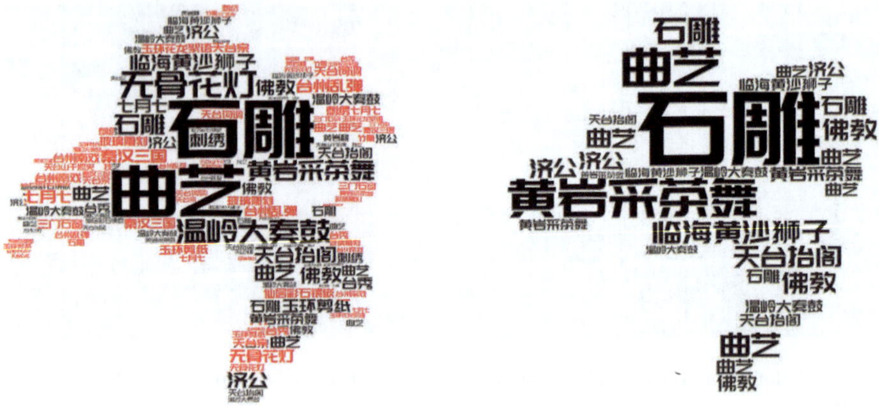

Fig. 10 Historical and cultural element perception diagnosis of Taizhou Bay New Area. *Source* by the author

(7) Diagnostic Dimension 7: Flow Diagnosis

Flow diagnosis is primarily based on the concept of "proximity" in space syntax. It can analyze the connectivity of roads within a target city and provide recommendations for optimizing road layouts and accessibility. Additionally, by overlaying the concept of "betweenness" from space syntax with major infrastructure like transportation hubs and public service facilities, it can evaluate the rationality of the current infrastructure placement. Using open-source transportation big data, flow diagnosis can assess traffic flows, networks, sources, and their effective coverage areas, evaluating the performance of the existing road network. This provides scientific support for assessing both the current state and future proposals for the city's transportation system (Fig. 11).

(8) Diagnostic Dimension 8: Spatial Diagnosis

Accessibility coverage area of car traffic in Jiakuntai area

District and county	15 minutes coverage	15 minute coverage average	30 minutes coverage	30 minute coverage average	45 minutes coverage	45 minute coverage average	60 minutes coverage	60 minute coverage average
Jiading	58km²	—	785km²	—	2248km²	—	5196km²	—
Kunshan	47km² ↓	52.33km²	351km² ↓	536.67km²	1112km² ↓	1724.67km²	2914km² ↓	4089.67km²
Taicang	52km²	—	474km²	—	1814km²	—	4159km²	—

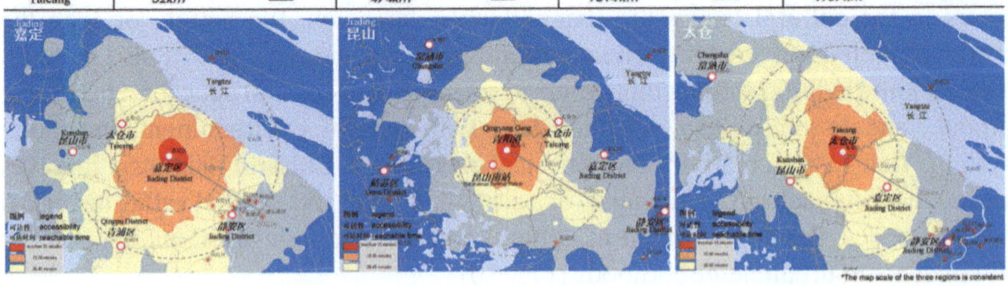

Fig. 11 Comparative study of car accessibility coverage in the Jiading-Kunshan-Taicang Region. *Source* by the author

The use and governance of urban space should not be solely focused on economics or ecology. Urban space, regardless of its scale, should be viewed as a part of the ecological, economic, and social benefits that contribute to the development of land and civilization. Spatial diagnosis uses big data technologies to assess the spatial efficiency of a city from ecological, economic, and social perspectives. It takes into account factors such as biodiversity, plant diversity, the importance of water conservation, human activities, and land productivity per unit area. Machine learning algorithms are employed to conduct an integrated evaluation of a city's ecological, economic, and social benefits, providing recommendations for optimizing land use structure and enhancing the efficiency of land utilization (Fig. 12).

(9) **Diagnostic Dimension 9: Safety and Health Elements Diagnosis**

Modern urban planning has its origins closely tied to public health. An important link between urban planning and public health is the built environment. The built environment, as a spatial reflection of urban planning and development, plays a crucial role in influencing residents' public health. Health is affected not only by genetic factors, lifestyle, and socioeconomic conditions but also by the built environment. Health can be understood at three levels: surviving, living long, and living well. Therefore, safety and health element diagnosis explores the impact of built environment elements on factors such as the sense of safety residents feel in a city, life expectancy, and happiness. This provides precise support for the development of healthy cities and includes assessments such as urban safety diagnosis, community health diagnosis, urban emotional intelligence diagnosis, and urban happiness diagnosis.

Urban diagnostics is an ancient discipline, its development tracing back to the establishment of human settlements and the birth of civilization, rooted in instinct and intuition. After the advent of modern science, humanity has continuously sought

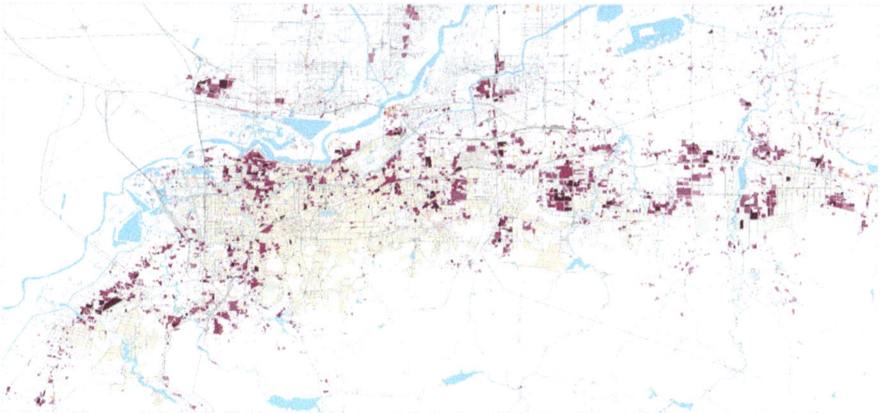

Fig. 12 Dynamic monitoring platform for land spatial efficiency in Jinan. *Source* by the author

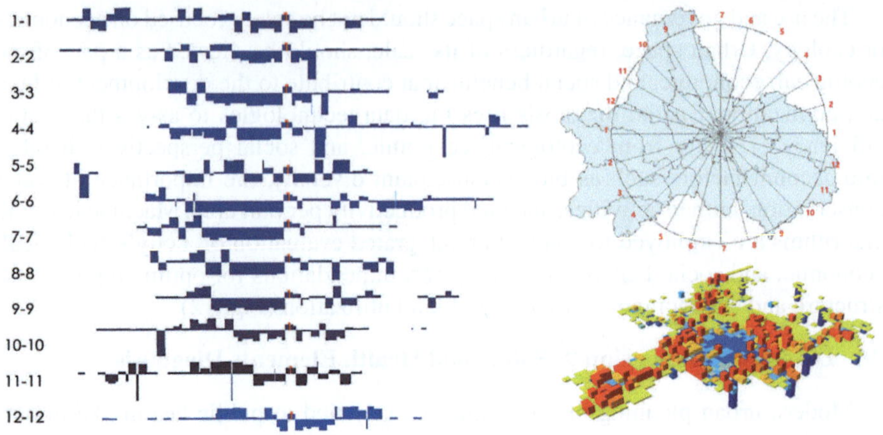

Fig. 13 AI prediction and spatial distribution laws of healthy communities in Chengdu. *Source* by the author

to identify the fundamental patterns of collective behavior and the deviations in individual cases, as well as the thresholds for reasonable deviations. However, achieving precise diagnostics has remained elusive.

The birth of Big Data, Artificial Intelligence, Mobile Internet, Cloud Computing, and Blockchain has provided entirely new perspectives and methods for observing, understanding, and analyzing urban issues. Tools for urban diagnostics, like those in medicine, "thermometer," "blood pressure monitor," and even "X-ray" or "MRI," have already entered into practice. The construction and development of urban diagnostics are now entering an unprecedentedly favorable environment. Only with the advent of "Big Data-AI-Cloud-Chain" has it become possible to conduct systematic, precise, comprehensive, and dynamic diagnostics of cities, which serve as carriers of human civilization. The introduction of artificial intelligence technology will enable learning of laws through big data, establishing a theoretical and methodological framework for precise planning and precise governance (Fig. 13).

10 Issues in Urban Governance and Their Philosophical Essence

10.1 Urban Development and Governance Issues in Chinese Cities

With the rapid development of urbanization in China, urban governance issues have continually emerged, making the modernization of urban governance systems and capabilities increasingly urgent. The development of Chinese cities and their governance face numerous challenges, specifically reflected in ecological imbalances, a large population combined with resource wastage, the destruction of historical and cultural heritage and the loss of urban identity, high energy consumption, deteriorating living environments, and frequent urban disasters.[12]

Urban development and governance issues in Chinese cities also include the scarcity of livable land and water resources, leading to sharp conflicts between people and land; a large population of migratory workers with imbalanced distribution patterns; an unbalanced energy structure with rapid growth in building energy consumption; the simultaneous occurrence of motorization and urbanization, contributing to urban sprawl; misaligned drivers of urbanization leading to uncontrolled environmental pollution; destruction of natural and cultural heritage, resulting in the homogenization of urban landscapes; an increasing income gap between urban and rural residents, leading to heightened social conflict; and intensified vicious competition between cities.[13]

In the National New-type Urbanization Plan (2014–2020), the urban issues faced by China, often referred to as "urban diseases," are described in detail. These include problems such as disorderly urban spatial development, excessive population concentration in cities, traffic congestion, public safety concerns, environmental pollution, inefficiencies in management, inadequate public service provision, and poor living environments.

Overall, during the rapid urban development process, the lack of a scientifically sound value system and the insufficient collaboration among various engineering disciplines have led to a series of serious urban diseases, posing a significant threat to China's sustainable development.

10.2 The Philosophical Essence of Urban Operations

Using Chinese ecological rationality to reflect on the complex issues facing cities today, the philosophical essence can be summarized into three levels: disharmony, disalignment, and discontinuity.

External Disharmony: This is evident in the lack of harmony between the development of cities in different geographic locations, between urban and rural areas, and between cities and nature.

Internal Disalignment: Within cities, there are mismatches in several aspects, including the lack of coherence in urban spatial development, the misalignment between urban resources and environmental capacities, the disparity between urban engineering and development, and the disconnect between urban transportation and overall urban growth. These represent the internal discord among different systems and elements within a city.

[12] Wu Liangyong (2004).

[13] Qiu Baoxing (2006).

Temporal Discontinuity: This is reflected in the growing strain on the ecological environment due to rapid population growth, the unsustainability of short-term development versus long-term planning, and the depletion of resources for future generations by contemporary development. Furthermore, there is a loss of sustainability between historical heritage and future innovation across generations.[14]

"Good living, good functioning, and good looking" are essential requirements for the development of city-being, as they are for any life system. Understanding harmony and sustainability remains the fundamental guiding principle for urban development.

References

Qiu Baoxing.(2006).*Outline of Urban Planning Reform in the Transition Period*. Planners, 2006(03): 5-14.

Wu Zhiqiang, He Rui, Xu Haowen, et al. *Discussion on the Iterative Patterns of New Infrastructure Construction* [J]. *Urban Planning*, 2021, 45(03): 9-14.

Wu Zhiqiang. *Artificial Intelligence-Assisted Urban Planning* [J]. Time+Architecture, 2018(01): 6–11.

Wu Zhiqiang, Li Xin. *The Sustainable Rationality of Urban Planning and Design* [J]. Southern Architecture, 2016(05): 4-9.

Wu Zhiqiang, Li Xiang, Zhou Xingang, et al. *Urban Diagnosis Based on the Intelligent City Evaluation Index System* [J]. Urban Planning Forum, 2020(02): 12-18.

Wu Liangyong. *Lewis Mumford's Academic Thought and Its Implications for the Construction of Human Settlements Science* [J]. Urban Planning, 1996(01): 35–41+48.

Wu Liangyong. (2004). *Scientific Issues in the Development of Chinese Cities*. Urban Development Studies(01),9–13.

Wu Zhiqiang. *On New Era Urban Planning and Its Ecological Rational Core* [J]. Urban Planning Forum, 2018(03): 19-23.

[14] Wu Zhiqiang (2018).

The Construction Logic and Operation Strategies of the Territorial Spatial Planning System of China

Zhao Min

The "Instructions on Establishing the Spatial Planning System and Supervising its Implementation" (hereinafter referred to as the "Instructions") by CPC Central Committee and State Council is of great significance and far-reaching meaning, marking that the overall framework of the "Spatial Planning System" has been clarified by integrating the existing Major Functional Zone Plan, Land Use Plan, Urban and Rural Plan and other spatial plans into one. This is an important reform progress and institutional innovation.

The construction of a unified national spatial planning system can be said to have created a new history. Previously, China had formed various types of spatial relevant plans, they have played active roles in serving the economic and social development and urbanization development since China adapted reform and opening policy, and promoting the rational use as well as effective protection of land and other territorial resources. At the same time, as pointed out in the Instructions, there were shortcomings in past such as excessive planning types, overlapping or conflicting contents, and frequent decision changes, etc. To overcome these problems, the key is to hold the concept of ecological civilization and practice in scientific approach on development, and eliminate the institutional and mechanism defects. Since the 18th National Congress of the CPC, the central authority has been promoting reforms in those concerns.[1,2]

At the First Session of the 13th National People's Congress of the People's Republic of China in March 2018, the institutional reform program of the State Council was approved, including the formation of a new "Ministry of Natural

[1] Lin Jian et al. (2018).

[2] Zhao Min (2018).

This paper was published in *Urban Planning Forum* (4, 2019).

Z. Min (✉)
College of Architecture and Urban Planning, Tongji University, Shanghai, China
e-mail: tjup@sh163.net

Z. Peng (ed.), *Theories and Frontiers in Territorial Spatial Planning*, Urban Sustainability, https://doi.org/10.1007/978-981-96-9825-7_3

Resources". According to the published institutional adjustment program, the national land and sea resource management and the preparation and management functions of spatial planning have been virtually integrated into the Ministry of Natural Resources. After solving the system and mechanism of "the multiple leaderships" in spatial resource management, the establishment of a unified territorial spatial planning system with clear responsibilities and rights, and operating in scientific and efficient way, has been put on the agenda.

There are many kinds of spatial planning that take land resources as the basic object at home and abroad, such as territorial planning, regional planning, urban and rural planning, urban cluster planning, metropolitan planning, and nature reserve planning, etc. Horizontally, these planning involve the functions of multiple government ministries and commissions; vertically, they are divided into multiple levels including the state, province, prefecture and city, county, town and township as in China. Under this circumstance, to construct a unified spatial planning framework, it is necessary to straighten out the relationship between various planning in the first place.

According to international experience, there are two patterns of spatial planning system: one is the simultaneous existence of multiple types of spatial planning under a unified framework; the other is the integration of multiple plannings. "Pattern one is a coordination strategy worked out on the basis of retaining the existing departmental planning functions, which is relatively mild, gradual and structural; while pattern two is a radical reform of the planning formulation based on the integration of departmental responsibilities, which is relatively revolutionary, thorough and substantial".[3] After several years of pilot practice, China has chosen the pattern of "multi-planning integration", that is, integrating China's unique Major Functional Zoning Planning with the Land Use Planning, Urban and Rural Planning, etc. into a unified "Spatial Planning", while the "Spatial planning System" is composed of "five levels and three categories" plannings, including Comprehensive Spatial planning at the national, provincial, municipal, county and township levels, Detailed Planning (regulatory planning) at the municipal and county levels and below, and theme planning related to some specific zones and various fields relevant with spatial matters. Besides, the new planning system also includes the planning making and approval system, implementation supervision system, law and policy system, and technical standard system.

After the central government made the decision and issued the "Instructions" to formulate and implement spatial planning, the Ministry of Natural Resources and local governments at all levels have quickly initiated the formulation of general plans at the national and local administrative levels. This is a huge and systematic project, and there are still many issues need to be solved. Therefore, it is necessary to probe into the theoretical and technical fields of the new planning system while carrying out the professional work. This article analyzes the construction logic of China's Spatial planning System from the historical dimension and discusses some related implementation concepts and strategies.

[3] Luo and Song (2018).

1 The Formation and Evolution of Spatial Planning in China

1.1 Operation in the Planned Economy Era and Reflection After Reform and Opening up

The urban planning of the People's Republic of China started in the era of planned economy, with a weak foundation. Early planning was mainly a tool to support the country's capital construction, focusing on physical planning and design. Taking the Urban Master Planning Schemes of Baotou, Taiyuan, Luoyang, Xi'an, Lanzhou, Wuhan, Chengdu and Datong in the 1950s as examples, it was mainly to implement the site selection task of the important industrial projects supported by the former Soviet Union during the First Five-Year Plan period.[4] At that time, the urban land use was realized by appling for site selection and land acquisition according to the procedures of capital construction plans. In January 1958, the State Council promulgated *the Measures for Land Acquisition for National Construction*, which clearly stipulated the procedures.

In the era of planned economy, urban planning and land management are essentially the continuation and materialization of national economic plan; urban planning has its ups and downs greatly due to the ups and downs of economic development. There has been "The Great Leap Forward"—style "rapid planning", as well as "not undertake urban planning for three years" and the retrograde step of canceling the major of urban planning in colleges and universities[5,6]. It is clear, if the scientific nature of urban development and planning is not respected, the discipline of planning and the profession of planning will not have a normal status and independence.

After the reform and opening up in 1978, the development of the market-led economy and the unprecedented demand for spatial development and stipulation promoted the development of planning and its legislation. Based on the reflection of historical experiences, the Minutes of the National Conference on Urban Planning issued in 1980 pointed out that "China's urban planning has been neglected for a long time, causing serious consequences", and "in order to completely change the situation of 'only rule by man and no legal system' formed over the years, it is necessary to make relevant laws to ensure the stable, continuous and effective implementation of urban planning".

---

[4] Li (2017).

[5] Zhao (2004).

[6] Li Hao (2012).

1.2 Urban Planning and Land Administration in the 1980 and 1990s

In 1984, the State Council issued the administrative by-law *City Planning Regulation*, which stipulated that "urban planning is composed of two stages: general planning and detailed planning". It made clear that local government departments shall formulate urban plans, and also established a hierarchical approval system for urban planning schemes. At the same time, it established the management system of "land use permit" and "building permit" in urban planning areas. In 1989, The National People's Congress of the People's Republic of China passed *the Urban Planning Law*. This is the first urban planning law in the history of New China, which establishes the examination and approval system for the preparation of Urban General Planning, Detailed Planning and Urban System Planning Schemes. In terms of the land use control, a clearer and more perfect system of "one notice and two permits" (i.e., construction project site selection notice, development land planning permit and construction project planning permit) has been established. So far, urban planning regulations and laws have given the local governments clear administrative responsibilities to manage urban planning and construction matters.

As for land use system, in 1982, the State Council issued a new edition of *the Rgulations on Land Acquisition for National Construction*, which still followed the management mode in the planned economy period. In order to adapt to the new situation of reform and opening up, in 1986, the National People's Congress of the People's Republic of China formulated *the Land Administration Law* (at the same time abolishing *the Regulations on Land Acquisition for National Construction*), which raised the issue of land use and protection to the national legislative level; In the same year, the State Land Administration was established. The law stipulates that governments at all levels shall formulate the general land use plan, and the general land use plan of the local government shall be implemented after approval by the higher-level government (Article 15). In the urban planning area, land use shall conform to the urban planning schemes (Article 16). In addition, the law also establishes the system of "grading quota check and approval" for land acquisition for national construction, namely: the acquisition of more than 1000mu(164 acres) of farm land and more than 2000mu(329 acres) of other land for national construction shall be approved by the State Council, and the acquisition of less than 3mu(0.49 acres) of farm land and less than 10mu(1.64 acres) of other land shall be approved by the government at the county level (Article 25).

It can be seen that the land use planning in the early stage of reform and opening up, like urban planning, was organized by local governments and submitted to higher-level governments for approval. Generally speaking, the formulation and implementation of urban planning and land use planning at that time were not strict and strict. Urban spatial development is basically dominated by local governments; land supply for construction is mainly administrative allocation, and the mode of "planning approval-planning permission" is adopted in the control of territorial space use.

At that time, urban planning and land management served the new situation of reform and opening up, and played an indispensable role in China's economic rise and social development. However, the limitations of bottom-up planning and the use of land and spatial resources have become increasingly prominent; as planning focuses on the expansion of construction space and only covers parts of the local administrative areas, the control mechanism of land allocation for new construction tends to be ineffective. Although land use is restricted by quota approval, local governments often use "break up the whole into parts" and other means to evade it. The spatial appearance is that many kinds of new districts and development zones blossom everywhere, which led to excessive occupation of farm land and excessive consumption of land resources, and in general, low development performance.

In the mid-1990s, in order to control the disorderly spread of development land and to change the situation of excessive consumption of land resources, the central government proposed the implementation of "the most stringent farm land management system". The National People's Congress revised the land administration law in 1998. In the same year, the Ministry of Land and Resources was established, and the land management power was only designated to the central and provincial governments. Subsequently, major reforms of the land management system such as vertical land management were implemented below the provincial levels. The new legislation stipulates that the general land use plan at a lower level should be prepared in accordance with the general land use planning at a higher level (Article 18). The scale of development land in Urban General Plans, Village and Town Plans shall not exceed the scale of development land for cities, villages and towns determined in the general land use plan (Article 22). And it also stipulates that the management of land use should be carried out in accordance with the plans, by controlling the total amount and the annual supply of construction land.

At this point, the bottom-up "urban planning" and the top-down "land use planning" have entered a cycle of contradiction-coordination-conflict due to their different logics, administrative systems and controlling mechanisms.

1.3 Legislation of Urban and Rural Planning After Entering the 21st Century

After entering the twenty-first century, the central government put forward a series of new development concepts, including "coordinated urban and rural development". In this concern, the National People's Congress revised the *Urban Planning Law* and formed the *Urban and Rural Planning Law*. The new legislation was passed in October 2007 and took effect in January 2008. *The urban and rural planning law* establishes a five-level urban and rural planning making system including urban system planning, urban planning, town planning, township planning and village planning. So far, "urban planning" has been developed into "urban and rural planning", and urban and rural planning and construction has been brought into a unified legal system and management framework. Apart from that, the *urban and rural planning law* also clarifies the binding relationship between regulatory detailed plan and the management of land use planning permission and construction project planning permission. The implementation of the *urban and rural planning law* has played an important role in upgrading the standards of urban and rural planning making, as well as in strengthening land use control in urban and rural development.

At the same time, due to the fact that different spatial plans were under the administration of different central departments, the lack of top-level framework design, and the delay of institutional reform, the overlap and contradiction between Urban and Rural Plans and other departmental Plans were inevitable. Although *the urban and rural planning law* stipulates that "the preparation of Urban General Plan, Town and Village Plans should be based on national economic and social development plan and in connection with Land Use Plan", it is difficult to implement it in fact. The Ministry of Housing and Urban–Rural Development, the Ministry of Land and Resources and other departments continue to make their own plans, with their own planning categories, planning effective periods and technical standards, resulting in hundreds of thousands of differences between urban and rural plan (referred to as "urban plan") and territorial plan (referred to as "land use plan") on the "land use map" of the same territorial object. "Taking Xiamen as an example, through the work of 'multi-plan integration', it is found that there are 124,000 different pots between urban planning map and land use planning map, which leads to the ineffective use of about 55 km^2 land quota." [7] There are contradictions between "urban plan" and "land use plan", but also between "forest plan", "grass plan" and "land use plan" (Figs. 3.1 and 3.2). At the same time, some cities urgently need development space, but there is no development land quota. In some cities, the actual development land used has broken through the development land quota of their Land Use Plan. This would inevitably reduce administrative efficiency and affect the healthy development of the economy and society.

1.4 Emergence and Analysis of the National Main Functional Zone Plan

In order to unify the order of spatial development and establish a reasonable spatial development structure, the *Eleventh Five-Year Plan for National Economic and Social Development*, adopted by the National People's Congress in 2006 proposed to promote the preparation of the Main Functional Zone Plan, i.e. "according to the carrying capacity of resources and the environment, the existing development density and development potential, taking into account the future population distribution, economic layout, land use and urbanization pattern of China, the territory space is divided into Optimized Development Zone, Key Development Zone, Development Restricted Zone, and Development Prohibited Zone. The regional policies and

[7] Wang Menghui (2015).

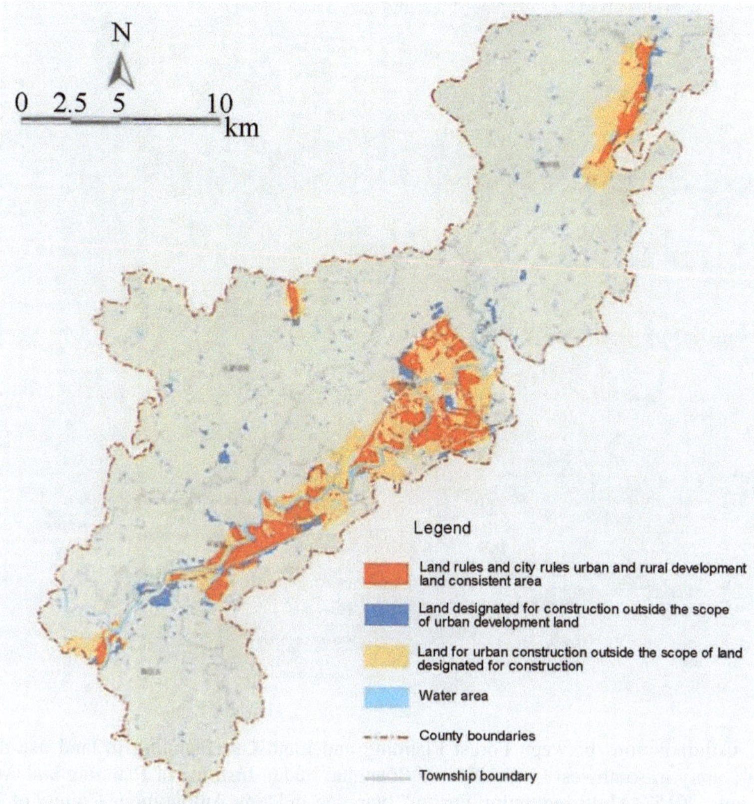

Fig. 3.1 Contradictions between Urban Planning and Land Use Planning in land use designation in a county in southwest China. *Source* Shanghai Fudan Planning and Architectural Design Institute. (2015) "Multi-regulation Fusion" Planning in yuping county, Guizhou Province

performance evaluation should be made or adjusted according to the main functional orientations." Accordingly, the Main Functional Zone Plan should be positioned as a strategic plan that incorporates "development" with "space".

In 2007, the State Council issued the "*Instructions on Formulating National Main Functional Zone Plan," stating that* "The National Main Functional Zone Plan is a strategic, fundamental and binding plan, which is the basic basis for the spatial development and layout of the general planning of national economic and social development, as well as population planning, regional planning, urban planning, land use Planning, environmental protection planning, ecological construction planning, comprehensive watershed planning, comprehensive water resources planning, marine functional zoning, sea area use planning, food production planning, transportation planning, disaster prevention and mitigation planning, etc." According to the documents of the State Council, the National Main Functional Zone Plan consists of the National Main Functional Zone Plan and the Provincial Main Functional

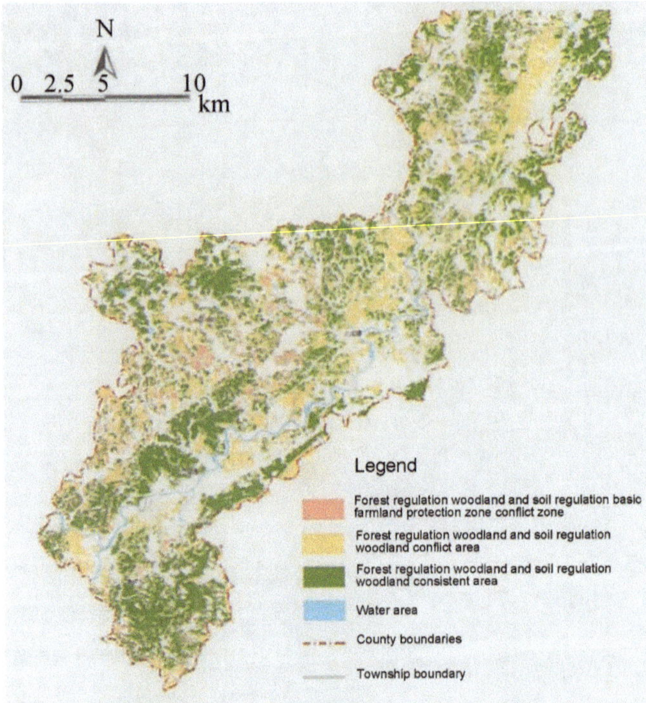

Fig. 3.2 Contradictions between Forest Planning and Land Use Planning in land use designation in a county in southwest China. *Source* Shanghai Fudan Institute of Planning and Architecture Design. (2015) "Multi-regulation Fusion" planning in Dong Autonomous County of Yuping, Guizhou Province

Zone Plan, which are prepared at the national and provincial levels. The National Main Functional Zone Plan is prepared by the National Main Functional Zone Plan Leading Group in conjunction with the governments of all provinces (provincial level autonomous regions and municipalities), and the planning period is until 2020. The Provincial Main Functional Zone Plan is organized by the governments of all provinces (provincial level autonomous regions and municipalities) and prepared by the governments at the municipal and county levels, and the planning period is until 2020.

In December 2010, the National Main Functional Zone Plan was issued by the State Council (*Notice of the State Council on printing and distributing the national main functional zone plan*). According to this document: "China's territorial space is designated to the following main functional zones: According to development approach, there are Optimizing Development Zones, Key Development Zones, Development Restricted Zones, and Development Prohibited Zones; according to development content, there are Urbanization Zones, Main Agricultural Production Zones, and Key Ecological Function Zones; according to the hierarchy, there are two levels, National and Provincial ones." Concerning definition, "the term of 'development' in the Optimizing Development Zones, Key Development Zones, Development Restricted Zones, and Development Prohibited Zones in this plan specifically refer to largescale and high-intensity industrialization and urbanization development." Therefore, if "development restricted" in a zone, it only specifically refers to restricting large-scale and high-intensity industrial and urban development, rather than restricting all development activities.

It can be seen that, on the one hand, the planning of Main Functional Zone Plan, as a strategic, fundamental and binding planning of China's territorial space development, is undoubtedly of great significance. On the other hand, as a new plan, its rigor seems insufficient, for example, how to deal with the relationship between government choice and market choice? How to define "large-scale and high-intensity industrialization and urbanization development"? In fact, the "Optimizing development zone" identified in the plan, such as "Bohai Rim", "Yangtze River Delta" and "Pearl River Delta", still need to have certain large-scale and high-intensity development and construction at this stage. From its planning diagram, it is too abstract and simplified. According to Professor Lin Jian's analysis, Main Functional Zone Plan is only functional zoning maps, and there is no way to manage specific plots. In terms of the maturity of management implementation means, the Land Use Plan and Urban and Rural Plan is high, while the Forest Land Protection and Use Plan, Water Functional Zoning Plan and Marine Functional zoning Plan is good. But there is no clear means for Main Functional Zone Plan and Ecological Function Zoning Plan[8].

Therefore, it can be assumed that Main Functional Zone Plan must form a close transmission and coordination relationship with other spatial plans in order to play its expected role.

2 Exploration and Construction Logic of the New Planning System

2.1 From "Multi-Plan Integration" Pilot in Cities and Counties to the "Provincial Spatial Planning" Pilot

In response to "the multiple leaderships" of the growing spatial planning inconsistency and even "fighting" each other, the National Development and Reform Commission, together with the former Ministry of Land and Resources, the Ministry of Environmental Protection and the Ministry of Housing and Urban–Rural Development (hereinafter referred to as the "four Ministries"), issued the Notice on the Pilot Work of "multi-plan integration" in Cities and Counties in August 2014.

According to the notice, "carrying out the pilot reform of city and county spatial planning, promoting 'multi-plan integration' of economic and social development

[8] Lin Jian et al. (2018).

planning, urban and rural planning, land use planning and ecological environmental protection planning, and forming a unified blueprint for each city and county is an important task in the central government's comprehensive deepening reform agenda in 2014"; The main task of carrying out the pilot work is to "explore the specific ideas of 'multi-plan integration' in economic and social development planning, urban and rural planning, land use planning, ecological environment protection and other planning, study and put forward 'multi-plan integration' pilot project that can be replicated and promoted, and form a city and county with one plan and one blueprint. At the same time, explore and improve the spatial planning system of cities and counties, and establish a coordination mechanism for relevant planning."

In terms of actual impact, this "bottom-up" experiment has its value, but it also has obvious limitations. If the current "multiple leaderships" is necessary, the meaning of "multi-plan integration" should be to integrate "multi-regulation within one framework", so as to coordinate the operation[9]. However, if the problem of "multiple leadership" is not solved, even if the contradiction spots are eliminated this time and "multi-plan integration" in a specific time and space is realized, the same dilemma may be repeated in the future. As for the pilot projects carried out by the four Ministries concerned, three schemes have actually been put forward. It can be seen that the department's understanding of "multi-plan integration" is not exactly the same[10]. In this case, it is difficult to popularize the results of pilots.

Subsequently, the central government upgraded the pilot level. In January 2017, the General Office of the CPC Central Committee and the General Office of the State Council issued the p*ilot program of provincial spatial planning.* Its purpose is:" In order to implement Fifth Plenary Session of the 18th Central Committee of the CPC to coordinate all kinds of spatial planning based on the Main Functional Zone Planning, promote the strategic use of 'multi-plan integration', deepen the reform and innovation of the planning system, establish and improve a unified spatial planning system, and enhance the territorial space governance capacity and efficiency, a provincial spatial planning pilot scheme is formulated on the basis of the pilot work of the 'multi-plan integratuion' in cities and counties." The pilot covers 9 provinces including Hainan, Ningxia, Jilin, Zhejiang, Fujian, Jiangxi, Henan, Guangxi and Guizhou. The basic principle of the pilot work is to "strengthen the top-level design of system, mechanism, laws and regulations in view of the problems existing in various spatial plans, study and propose a package solution to systematically solve key and difficult problems, and break the fragmented situation of various plans".

The pilot work of provincial spatial planning is very difficult. As far as land use control is concerned, the objectives and means of spatial planning and other land use planning are basically the same, but it is technically difficult to define the functional land use and draw control lines at the provincial scale. In terms of the relationship between spatial planning and urban and rural Planning, the levels of national urban system plan, provincial urban system plan and urban general plan coincide with the proposed spatial planning levels. It is evident that the exploration

9 Zhang Jie and Zhao Min (2015).

10 Sun Anjun (2018).

of spatial planning is intricately connected to the restructure of urban and rural planning and land use planning. The reform, which primarily centers on top-level design, inevitably confronts issues pertaining to the system and mechanism.

2.2 *Construction Logic of the Spatial Planning System*

For a large country with a vast territory like China, the division of responsibilities in the government management system to some extent is inevitable; the creation and development of China's various spatial plans are also necessary and reasonable, but the plans must be under a unified framework of territory development and protection. In view of the existing problems in China's land development and protection, and in order to solve the contradictions between spatial plans, the *General Plan of Reform for Promoting Ecological System* by CPC Central Committee in 2015 clearly proposed to "build a territorial space development and protection system based on spatial planning, with the main means of land use control", and "build a national unified, interconnected and hierarchical administrated spatial planning system, with spatial governance and spatial structural optimization as the main contents".

Years of exploration and pilot projects show that under the current conditions, it is often difficult to achieve the expected goals by exploring spatial planning technology alone; The optimization and adjustment of institutions and corresponding administrative management system would generally produce the required governance effects. If we probe in depth into the incoordination of departmental planning and management in the allocation of spatial resources, we could find that the most fundamental reason was the internal conflicts of the departmental policies, behind were different planning logics as well as the competition for "dominance" among the departments. These problems were manifested externally as each department formulating its own plan, with different planning scope, timing, technical standards and so on, that inevitably led to dilemmas and conflicts.

If only solve the problem and achieve "multi-plan integration" at the technical level, it would inevitably be "only touch the surface". Through the institutional reform, the central government concentrated the responsibilities of spatial planning and management related to the protection and use of natural resources under the newly established Ministry of Natural Resources, hoping to effectively resolve the contradictions arising from the previous system and mechanism[11]. On this basis, the construction of territorial spatial planning system with "multi-plan integration" can be described as a matter of course.

Viewing from the origin and development of all kinds of spatial planning in China, they all have specific objectives, and all of them are based on sufficient technical rationalities. Therefore, the reform and construction of the new planning system should be a process of "sublation", rather than reinventing them all. That is, after the completion of institutional reform and top-level design, the frictions caused by the

[11] Zhao Min (2018).

old systems are eliminated, and the reasonable connotation of the original plans at all levels can and should be retained and integrated into the new system. All effective planning tools, developed in past for the management of territorial space and the development control, should be integrated into the new framework and give full play to their effectiveness. This can be described as the technical logic of the construction of the new planning system.

The main body of the new planning system is the "General Plans" of "multi-plan integration", which is compiled at different levels according to the level of government. In cities, counties and below, Detailed Plans are prepared for both urban and rural areas to elaborate and specify General Plans as well as to implement land use control. As an integral part of the formulation system of spatial panning, there are also Theme Plans of some specific zones and various fields relevant with land use matters. It can be said that this planning system has both characteristics of integration of "multi-plan in one" and openness in certain degree, but it should be also coordinated and unified with each other. The relationship between them is clearly stated in Instructions: First, it is necessary to "strengthen the guiding and binding role of Territorial Spatial Plans on Theme Plans"; second, each theme plan "shall not violate the mandatory contents of the General Plans, and its main contents shall be included in the Detailed Plans".

3 Legal Construction of the New Planning System and Implementation Strategies

3.1 Strengthen the Legal System Construction in Line with the New Planning System

It is necessary to establish a spatial planning system and supervise its implementation based on the perception of rule of law while attaching importance to legal construction. As the reform and the construction of the new planning system still continue and inherit the previous. Planning system to a certain extent, the reasonable con- notations of the previous planning of all types and at all levels should be retained. Therefore, the construction of the new planning system is inherently consistent with the legislative spirit of existing relevant laws such as the Urban and Rural Planning Law of the People's Republic of China, the Land Administration Law of the People's Republic of China and the basic requirements of the State Council's policy documents. From the viewpoints of judicial logic and administrative rules, the formulation and supervision of the spatial planning system do have some legal sources. But in order to better implement the law-based governance strategy, the legal construction must also be coherent with major reforms, timely institutionalize the results of reform and innovation, and pro- vide accurate and systematic legal norms and technical standards.

At the same time, we must also aware that new legislation will take some time. The formulation of new laws such as the Territory Development and Protection Law and the Spatial planning Law cannot be accomplished overnight. However, it is also unrealistic to wait until all legislation and technical standards are complete before starting the formulation, approval and implementation of the Spatial Plan. Therefore, there is a problem called "legal linkage during the transitional period," which requires the attention and proper handling by the People's Congresses and Governments at all levels. As the "multi-plan integration" reform that integrates Land Use Plan, Urban and Rural Plan, etc., the new Planning making practice will inevitably go beyond the content and provisions of the current laws and regulations. In this regard, the "Instructions" clearly requires "submitting for approval in accordance with the procedures first, and proceeding after obtaining authorization." This requirement embodies the spirit of law-based governance.

While the National People's Congress and the central government make planning-related laws, administrative regulations and rules, local people's congresses and governments with legislative authority can also take the initiative to formulate local planning regulations and make government decrees. Since the Spatial planning, corresponding Theme Planning and Detailed Planning preparations involve various aspects of relationships and technical issues, the formulation of technical standards and specifications must also be advanced in a timely and orderly manner. Furthermore, according to the Standardization Law of the People's Republic of China revised in 2017, apart from making National Standards by central authorities, the local governments, organizations, professional and social groups are all delegated rights to make planning related standards, that would help to fully meet the demands of the spatial planning in the new era.

3.2 The Hierarchical and Classified Formulation and the Transmission Mechanism of the Spatial Planning

The construction and supervision of the spatial planning system are the specific measures to improve national governance. Therefore, the Comprehensive Spatial Plans need to have a transmission mechanism between the upper and lower levels, and then play a guiding and restrictive role on Theme Plans, and finally be translated and elaborated into Detailed Plans and rules for land use control, to realize the goals of territorial functional zoning as well as urban and rural planning administration.

The new planning system highlights the different role setting and mechanism design of the Comprehensive Spatial Plans at all levels, and emphasizes the principle of "top-down, and upper and lower-level coordination." According to the "Instructions", the National Comprehensive Spatial Plans is the general outline for the protection and development of nation-wide territory, focusing on the strategic issues; Provincial Comprehensive Spatial Plans implement National Comprehensive

Spatial Plans and guides the formulation of municipal and county-level Comprehensive Spatial Plan, focusing on the coordination role; and the Comprehensive Spatial Plans in municipal and county-levels are the detailed implementation of the planning requirements of the higher-level governments, and the specific arrangements for the development and protection of their administrative areas, focusing on implementation issues.

Viewing the general situation, the transmission mechanism between the upper and lower levels and the assigned focuses of the planning at different levels are undoubtedly necessary, and would be of help to over- come the blindness of the local governments in their planning making. In terms of current planning tasks at the local level, the provincial, municipal and county levels will formulate their Comprehensive Spatial Plans up to 2035 and looking forward to 2050, in a top-down manner. This requires lower-level authorities to follow and elaborate the higher-level planning requirements, at the same time, carry out various researches concerning local conditions, in order to make "strategic, basic and restrictive" arrangements for their administrative areas of different levels. The local Comprehensive Spatial planning making should give feedback to the upper-level planning making and provide terms to the lower levels' planning making, as well as to exercise guiding and constraining functions on theme planning processes.

The new planning system still follows the hierarchical system with division of administrative powers in formulation and approval procedures, but pays more attention to "implementation and supervision." It emphasizes the compliance of the lower-level plan with the upper- level plan, and of the Theme Plan and Detailed Plan with the Comprehensive Spatial Plan. At the same time, feedback should be given by the lower-level planning making to the upper-level planning making, and by the theme planning and detailed planning making to the general planning making. And reasonable interaction and corresponding confirming mechanisms are also required. Spatial planning making and the corresponding supervision and implementation are a systematic project, the systematic approaches are the premise that will ensure the smooth and effective operation of the system.

3.3 The Strategic Guidance and Baseline Control Roles of Municipal and County-Level General Plans

Compared with the national and provincial-level Comprehensive Spatial planning, the municipal and county-level Comprehensive Spatial planning should lay particular emphasis on implementation issues. It is also true, from international and domestic experience, the formulation of municipal and county-level Comprehensive Spatial planning will definitely involve the concerns of the regional and socioeconomic development issues, rather than to be limited on land use allocation and control matters.

For example, regarding the Beijing's new Comprehensive Urban Planning Scheme, General Secretary Xi Jinping once clear requested that Beijing's urban planning should have in-depth thinking on what kind of Capital City to build and how to build it, grasp the strategic positioning, spatial pattern, and element configuration, persist in urban and rural overall development, and implement "multi-plan integration," to form one plan and one blueprint. Regarding the Shanghai's new Comprehensive Urban Planning Scheme, the State Council pointed out in its ratification document: under the guidance of the Shanghai's new plan, efforts should be made to enhance the city's functions, shape its distinctive features, improve environmental quality, optimize urban management and services, and strive to build Shanghai into an admirable city of innovation, humanity and sustainability, an excellent global city, and a modern socialist international metropolis with world influence. It can be seen that a municipal spatial comprehensive or master plan is a strategic plan that combines economic and social development goals with spatial arrangements, and its formulation and implementation should be the joint task of all competent departments under the leader- ship of the municipal government. The Comprehensive Spatial Plans at the national, provincial, municipal and county levels, with the planning period to 2035 and prospects to 2050, is both basic and strategic. It should be a joint master plan that guiding the making of five-year local "economic and social development plan" and the five-year "urban and rural development plan" as well as other action plans.

In short, taking into account the function of municipal and county-level Comprehensive Spatial Plans, as an integration of former Major Functional Zoning Plan, Land Use Plan, Urban and Rural Plan and other spatial and sector plans, it should continue and further exert its original roles of "strategic guidance" and "base- line control" (Dong & Zhang, 2018). The content of its formulation should cover the aspects of economy, society, ecology, and social service and management. Planners should not limit its content to spatial matter itself, and should never artificially create new conflicts between development, spatial and other plans.

3.4 Formulation Method of Detailed Plans Under the Spacial Planning System

In the formulation of municipal and county-level Comprehensive Spatial Plans, "multi-plan integration" must be realized as the full coverage of urban and rural areas in the functional and policy zoning maps of a governmental spatial plan; its vertical transmission and elaboration should also achieve full coverage of urban and rural areas in a city or county area. Therefore, the Detailed Plan should be adopted at the municipal, county and below levels.

Detailed Plan is an implementation arrangement for the use of specific land plots. It is the legal basis for the administration of spatial development and protection, including the implementation of land use control, the grant of building permits for urban and rural construction projects, and carrying out various constructions. Specifically, it is divided into detailed plan within the urban development boundary and detailed plan for rural areas outside the urban development boundary. Through two types of Detailed Planning, full coverage of urban and rural areas will be achieved. The Detailed Plan under the spatial planning formulation system seems to be the same as the current "Regulatory Plan." But under the new system, detailed planning making needs to be innovated and improved on the basis of the past experience and new demands.

According to the observation of Zoning By-Law and Local Planning in North America and European countries, the planning formulation at the development control level generally covers the entire administrative area. There is great flexibility in the setting of the scale of the planning maps, land use classification and other planning control elements to suit different planning objects such as densely constructed areas and agricultural areas. In contrast, the formulation of China's previous "Regulatory Plan" basically followed the same pattern, which ignored the diverse characteristics of spatial objects. Under the new "multi-plan integration" planning system, the formulation and implementation of the Detailed Plan should be unified for urban and rural areas, with classified guidance provided. Unified Detailed Plan should be formulated for urban construction areas and rural areas, but differentiated formulation methods and technical standards should be developed, so the drawing scale, land use classification, setting of control elements, etc. could be different according to the real needs. In the large-scale drawings of cities and counties, aiming at the local dense construction areas, we can learn from the local "insert" method of British local planning to meet the requirements of the formulation depth of the Detailed Plan (Fig. 3.3).

In fact, as far as the methodology of Detailed Plan is concerned, even in the planned built-up areas, different functional areas (such as industrial areas, commercial areas and residential areas), different characteristic areas (such as new urban areas, old urban areas and historical blocks) and different stages of development (such as deterministic development and uncertain development) should adopt differentiated strategies-the choice of control elements and "strict" and "loose". Only by ensuring the applicability of the "control power" of the Detailed Plan can we really contribute to strict implementation of the plan.

3.5 *The Division of Planning Administration and the Innovation of Planning Documentation*

In order to improve the efficiency of planning administration, it is necessary to reasonably divide the administrative responsibilities of planning. There have been many discussions on this subject in the planning academic circle. It is generally believed that a higher-level government should mainly to pay attention to the functional orientation, development scale, environment and resources, important historical and cultural

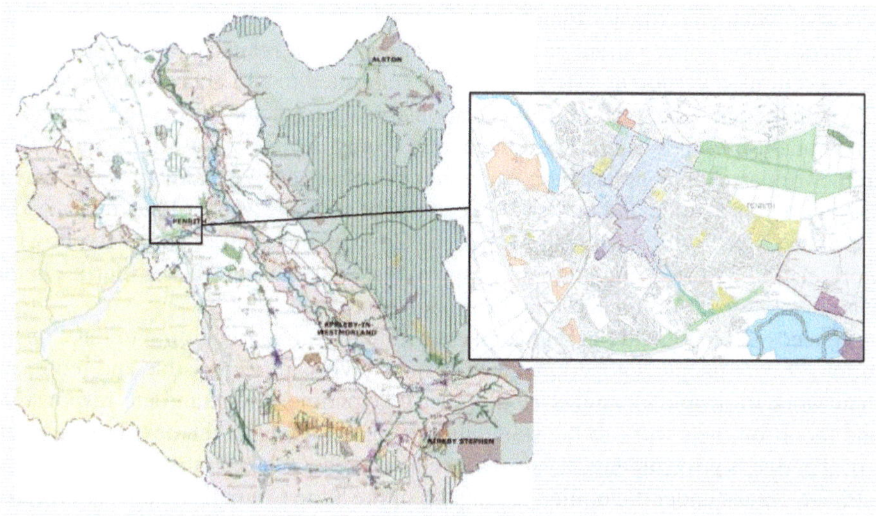

Fig. 3.3 The global and local "insertion" relationships of local planning in the United Kingdom. *Source* The author redrawed by local government website: https://www.eden.gov.uk/planning-and-building/planning-policy/

heritage, regional infrastructure, ecological protection and other planning content of lower-level cities and counties that involves national and regional interests. The layout of urban and rural land use, public service facilities, commercial service facilities, road transportation facilities, municipal public facilities, green space and squares, urban landscape construction, etc. could be classified as affairs of local governments.

Taking the lessons of previous practice, the principle of division of planning powers in the Spatial planning System has set up. The approval, and supervision will be that "the higher-level governments only verify their concerned contents of lower-level governments' plans". And based on the principle of "The government approving a plan shall also be responsible for the supervision," a review and record system for the Spatial Planning at different levels will be established. In order to implement this reform, the content of planning formulation, the documentation of written plan, and the method of sub-mission for approval must also be adjusted or innovated.

The innovation of the planning document system can not only better fit the division of administrative powers, but also help highlight the strategic roles and public policy attributes of the governmental Comprehensive Spatial planning. Planning document must be comprehensible and readable to the public. To this end, the previous boring panning texts and maps that focused on "technical" clauses and drawings must be changed intopublic-friendly booklets, including the use of illustrated typesetting and well-founded explanations.

In this regard, the new Shanghai Master Plan is a good example worthy of reference. Shanghai has adopted the "1 + 3" documentation, of which "1" is the general report of "Shanghai Master Plan (2017–2035)," a framework document guiding the urban development at a strategic level.[1] In general, "1" is related to the responsibility of the central government. After the plan is submitted to and approved by the State Council, the Shanghai government is responsible for implementation and the central government for supervision. The content of "3" only involves local affairs, and only a framework arrangement has been made in the comprehensive planning stage, which serves only as a guide and outline. In accordance with the Master Plan approved by the State Council, the Shanghai Municipal Government shall organize the flowing work, including the formulation and approval of district-level Master Plans, Theme Plans and various Action and major Construction Plans.

It needs to be pointed out that China is a Unitary State. According to the Constitution, "Local people's governments at various levels are responsible and report on their work to people's congresses at the corresponding levels, and are responsible and report on their work to the state administrative organs at the next higher level. Local people's governments at various levels throughout the country are state administrative organs under the unified leadership of the State Council and are subordinate to it" (Article 110 of the Constitution of the People's Republic of China). Therefore, the division of planning administrative powers is not absolute. Even if it is a "local responsibility," the higher-level governments and the central government can still exercise supervisory powers and hold the local government accountable.

4 Conclusion

Spatial governance is an essential part of national governance. The major decision made by the CPC Central Committee and the State Council to establish a Spatial planning System and supervise its implementation is the specific measures to improve China's governance system and governance capabilities, which is also a response to the requirement of the ecological civilization in the new era.

Research has revealed that all spatial relevant planning were created for some specific reasons and all have their specific evolution courses; behind the establishment of a unified Spatial planning System, there are administrative and technical logics. As far as the professional practice of planning is concerned, both the planning formulation system and the planning supervision and implementation mechanism must conform to the new development concept so as to promote the modernization of national governance.[1]

With the continuous development of China's economy, society and urbanization, and the constant changes in natural resources and environmental conditions, the spatial planning system, the professional practice, and the disciplinary connotation behind will also inevitably change. As far as China's urban planning is concerned, space and scope have been continuously expanded over the years: from planning design to permit management; from urban planning to urban system planning; from physical planning and design to economic and social development strategies; from urban planning to urban and rural planning and "city-region" planning. It can be said that urban planning has evolved into comprehensive planning, and needs to deal with issues at multiple levels and of various natures such as "urban area" and "region," "city" and "rural," "development" and "protection," etc. Therefore, the intersection and integration of the urban planning profession and other planning professions become inevitable, and the urban planning discipline is constantly developing and changing because of the continuous absorption of knowledge and methods from other disciplines. However, the discipline of urban and rural planning still needs to define the boundaries of its own field as it cannot expand indefinitely and cover everything.

Besides, there are also aspects that remain unchanged during the tremendous changes, such as the original aspiration of planning. Original aspirations should be maintained and new advancements should also be made, whether it is Ebenezer Howard's "Garden City" vision, the concepts advocated by the "Garden City and Town Planning Association," or the public intervention theory that emerged in the desire of overcoming the defects of the market mechanism, or the development concept of protecting the ecological environment, historical culture, and pursuing efficiency and fairness of resource and welfare allocation.

References

Lin Jian, Wu Yuxiang, Wu Jiayu & Liu Shiyi. (2018). On the construction of spatial planning system: An analysis of the relationship between spatial planning, territorial space use regulation and natural resources regulation. *City Planning Review* (05),9-17.

Zhao Min. (2018). On innovative practice and policy orientation of urban master planning in the new era. *Urban & Rural Planning* (02),8-18.

Luo, Z. D. and Song, Y. (2018). Between Change and Immutability: The evolution of Land Use Plan in the United States since World War II and its implications for China. *Urban Planning International* (04),4-11.

Li, H. (2017). Establishment of Strategic Planning Urban Design Institute and eight key urban planning. *City Planning Review* (10),122-124.

Zhao, M. (2004). *Further promoting the development of urban planning in China under the market economy.* Urban Planning Forum (05), 29–30+95.

Li Hao. (2012). Historical Review and Reflection -- Written on the 50th anniversary of the proposal of "No urban planning for Three Years". City Planning Review (01), 73-79.

Wang Menghui. (2015). *Promoting the transformation of government functions and realizing the overall development of resources and environment in urban and rural areas -- Thinking and practice of carrying out the reform of "combining multiple regulations" in Xiamen City.* City Planning Review (06), 9–13+42.

Zhang Jie & Zhao Min. (2015). Discussion on China's overall urban planning reform from the perspective of "multi-planning integration". Shanghai Urban Planning Review (06), 8-13.

Sun Anjun. (2018). *Thoughts on spatial planning reform. Urban Planning Forum* (01), 10-17.

Content and Structure of Territorial Spatial Planning

Sun Shiwen

In accordance with the overall deployment of *Several opinions on building a national territorial spatial planning system and supervising the implementation by CPC Central Committee and State Council* (hereinafter referred to as "*Several Opinions*"), the construction of relevant systems of territory spatial planning is progressing in an orderly manner, and the preparation and implementation of plans at different levels is also underway with great intensity. Meanwhile, various academic studies on territory spatial planning are flourishing in different directions, across different directions, disciplines, fields, and levels. However, these discussions and research primarily focus on what territory spatial planning does, how it is done, and what it achieves, etc.,[1,2] with little attention paid to the knowledge foundation supporting the implementation of territorial spatial planning and the interrelationships among this knowledge. From an academic perspective, exploring the substantive content and composition of territory spatial planning, as well as the interrelationships with various related knowledge, has positive implications for gaining a deeper understanding of territorial spatial planning, improving planning content, and better carrying out related work.

[1] Pan Haixia and Zhao Min (2020).

[2] Wang Xinzhe (2019).

This article was published in Urban Planning Forum, No.6, 2020.

S. Shiwen (✉)
College of Architecture and Urban Planning, Tongji University, Shanghai, China
e-mail: shsun@tongji.edu.cn

Z. Peng (ed.), *Theories and Frontiers in Territorial Spatial Planning*, Urban Sustainability, https://doi.org/10.1007/978-981-96-9825-7_4

1 The Substantive Contents of Territory Spatial Planning

Several Opinions clearly provide a clear definition of Territory Spatial Planning: "Territory Spatial Planning serves as the compass for national spatial development and the spatial blueprint for sustainable development. It provides the fundamental basis for various activities related to development, protection and construction. The main tasks of territory spatial planning include: "The Overall planning of territory spatial development and protection pattern in the new era takes into account various factors such as population distribution, economic layout, land use, and ecological environment protection, and the scientifically allocates production space, living space, and ecological space." It is evident that Territory Spatial Planning is aimed at future development, providing holistic overall planning and arrangements, as well as constraining various development and protection activities. Simultaneously, it takes into consideration social and economic environments to scientifically allocate space for future use, including production, living, and ecological spaces. Therefore, as an important part of the national planning system, Territory Spatial Planning is the first and foremost planning, that is, on the basis of coordinating and arranging all kinds of land space use in advance, the future development goals can be achieved through control of land space utilization activities. Therefore, based on the establishment of Territory Spatial Planning system, *Several Opinions* focus on "epitomizing strategic", "improving scientific, "strengthening coordination", "focusing on operability" in the "Compilation Requirements", These opinions clearly define the specific work content and requirements of Territory Spatial Planning. In the sections on "implementation and supervision", a series of institutional frameworks have been established to ensure the effective implementation of territorial spatial planning, thereby providing guidance for the development of this work.

"The overall planning of territory and spacial development and protection pattern in the new era" is the core task of Territory Spatial Planning. The pattern is formed by the overall structure and layout relationship that emerges from various utilization processes and it needs to be realized through various development, protection and construction activities in the future development process. Therefore, Territory Spatial Planning is not just about creating a blueprint, but also about achieving spatial outcomes through the control and management of various development, protection, and construction activities. This also reveals that the essential nature of planning is the control of various spacial usage behaviors. In a broader sense, the protection, development, utilization, restoration, and governance of Territorial Space are all human activities involving the use of territorial space. When it comes to the future state of the constituent elements of Territorial Space, in the final, it is ultimately the result of human activities (i.e. protection, development, utilization, restoration, governance, etc.) that it takes shape. The core of planning lies in achieving it through regulated processes. In that case, the object of Territorial Spatial Planning and control is the use of Territorial Space, which regulates various types of spatial use, rather than just arranging for specific elements of territorial space, which is also the essence of the existence of Territorial Spatial Planning as a social institution. The constituent elements of various types of territorial space are the specific objects of all kinds of social subjects, that is, the objects of spatial usage. For territorial spatial planning, it serves as the foundation for regulating various spatial usage behaviors and determines the outcomes. From this perspective, the essence of Territory Spatial Planning lies in the management of territorial spatial changes, which serves as a guiding force through planning and provides a regulatory basis through declaration, ultimately achieving the desired transformations.[3]

Based on this understanding, from the perspective of planning theory, the knowledge content of Territory Spatial Planning includes three levels: The first is about the the nature of planning and how it is carried out. The second aspect pertains to the objects of planning, which means the objects and contents of the overall arrangement and control of Territory Spatial Planning, which involve the coordinated arrangement and management of national territorial space, as well as the various types of Territorial Spatial Planning uses and their interrelationships; The third focuses on the elements that constitute the utilization of territorial space, namely the constituent factors of national territorial space. Therefore, as the knowledge type that supports the development of Territory Spatial Planning, it can also be roughly divided into three parts, namely, the knowledge about planning, the knowledge about Territorial Space Usage, and the elements that constitute territorial space. Based on the substantive content of Territory Spatial Planning, these three parts of knowledge are closely related, but there are also differences in terms of the core and the extension. More importantly, there are knowledge contents from different disciplines and specialties, which are intertwined with each other in the process of planning so that throughout the whole process of planning, these different levels, sources and various types of knowledge form a complex network to jointly support the development of Territory Spatial Planning.[4,5]

For the sake of narrative clarity, the following discussion begins with the content of physical entities, then moves on to the knowledge related to the use of physical space, and concludes with a discussion of the planning itself.

2 Types of Knowledge About the Elements of Territorial Space

The concept of territorial space in its entirety refers to the overall space including territory, airspace and territorial sea within the national boundaries, including the utilized and unutilized land surface and the space above and below the water surface. In terms of its composition, it is a complex of different landforms, soils, rocks, hydrology, vegetation and buildings; If it is distinguished by surface projection and

[3] Tewdwr-Jones (2012).

[4] Innes and Bother (2010).

[5] Sanyal et al. (2012).

its collection elements, it can be broadly summarized as mountains, water, forests, fields, lakes, grasses, beaches, deserts, seas, islands, cities, towns, villages, etc.

According to the spatial characteristics formed by these elements, they can be broadly divided into three categories: Natural space (environment), artificial space (environment) and artificial-natural hybrid space (environment). Nowadays, it is common to see that "production space, living space, ecological space" or "ecological space, agricultural space, urban space" used to distinguish the composition of territorial space. However, this kind of differentiation is essentially based on the external appearance without distinguishing functionality and basic attributes.[6] Moreover, some terms may bring about ambiguity in usage. For example, ecological space, production space and living space should permeate all three categories mentioned above. In terms of pure human production and life are concerned, even in artificial space environments such as cities and towns, there are still ecological spaces and spaces that serve ecological functions.

Strictly speaking, there are not many purely natural environments left on earth, and many of those that can be described as natural have already been altered or affected by human activities. However, there still remains a substantial portion that has not undergone complete transformation. Therefore, this natural environment refers to the space environment that has not been created by human beings, has not been massively modified by human beings, and remains undisturbed by human activities, as well as to areas that still maintain their fundamental natural habitat characteristics; The human significance attributed to these natural environments is bestowed by human beings, but without the prerequisite of altering their inherent qualities. The term "artificial environment" refers to a spatial environment created by manpower, which is formed through human beings' modification of the natural environment, some of which still maintain their natural characteristics, such as mountains, rivers, lakes, forests, etc., but they have usually been intervened by manpower in some aspects or are constantly disturbed by human beings; the humanistic nature of the artificial environment is shaped by altering the natural form and creating new non-natural characteristics based on human intentions and the cultural significance. The concept of artificial-natural mixed environment refers to an environment that is created through artificial modifications, but its usage and operation still rely on natural characteristics or are predominantly characterized by natural growth. The extensive agricultural production regions clearly possess such characteristics: These spaces have undergone human cultivation and have changed their natural features, but the way of production still need to follow the natural rhythm and rely on the natural growth force to obtain harvest. Of course, human disturbance and intervention are uninterrupted, so this interaction between man and nature is the most remarkable feature of this kind of area, and humanistic nature of their environment is also expressed through this interaction, whether it is be rice fields, "green curtains," or fields of flowers.

[6] Wu Cifang et al. (2019).

From the perspective of knowledge sources for environmental cognition, from natural environment to the man-made environment, natural elements and artificial entity elements are studied by various natural science and engineering technology disciplines, such as geology, geophysics, oceanography, water conservancy, agronomy, forestry, biology, civil engineering, etc. These disciplines provide fundamental knowledge for understanding the real world; On the other hand, the relationship between the man-made environment elements and entity elements influenced or determined by social factors, a series of social science disciplines and engineering disciplines provide corresponding knowledge, such as economics, sociology, history, psychology, architecture, transportation, safety engineering, etc. Additionally, there are some disciplines, such as geography, environmental science, ecology, surveying and mapping, which combine natural and artificial elements, cover the natural environment to the whole artificial environment, and provide foundational knowledge for understanding the spatial aspects of our land.

In terms of understanding the various constituent elements of territorial space, these disciplines have the main task of understanding real phenomena and exploring the mechanisms of things themselves. They aim to reveal the inherent laws of things by employing a clear conceptual system and consistent principles of interpretation. However, the same object often possesses different attributes and facets, so it has become the subject of research for multiple disciplines. For instance, the study of soil encompasses various disciplines such as soil physics, soil chemistry, soil biology, and soil geography, which are often carried out according to the disciplinary perspectives, problem orientation, conceptual framework with methods of physics, chemistry, biology, and geography. Similarly, there are subfields such as urban economics in economics, urban sociology in sociology, urban geography in geography, urban politics in political science, and so on. This also demonstrates that the distinction between disciplines is not determined by research objects but rather by discipline traditions formed by research problems, methods, and conceptual systems. Each discipline investigates a specific aspect of objective phenomena, and although these aspects are all interconnected within the same object, the research outcomes of these disciplines do not adhere to the same logic framework. This is what is referred to as the "incommensurability" of knowledge in the philosophy of science.[7,8]

When it comes to the physical composition of space, it is often composed of multiple elements. To understand the characteristics and evolutionary features of these elements within the space, even without considering human transformation, intervention, and usage activities, merely gaining knowledge about the objective objects themselves would still constitute a vast web of knowledge. For example, in the case of a particular type of geomorphic region, it may be related to geological structures, stratigraphy, soil composition, vegetation, climate, biodiversity, hydrology, minerals, etc., and these factors interact and mutually influence each other. Therefore, in the process of practice, it is necessary to transform and synthesize various types of knowledge that originate from different disciplines and lack compatibility

---

[7] Kahn et al. (2009).

[8] Cahn et al. (2003).

when dealing with the work of the same object. This will allow the various types of knowledge to intertwine and work together, rather than simply being pieced together.[9,10,11,12]

3 Types of Knowledge About the Utilization of Territorial Space

The utilization of territorial space refers to the actions and methods through which human beings transform and utilize the natural elements of the land. The current pattern of territorial space is not only an objective existence but also a consequence of human utilization of the land in the past. All utilization of space is a purposeful act by human beings, driven by their own needs, values, and ability to transform the natural environment. Towns, agricultural and rural areas, etc., are all the outcomes of human beings transforming nature to varying degrees and in different ways, continuously refining their utilization methods. Even in terms of the protection of the natural environment, it is still a way for human to utilize natural space. The ability of forests, grasslands, deserts, mountains, rivers, lakes, and uninhabited islands to persist in a relatively natural state is also a result of active or passive choices made by humans. From this point of view, human spatial behavior includes protection, development, utilization, restoration, governance and other main types. Each type involves different ways of utilizing the various components of the national space.

The use of space is usually expressed through land use or land utilization.[13] After all, land is the carrier of all human activities and the substrate of space, and space use is based on land use. However, there are still some distinctions between the two. Land use often refers to the concrete entities carried on the plot, emphasizing the use of things and their material existence; Space use reflects the ways in which people utilize and behave within a space, acting as a manifestation of human spatial activities. From another perspective, land use emphasizes functionality, while space utilization more emphasis on relationship, much like the distinction between "residing" and "dwelling". In the system of Territory Spatial Planning, the overall planning level should be the main focus at the organization of space use, while the detailed planning level should implement space use to the arrangement of land use; In terms of the use

---

[9] Wilson (2016).

[10] Wilson (1999).

[11] Gibbons et al. (2011).

[12] Gibbons et al. (1994).

[13] The basic meanings of "land use" and "land utilization" used in different plans should be consistent. The reason why there is such a distinction and it is solidified separately stems from the past departmental consciousness and the regulations of different planning categories. Linguistically speaking, "use" is positive, while "utilization" has the meaning of "being used", which usually has some derogatory meanings in Chinese; In addition, "utilization" has the meaning of one-time effect or single use, while "use" can be continuous and changing, that is, different uses. Therefore, "land use" and "space use" should be used in the context of "multi-regulation integration".

control, the regional use control should be based on space use (the spatial relationship of the realization of the main function should be controlled, instead of formulating a large number of access rules by function categories), and the planning permission can take land use as the control basis, but the classification of land use should be carried out according to the characteristics of space use in order to meet the requirement of land use compatibility.[14,15,16]

Due to the fact that space use is a human behavior and a purposeful activity, national spatial planning is not only concerned with the physical state of national space. These physical states are the result of deliberate and conscious choices made by humans in their interactions with natural space. Therefore, in Territory Spatial Planning, it is necessary to study a wide range of knowledge about human beings, including various types of needs, ways of activities, production and lifestyle, as well as their relationship with space and the ability and methods to change space usage. This involves understanding the spatial needs of individuals, families, social groups, organizations, and the whole society and various kinds of knowledge related to spatial decision-making, which derived from the research results of various social sciences, humanities, and natural sciences such as physiology and anthropology. Human behavior and social development have their own patterns, but it is evident that they are not natural laws. Instead, they are general characteristics formed under influence of human agency, collective cultural consciousness, social norms, and other factors. These general characteristics possess a certain level of objectivity, which serves as the foundation for a series of social science studies.[17,18] In the use of space, there will also be the need to change and transform space in order to meet the needs of space use activities, which involves the knowledge of natural conditions and the corresponding engineering requirements.[19,20] This section will continue the discussion in relation to space change management in the following content.

Any use of space has externalities, both positive and negative. The utilization of a piece of land, regardless of its purpose or manner of use, is not solely determined by its own capabilities but rather requires assessment in relation to its surrounding environment. This is known as spatial relationship, which should be a key focus in planning. In terms of various applications, it also relies on the externalities resulting from the use in the surrounding areas. However, different types of usage methods, require different conditions and approaches, as they face different spatial objects. When it comes to

[14] In the past, many land use classifications were divided according to management institutions or management methods, rather than according to the characteristics of space use and its external effects. For example, hotel land in cities was included in commercial land instead of residential land.

[15] Silva et al. (2016).

[16] Silva et al. (2015).

[17] Kahn et al. (2009).

[18] Cahn et al. (2003).

[19] Burke et al. (2009).

[20] Berke et al. (2006).

any space use mode of protection, development, utilization, restoration, and governance, it is a systematic project undertaking closely associated with areas beyond the specific region. Even the protection of virgin forests involves the prevention and control of various surrounding pollution and its impact, such as air pollution, acid rain, groundwater level drops caused by sewage inflow or over-extraction of groundwater from the outside, alien species and pests, etc. For ecological restoration, cultivated land restoration, land administration, watershed management, etc., not only their own restoration and management requirements and methods are different but also their external requirements and impacts, even more.

In terms of development and utilization, it is not only the changes of geomorphology and features but also the change of the relationship with the outside world. For example, the cultivation of agricultural land or the construction on farmland involves the introduction of irrigation water for farm land, the introduction of infrastructure pipelines and transportation for urban construction, and the relationship between farming activities and the use of built-up space and various types of production and living activities will also be considered. It can be inferred from this that all kinds of use activities are not only determined by the assessment of natural conditions, but also involve people's purposes, wishes, and specific use behaviors, and also involve specific operations. It is a complete process from planning to actual landing, in which a lot of work, especially engineering work, can achieve better use purposes. These works are carried out by relevant decision makers, project undertakers, or users, but their external needs and possible external effects in the project process are often less taken into account. This needs to be carefully considered and monitored in the process of Territory Spatial Planning.

It is worth noting that space utilization activities are carried out by specific users rather than by planners or planning management institutions. Some changes to the space are commissioned to other institutions or acquired and used through lease or property rights conversion after completion. Moreover, many activities themselves are managed by other public departments.[21] For the implementation of these activities and management, the owners, users, builders, or managers of physical space activities have their own starting points, concerns, and logical systems for thinking about these problems. Therefore, their utilization of space is influenced by their individual perspectives and values. In terms of planning, decision-making, and implementation of different space use changes, also involve the organizational ability and mode of society and groups, national and local laws and regulations, engineering technology, etc. This kind of knowledge is not only distributed across various natural and social sciences, engineering technology and related applied disciplines but also integrated into the practice processes of specific contexts.

Therefore, it is evidently impractical to have a thorough understanding of all spatial arrangement organizations and make clear arrangements on a case-by-case basis in planning. All spatial organizations are strategiclly and directionally arranged based on aspirations and certain values and the determination of specific space use needs to be reevaluated and reweighed during implementation, which is also the basic

[21] Sun Shiwen (2000).

concept that Territory Spatial Planning must have. Regarding the arrangement of space utilization, not to mentioned the anticipatory arrangement, even the empirical research afterward can only be analyzed and deduced according to the review of the implementation results and the process. It is difficult to fully capture the original intentions of individuals and institutions, their understanding at that time, and the specific strategy adjustment, etc. This is because all initial understanding needs to compromise in coordinating the relationship between all parties and bargaining with those involved in the current situation, which is an important feature of practical activities.[22,23]

In this regard, the compilers of Territory Spatial Planning should be fully aware of the uncertainty of the future, and the planning cannot determine the body of language and value judgment of future spacial users who are a different group of people. And compared with the planners, they are people in different times and groups, so the arrangement in the planning results should be quite inclusive; Managers of Territory Spatial Planning should also fully realize that planning implementation management is also an important transformation process, and the planning results determine the basic direction. The specific use still needs strategic evaluation and comprehensive trade-off, rather than just a process of following the map and applying for a permit according to the map. In summary, it is essential to establish a planning concept and institutional framework that covers the entire process from planning formulation to planning implementation.[24]

4 Types of Knowledge Related to Planning

Planning is the concentrated expression of human ability to anticipate and organize future events. Territory Spatial Planning arranges and organizes the future development of territorial space, which is based on the spatial usage behavior of various social groups and achieved by the change in the use of territorial space. Therefore, the essence of Territory Spatial Planning is to manage the changes in the future use of territorial space.[25]

Under modern scientific conditions, the use of Territorial Space resources is characterized by its versatility, but the total amount of territorial space resources is fixed, and the expansion of built-up space will inevitably encroach on natural space or agricultural production space, which will affect the supply of fresh air, water, food and the purification of waste; Conversely, if the built-up space is too small, it will result in unmet human living needs and a decline in quality of life. Similarly, in urban areas, with the improvement of living standards, people want to increase the residential area, which requires increasing the residential land. Under the condition that other

[22] Baldwin et al. (2017).

[23] Baldwin et al. (2010).

[24] Morphet (2011).

[25] Tewdwr-Jones (2012).

land remains unchanged, the city needs to expand and occupy more natural space or agricultural production space; In order to ensure that the natural space and agricultural production space do not change and the city does not expand, it is necessary to reduce the land use of other facilities in the city, which will lead to insufficient allocation of land and facilities such as schools, offices, infrastructure and roads, and will also affect the overall quality of production and life. If the total amount of residential land is reduced as a result, it will lead to an increase in housing prices and a decrease in the quality of life under the same income.

In summary, the essence of resource allocation lies in striking a balance between the demands for different types of space utilization, thus addressing the issue of competing spatial needs. In this regard, the allocation of territorial space use resources is based on the output capacity of various resources and the allocation and combination according to the future development goals. The output capacity of resource use is not only economic (such as GDP or per capita output value). It also encompasses social and environmental outputs. What is more important is to take into account all kinds of factors that adapt to the future society, people's survival, and activities, such as oxygen, clean water, food, energy production, waste and pollution purification, etc. Of course, it also includes the needs of human life, production, and recreation. Fundamentally, it is about fulfilling a wide range of needs for the future survival, security, enjoyment, and development of humanity.

Resource allocation is based on the scientific and rational relationships between different spatial uses, and the relationships between various types of resource use need to be continuously adjusted as needs, goals and output capacities change, which requires continuous support from corresponding socio-economic and ecological knowledge. Meanwhile, due to the influence of regional scale and the existing pattern of Territorial Space, it is difficult to coordinate in a closed way in a small region, which makes it necessary to establish a system of allocation from state to local, from region to city and from top to bottom. Of course, in small regions or cities, there needs to be corresponding equilibrium, but it is definitely not self-sufficient local equilibrium. This also requires corresponding ways and means of resource allocation of Territorial Space use based on human survival and healthy development, which are compatible with the local native environment.[26] However, the allocation of resources based on countries and large regions will inevitably lead to differentiation of interests, especially for regions that provide a large number of public goods outputs, thus there may be issues of regional fairness that need to be addressed through the establishment of specialized systems.

As demands and objectives evolve, along with technological advancements, there is a continuous need to harmonize the relationship between various spatial uses and undertake purposeful reorganization in order to achieve high-quality development and a high standard of living. Of course, this kind of tuning and reorganization is not a holistic reconstruction, but an adjustment and perfection based on the existing pattern, but there is also the possibility and need for reconstruction in some areas

[26] Urban Planning Society of China (2020).

and some aspects.[27] Changes of territorial space use is always taking place, and any change may lead the change of surrounding areas or produce the demand for change, which is determined by the externality of space use. There are positive and negative externalities, and for each type of use, users pay attention to the influence of peripheral use on their use, which is also the basic content considered in their use decision-making; However, the impact of this use on the surrounding areas is usually less considered by users, and both positive and negative are not considered, while the surrounding areas just change like shocks under its positive or negative external effects. Planners should pay focus on this kind of external effect, pay attention to the external effect of different uses, and combine various uses on this basis to coordinate the relationship between them, from the macro-regional scale to the block scale. The use of any Territorial Space is not determined solely by the space itself, but rather by the external association, and one could even say the use of surrounding areas.

Using coordinated land and marine development as an example, when using different space resources, the decision-making basis will be chosen from the aspects of scarcity, strategic significance, and the difficulty of creating similar space types. Therefore, the essence of prioritizing land based on sea is to give prominence to the comparison between marine and terrestrial resources. It is necessary to determine the use of land and coast from the perspective of marine use. For example, in the protected zones of specific marine resources, fishery resources, mangroves, wetlands, etc., no construction projects should be arranged, either should the leisure and entertainment facilities to attract people be set up. Even if favorable conditions exist in terms of water depth and coastal infrastructure, large-tonnage docks, industrial projects using the sea, or relying on marine transportation should not be set up, because these uses may bring interference or even harm to marine resources and bring marine pollution. Similarly, according to the distribution of marine use functions, routes, anchorages, and water depth conditions, it is possible to determine the areas where container terminals can be established. These areas may have unsatisfactory onshore construction conditions. On the basis of regional port layout, only such areas can be selected for construction without other choices, instead of areas that may have negative impacts on marine ecological space.

Similarly, different organizational principles need to be considered for different space uses. For example, in natural environments, it is crucial to fully consider the natural habitat requirements of different biological communities; in rural areas, it is necessary to consider the agricultural production mode and its matching with living space, and in urban areas, it is even more necessary to consider the contact requirements of human activities. When dealing with the interrelationships, it is necessary to analyze the use of specific arrangements. However, it is clear that spatial differentiation can never be solved by drawing a line.

For example, the juxtaposition of permanent prime farmland and high-tech development zone can never be separated by a red line, a ditch, a road, or even a physical fence. In fact, in actual use, they have mutual interference, which is not only the problem of space expansion between them, but the use of either party may have

[27] Sun Shiwen (2020).

adverse effects on the other party. Similarly, if permanent prime farmland is not effectively integrated with villages, it will also impact the practical utilization and production efficiency of the farmland. In protected areas, the setting of fences not only isolates the interaction between environments but also interferes with the exchange of species and the migration of animals. In cities, because the population and activities are more concentrated, this mutual influence is more significant, which is also an important reason why spatial planning, both at home and abroad, starts from cities first. Not to mention the relationship between industry and residence, residence and office, and setting up facilities of Nimbyism, even residential areas and hospitals that are both living facilities, even if they are separated by an urban road layout, they will bring a lot of troubles to residents' lives; Although secondary schools, primary schools, kindergarten and nursery facilities are all cultural and educational facilities, the adjacent layouts will interfere with each other; Even the adjacent layout of schools and residential areas will cause interference. When these facilities need to be arranged adjacent to each other, it is necessary to analyze the actual use situation, make combined arrangements, and revise and improve the arrangement of entrances and exits and the layout of buildings.

Changes in space usage are relative to the current situation, so it is necessary we need to pay attention to the influence and problems brought by the change of reality and also need to pay attention to the possible influence on the actual use after the change. This includes not only the reduction and compensation of agricultural land caused by the transformation of agricultural land into urban construction land, but also the changes in agricultural production units, the adjustment to drainage in agricultural areas, the matching degree between rural production and life, and the impact of pollution caused by urban land on agricultural production. Similarly, reclaiming woodland or barren hills into cultivated land will not only change the topography and vegetation conditions but also change the microclimate, groundwater, and even geological conditions of the region because of the water introduced by agricultural production. The increase in human activities and chemical fertilizers used in crop production will interfere with the surrounding woodland or protected areas. However, measures such as Converting Farmland to Forest and Sea will have the need to adjust the farm land system and ditch water flow, and will also interfere with and influence forest and sea areas that have undergone ecological adaptation for many years.

In the city, not only will new construction change the spatial structure of the city or a specific area of the city, but even the urban renewal, such as renewing the original residential area into a commercial area, or renewing the original residential-commercial area for a purely commercial area or a residential area, will have an impact on the indigenous residents on the plot. The introduction of new residents, facilities, and users has also altered the life trajectories of citizens in other areas who pass through this place and use the facilities in this area on a daily basis, resulting in the reshaping of the spatial structure of cities or specific areas. For this reason, in the process of planning and implementation, it is necessary to fully study these changes and evaluate the possible impacts of these changes in order to better organize the space use. In a certain way, it is precisely because of such external impacts that advance planning studies and arrangements and controls are needed. Certainly, to change the status quo and reshape the interrelationships between various types of space use, it is necessary to have sufficient reasons, consider from the perspective of the common well-being of the whole region, give full play to the external positive effects, reduce or eliminate the external negative effects for other people, other communities, other species, and adopt feasible ways and means to minimize the inappropriateness and adjustment period of the transformation process.

It is important to emphasized that during the planning implementation and planning permission for use control, it is not only a matter of check whether the land uses nature and development intensity of the project to be implemented are consistent with the planning map, but also to make an actual evaluation and comprehensive decision on these contents. The results of planning preparation serve as guidance for the intentions and direction under uncertain circumstances in the future. However, planning implementation management and planning permission are conducted under relatively certain conditions regarding the project and its surroundings. Therefore, it is neither appropriate nor advisable to simply and mechanically apply the results of planning preparation as the sole basis for decision-making.

5 Knowledge Ontology and Integration of Various Types of Knowledge in Territory Spatial Planning

Essentially, Territory Spatial Planning is a public intervention in the development and protection of territorial space based on a specific governance structure. The purpose of this intervention is to solve the existing problems and ensure the realization of future goals. Therefore, it aims to change the existing conditions or existing usage patterns and promote the occurrence of desired changes through the allocation of land and spatial resources and the adjustment of various spatial relationships. It also aims to control undesirable changes that are inconsistent with the intended objectives or intentions. Whether it is to solve the existing problems or achieve future goals, it is necessary to make the future development of territorial space higher quality and higher quality than the present situation, which can only be achieved through certain human efforts, and these spatial changes will be carried out in the future and have not yet happened. Therefore, it is very important to recognize the future and future changes. Most human knowledge is derived from past experience, and there are a lot of judgments and choices accompanying values, which are a process of unifying world outlook, epistemology, and methodology, and the result of planning is the product of continuous tuning in this process.[28]

Equally important is the fact that planning is a public intervention, conducted in response to decentralized decisions about space use. Therefore, in addition to the above-mentioned understanding of the cause-and-effect relationship and logic of space use, the understanding and judgment of the preconditions for intervention

[28] Weberr Crane (2012).

implementation, such as system, policy, and regulation, the mechanism of the implementation process, and the external effects of possible consequences, directly determine the effectiveness of planning. This is the characteristic shared by all public policies. The implementation of public policy is not only related to the effectiveness of policy implementation but also an important determinant of policy content, including policy objectives and measures. Therefore, the content and process of Territory Spatial Planning should be genuinely integrated into the goal and process of governance.

Planning is evidently more than just setting of goals or the determination of the future goals. It is not simply about sketching a vision of future goals. It is important to achieve goals or provide a basis for the realization of goals through the use of national space, that is to say, to establish an action program to achieve goals and put them into practice continuously. Planning objectives play an pivotal role in the formulation and implementation of planning, serving as the foundation for all arrangements and decisions. The development of various territorial space elements and the organization and management of various space uses need to be evaluated and weighed according to future development objectives before appropriate decisions can be made. As for Territory Spatial Planning, which is a work involving multiple disciplines and specialties, there are great differences in the concerns, conceptual systems, cognitive methods, and logical basis of disciplines and specialties, so there will be different understandings of territorial space resources and elements, the space use, possible results, the weight of their relations and the priority of their arrangements. It is normal for different requirements and contradictions to arise in the process of mutual cooperation and cooperation. To solve these contradictions, it is necessary to judge and reach a consensus under a unified goal, which is an inevitable way for planning to achieve a unified whole.

Of course, the planning goal itself serves as both the decomposition of the higher-level planning goal, but also the component of the action strategy for implementing the upper-level planning goal, and the result of the combination of the implementation of the upper-level strategic requirements and the action direction of this level.[29] Therefore, the determination of the planning goal is the synthesis of future forecast, strategic research, social vision, and upper-level planning arrangements. This means that no matter whether in terms of socio-economic environment or in the protection, development, utilization, restoration, and governance of territorial space, the planning goal is formed by the results of compound prediction research, countermeasure research, and practical work and their interaction. Furthermore, the results of planning goal formulation are to reach a social consensus, which is an important political process. Therefore, social mobilization and social work are also very important work foundations. Similarly, the requirements of human beings for space use or the goals of space use are diverse, and there is no inevitable logical relationship between these diverse requirements.[30] Take building a house as an example: it should be safe and sturdy, economical and applicable, and beautiful. There are differences in values in

[29] Zhao Min (2019).

[30] Xu Changfu (2002).

every project requirement, and these requirements are parallel, so it is impossible to deduce another requirement from the quality of one requirement; At the same time, they can't be superimposed, which is not determined by the required proportion. In the process of practice, with the participation of different groups, specific trade-offs should be made according to specific target requirements. It is necessary to establish a set of evaluation methods, comprehensively evaluate from the perspective of the overall well-being of society, and make choices according to the target requirements.[31,32,33] This point will be further magnified during the planning process, not only in the specific space use but also among the objectives of the whole planning. It can even be said that there will be a key problem in all levels and aspects of the planning: Any choice and decision must simultaneously satisfy these parallel and discrete goals at the same time. This is a problem that planning, as a practical activity actually faces and must be addressed.

Therefore, a significant amount of comprehensive knowledge about planning comes from the analysis of specific space usage and the process of solving arising issues. Historical experience and precedents can provide partial knowledge, but in the solution of specific problems, they often only play a leading role in defining and solving problems, and the actual application of the process still requires constant revision, enrichment and improvement. That is why a single discipline can't provide comprehensive knowledge, it needs multi-disciplinary cooperation, and there is no fixed mode and general way of this cooperation. Even the combination mode of different disciplines needs to be determined according to the analysis of problems in specific scenarios, and knowledge needs to be established and improved constantly during the implementation of the Territory Spatial Planning system. In the realm of Territory Spatial Planning, the demand and decision-making of territorial space use come from multi-type, multi-group, and multi-dimensional considerations, the outcomes and impacts of various territorial space uses are also related to society, economy, environment, ecology, and other aspects, and multidisciplinary collaborative work is particularly important. As a platform for collaborative work, it is necessary to have a professional perspective and basic consensus, which is based on common values, consensus on future development goals, basic understanding of good territorial space patterns and territorial space use, so as to stand on different disciplinary positions, use their own disciplinary knowledge and analysis tools, and reach a comprehensive solution through common consultation and perspective taking.[34] The establishment of this process requires each discipline to analyze the internal and external conditions and requirements of various space uses, as well as the ways and results of quality improvement and its results, and to integrate the understanding and development prospects of natural laws and social laws, forecast and the application conditions of various laws, in order to establish the evaluation criteria of interaction and its results, which is also a multidisciplinary collaborative process.

---

[31] Dali and Cobb (2017).

[32] Daly He and Cobb (1994).

[33] Urban Planning Society of China (2020).

[34] Gu Chaolin et al. (2019).

Mark Tewdwr-Jones believes that the planning work consists of four components: a regulatory process, a strategic assessment, a governance framework, and a future project. In the entire planning process, regulation is the most obvious power and approach in terms of external influence, as it involves the management of various development, protection, and construction activities. The scope and procedure of regulation implementation are defined and authorized by laws and regulations; The substantive content of regulation implementation is also defined by laws and regulations and specified by statutory planning. In the substantive content of planning regulation, resource allocation, and structural combination are carried out through strategic evaluation, and the management and control system is set in combination with the management and control powers at all levels, so as to form the future space use arrangement and program of action, which is implemented as a social contract. As mentioned earlier, the essence of Territory Spatial Planning is planning, and the direct object of planning is the use of territorial space and its changes, which are formed based on people's actions. Therefore, for the work of planning, the knowledge of organizing and controlling purposeful human behavior is at its core. It involves not only laws, regulations, and administration but also the comprehensive coordination among various use activities under the condition of the unity of space elements and use methods. Policy objectives, the logic of the relationship among various space use activities, the comprehensive evaluation under the goal orientation, and the coordination process of social organizations are the knowledge basis of this coordination.[35,36]

To understand the ways and trends of land use, it is essential to have a comprehensive understanding of the spatial use needs, wishes, implementation capabilities, and possible consequences and external effects of its use from various groups and individuals; It is also necessary to recognize the conditions under which these space use behaviors can be carried out, which requires a full understanding of the engineering and technical aspects related to the status and changes of various spaces, and the knowledge base of Territory Spatial Planning is thus integrated hierarchically into a unified whole.

The institutional construction of Territorial Spatial Planning is still in its early stages, and the preparation of plans is being promoted in an experimental manner, with exploration being the main theme of this time period. In this process, the previously isolated elements of development and preservation, natural environment and artificial environment, urban construction, and non-urban construction are placed in the continuum from the natural environment to the artificial environment for comprehensive deployment and arrangement, thus constructing a new planning system, and all kinds of knowledge used in planning are faced with the need of adjusting and reorganizing their mutual relations. In the process of this structure, the first thing that needs to be cleaned up is the levels of various kinds of knowledge. This paper puts forward the types of knowledge and characteristics of planning ontology, planning object, and three levels as the basis of planning object, thus providing a discussion

[35] Heineman et al. (2011).

[36] Heinbman et al. (2001).

basis for the comprehensive knowledge system framework of Territory Spatial Planning and the operation of Territory Spatial Planning system. The application and production of planning knowledge entail a strong practical thinking and reactive learning from the past process, which requires that interdisciplinary collaborative research should be carried out to solve specific problems continuously in the planning process so that all kinds of professional knowledge can cooperate with each other in the same planning and produce new knowledge contents and types.[37] This also requires the organizers and operators of planning and implementation to recognize the hierarchical nature of the planning process and the various dimensions of planning problems. It is important to build a competent workforce that encompasses key knowledge domains and to integrate the decision-making process within a comprehensive workflow.

References

Baldwin R, Cave M, & Rocky. (2017). *Oxford Regulation Manual*. Song Hualin, Li, & An Yongkang, et al., translated. Shanghai: Shanghai Joint Publishing Company.

Baldwin R, Cave M, & Lodge M. (2010). *The Oxford handbook of regulation*. New York: Oxford University Press.

Berke PR, Godschalk DR, Kaiser EJ, et al. (2006). *Urban land use planning*. 5th ed. Urbana: University of Illinois Press.

Burke, Goldshack, & Kaiser, et al. (2009). *Urban land use planning*. Wu Zhiqiang, translated. Beijing: China Architecture & Building Press.

Cahn SM, Turner SP, & Roth PA. (2003). *The blackwell guide to the philosophy of the social science*. London: Wiley-Blackwell.

Dali, Cobb. (2017). For the common good: Reshaping the economy for community, environment and sust*ainable future*. Wang Jun, Han Dongjun, translated. Beijing: Central Compilation and Translation Publishing House.

Daly He, Jr. Cobb JB. (1994). *For the common good: Redirecting the economy tower community, the environment, and a sustainable future*. Boston: Beacon Press.

Gibbons M, Limoges C, & Nowotny H, et al. (1994). *The new production of knowledge: the dynamics of science and research in contemporary societies*. London: SAGE Publications Ltd.

Gibbons, Limoges, & Novotzman, et al. (2011). *New Mode of Knowledge Production: Dynamics of contemporary social sciences and research*. Chen Hongjie, Shen Wenqin, et al., translated. Beijing: Peking University Press.

Gu Chaolin, Wu Tinghai, & Liu Wan. (2019). *Classics of Territory Spatial Planning*. Beijing: Commercial Press.

Heineman, Bloom, & Peterson, et al. (2011). *The World of Policy Analysts: Reason, Values and Politics*. Li Lingling, translated. Beijing: Peking University Press.

Heinbman RA, Bluhm WT, & Peterson SA, et al. (2001). *The world of the policy analysis: rationality, values, and politics*. 3rd ed. Thousand Oaks, CA: CQ Press.

Innes JE, Bother DE. (2010). *Planning with complexity: an introduction to collaborative rationality for public policy*. London and New York: Routledge.

Kahn, Turner, & Roth, *Philosophy of Social Sciences*. (2009). Yang Fubin, translated. Beijing: Beijing: China Remin University Press.

Morphet J. (2011). *Effective practice in spatial planning*. New York: Routledge.

[37] Gibbons et al. (2011).

Pan Haixia, Zhao Min. (2020). *Some Discrimination and Technical Difficulties on the Construction of Territory Spatial Planning System.* Urban Planning Forum, (01):17–22.
Sanyal B, Vale LJ, & Rosan CD. (2012). *Planning ideas that matter: livability, territoriality, governance, and reflective practice.* London: The MIT Press.
Silva EA, Healey YP, Harris N, et al. (2015). *The Routledge handbook of planning research methods.* London: Routledge.
Silva, Healy, & Harris, et al. (2016). *Handbook of planning research methods.* Gu Chaolin, Tian Li, & Wang Shifu, et al., translated. Beijing: China Building Industry Press.
Sun Shiwen. (2000). *Basic research on the implementation of urban planning.* Urban Planning, (07): 12–16.
Sun Shiwen. (2020). *From urban and rural planning to land spatial planning.* Urban Planning Forum, (04): 11–17.
Tewdwr-Jones M. (2012). *Spatial planning and governance: understanding UK planning.* Basingstoke and New York: Palgrave.
Urban Planning Society of China. (2020). *Technical roadmap of urban and rural planning discipline.* Beijing: China Science and Technology Press.
Wang Xinzhe. (2019). *Position and function of master plan of territory and space in prefecture-level cities.* Urban Planning Forum, (04): 31–36.
Weberr, Crane R. (2012). *The oxford handbook of urban planning.* New York: Oxford University Press.
Wilson. (2016). *Great integration of knowledge: Science and Humanities in the 21st Century.* Liang Jin, translated. Beijing: CITIC Publishing Group.
Wilson EO. (1999). *Consilience: the unity of knowledge.* London: Vintage.
Wu Cifang, Ye Yanmei, & Wu Yuzhe, et al. (2019). *Territory Spatial Planning.* Beijing: Geological Publishing House.
Xu Changfu. (2002). *Theoretical thinking and engineering thinking: The arrogation and demarcation of two ways of thinking.* Shanghai: Shanghai People's Publishing House.
Zhao Min. (2019). *Discussion on the logic and operation strategy of the construction of Territory Spatial Planning system.* Urban Planning Forum, (04): 8–15.

Topics and Prospects of Spatial Planning Reform: Thoughts on Planning and Discipline Development

Zhang Shangwu

Several Opinions on Establishing the National Territorial Spatial Planning System and Supervising Its Implementation by CPC Central Committee and State Council (hereinafter referred to as the "*Several Opinions*") was officially promulgated in May 2019, and the curtain of spatial planning reform has been opened. It is the basic requirement of this reform to integrate spatial planning such as main functional area planning, land use planning and urban and rural planning into a unified "territorial spatial planning system" and realize "multi-planning integration". Its great significance stems from its role in improving the institutional framework for territorial spatial development and protection. This encompasses strategic, scientific, authoritative, coordinated, and operational aspects, which in turn promote the modernization of space governance system and governance capacity. Furthermore, it facilitates the transformation of approaches to territorial spatial development and protection methods and leads to higher quality, greater efficiency, increased fairness, and more sustainable land development.

The establishment of territorial spatial planning is the top-level design of territorial space development and a major planning system reform promoted from top to bottom at the national level, emphasizing the basic role of territorial spatial planning in the national planning system; It reflects the structural reform requirements of the spatial supply side in the new era of ecological civilization, and the systematic, holistic and restructuring in accordance with the requirements of the modernization of national spatial governance[1]; It involves the reconsideration of the role of planning, the reconstruction of planning system, the reorganization of administrative institutions and other aspects.

[1] Zhuang Shaoqin (2018).

This paper was published in Urban Planning Forum (4, 2019).

Z. Shangwu (✉)
College of Architecture and Urban Planning, Tongji University, Shanghai, China
e-mail: zhshangwu@tongji.edu.cn

Z. Peng (ed.), *Theories and Frontiers in Territorial Spatial Planning*, Urban Sustainability, https://doi.org/10.1007/978-981-96-9825-7_5

The establishment of territorial spatial planning system will have a far-reaching impact on the planning practice and the development of urban and rural planning discipline.[2,3,4,5] At present, it is necessary to clarify the basic requirements, tasks and internal logic of the reform. In particular, as a technical undertaking, it is important to understand the objectives and crucial elements of planning in the context of reform. It is equally important to correctly manage the interplay between the "foundational role" and "strategic planning" aspects of planning, as well as to identify the relationship between spatial planning reform and the development of the urban planning discipline.

This paper argues that spatial planning reform should be understood in the macro environment of China's Reform and Opening-up and in the context of Chinese planning, and discusses spatial planning from the overarching themes of planning reform, the topics of planning reform, the development trend of spatial planning and the prospect of discipline development.

1 The Origin and Intrinsic Logic of Spatial Planning Reform

China's 40 years of reform and opening up is also a process of continuous development and evolution of various spatial plans. *Several Opinions* pointed out that "all kinds of spatial planning at all levels have played an active role in supporting the rapid development of urbanization and promoting the rational use and effective protection of territorial space, but there are also many problems, such as too many types of planning, overlapping and conflicting contents, complicated approval process, too long cycle, and constantly changing local planning". The requirements and intrinsic logic of spatial planning reform should be understood under the macro background of 40 years of reform and opening up.

1.1 Understanding the Overarching Themes of Spatial Planning Reform

(1) Urbanization transformation: scale, quality and challenges of spatial governance

The first main line is urbanization transformation. A basic characteristic of urbanization is its periodic transformations, which will inevitably bring problems and challenges of spatial management. On the one hand, urbanization, as a spatial

[2] Editorial Department of this Journal (2019).

[3] Sun Anjun (2018).

[4] Wu Weijia et al. (2019).

[5] Zhang Tingwei (2019).

phenomenon of unbalanced development and a more efficient economic form, is an essential path towards modernization, and will go through a process of speed change from slow to fast to relatively stable. On the other hand, urbanization is also a double-edged sword. While it promotes economic growth, it deepens the accumulated contradictions behind it. With regards to cross-cultural experiences, it can be noted that at the point of urbanization rates approaching 50% to 70%, a shift from quantitative expansion to qualitative improvement takes place. This transformation is not merely confined to the realm of economic growth drivers, but rather, it initiates various changes in social conditions and social governance issues. Wu Zhiqiang and others called it the transformation from "physical urbanization" to "intellectual urbanization"[6].

China pays great attention to urbanization especially after entering the twenty-first century. "During the Tenth Five-Year Plan period, urbanization was brought into the medium and long-term plan of national science and technology development". The Eleventh Five-Year Plan brings new urbanization into the national development strategy for the first time. In 2013, the Central Urbanization Work Conference was held, proposing to strengthen macro-management of urbanization in accordance with the law. In 2014, the *National Plan on Urbanization (2014–2020)* was promulgated, which clarified the connotation and basic principles of new urbanization, and put forward a people-oriented path of new urbanization with Chinese characteristics, emphasizing four modernization (industrialization, informatization, urbanization, and agricultural modernization), optimized spatial layouts, ecological civilization and cultural inheritance. The Central Urban Work Conference was held in 2015, which proposed that urban development should be people-centered, embodying five development concepts and the requirements of "one respect and five overall plans", which is a comprehensive exposition of urban governance capacity and modernization of governance system. In the report of the 19th National Congress of the Communist Party of China in 2017, it was comprehensively put forward that China's social and economic development stage is undergoing historic changes, and it is necessary to shift from high-speed development to high-quality development, and clarify the contents and requirements of "economic construction, political construction, social construction, cultural construction and ecological civilization construction" in the new era. From the perspective of urbanization development stage, adapting to the requirements of shifting from high-speed to high-quality development and improving spatial management capacity have become the key to China's modernization and sustainable development.

(2) Market transition: Contradiction between planning, project and multi-planning integration

The second main line is market transition. Reform and opening-up brings about market-oriented transition, which is also a process of changing from "plan" to "planning". With the increasing status and role of planning, the planning operation system adapted to the market-oriented environment has not been effectively established, and

[6] Wu Zhiqiang et al. (2015).

the contradiction of multi-planning integration comes into being. This process first appeared in the field of urban planning and construction in the 1980s, then land use planning, and expanded to development planning after 2000.

The "Eleventh Five-Year Plan" changed the past "development plan" into "development planning" for the first time. The "plan" has changed to "planning", and "space" is an important means to deal with the uncertainty of development. The National Development and Reform Commission took the lead in drawing up the "National Main Functional Area Plan" with spatial significance, and issued a series of "Economic Zone Plans" around 2010.

Space plays a fundamental role in development and protection, but the overlapping and conflict of various planning spaces are becoming increasingly apparent, not only because of the lack of overall planning, but also the differences in logic, basis, technology and methods of planning. Contradictions between central and local governments, governments and markets, different administrative bodies, and within planning and implementation and management systems are becoming more apparent.

In 2013, the Central Urbanization Work Conference proposed to establish a unified spatial planning system, promote the integration of various types of plans at city prefecture and county levels, and form "one plan and one blueprint". In 2014, four ministries (National Development and Reform Commission, Ministry of Land and Resources, Ministry of Environmental Protection, Ministry of Housing and Urban–Rural Development, etc.) jointly issued the *Notice on Launching the Pilot Work of the "Multi-planning Integration*" in Cities and Counties. The pilot work of provincial spatial planning started in 2015. In 2016, *Several Opinions* of the CPC Central Committee and the State Council on Further Strengthening the Management of Urban Planning and Construction" was issued. In 2018, the institutional reform at the national level was further promoted. This reflects that multi-planning integration is not only a problem of planning itself, but also a deep process of the reforming the planning operation system and administrative management system.

(3) Ecological civilization construction: Development, Protection and Reform of Ecological Civilization System

The third main line is the construction of ecological civilization. The rapid development of urban social economy can not be separated from the leading role played by local governments, but the contradiction between development and protection caused by the local development impulse is getting worse. China is a country with a fragile ecological environment, scarce natural resources and uneven distribution. It is necessary to follow an intensive and ecological development path based on the carrying capacity of ecological resources and the environment.

The 40 years of reform and opening up has been a process of constantly changing the concept of development, from focusing on the balance between efficiency and equity to emphasizing the balance between development and protection. Ecological construction was first proposed in the Report of the 17th National Congress of the Communist Party of China. The basic idea of the National Main Functional Area Plan issued in 2011 is to guide the population distribution to adapt to the economic development trend, and to guide the population and economic distribution to adapt to the carrying capacity of resources and the environment. The report of the 18th National Congress of the Communist Party of China further incorporated the construction of ecological civilization into the overall layout of "Five-in-One" overall layout, which integrates economic, political, cultural, social, and ecological development. In 2015, the *Overall Plan for the Reform of Ecological Civilization System* was issued, which put forward the requirements for building an ecological civilization system with the strengthening of natural resources management as the core, and established a spatial planning system as one of the eight reform tasks. The report of the 19th National Congress of the Communist Party of China put forward the general strategy of building a socialist modernization power with Chinese characteristics in an all-round way. The construction of Ecological civilization is an important guide to guide the development of the new era and the basic program of China's future modernization.

1.2 Tasks and Intrinsic Logic of Spatial Planning Reform

Analyzing the three main lines of spatial planning reform provides background context for the release of the "*Several Opinions*", which is helpful in understanding the tasks and intrinsic logic of the planning reform.

The aim of the spatial planning reform is to build a spatial governance model guided by ecological civilization and oriented toward high-quality development. Its basic tasks are divided into three aspects, namely, promoting the construction of ecological civilization, realizing multi-planning integration and leading high-quality development.

(1) Strictly adhere to the bottom line of development and emphasize green development. Strictly delineate the red line of ecological protection, the red line of permanent basic protection of farmland and the boundary of urban development, strictly abide by the binding requirements of natural resources protection, and take the bottom line constraint as the premise and foundation of green development.
(2) Coordinate spatial development and realize multi-planning integration. Emphasize the overall control of cities and towns, agricultural and ecological spaces, and change the situation of division of departments and conflicts among overlapping regulations.
(3) Adopt a people-centered approach to promote high-quality development. It is necessary to draw an insurmountable red line for development, which means not to restrict development, but to seek better development, "to achieve higher quality, more efficient, fairer and more sustainable development and protection of territorial space".

From the perspective of the relationship among the three, "realizing multi-planning integration" is the leverage of the reform, "promoting the construction of ecological civilization" is the content of reform, and "leading high-quality development" is the core task of reform, which is also the ultimate goal contained in the other two tasks. Its intrinsic logic lies in grasping two main relationships. The first is the relationship between protection and development. High-quality development must be based on protection, and protection is to achieve better development; second, the relationship between planning reform and high-quality development. Promoting high-quality urban development is the purpose of spatial planning system reform, which is the guarantee for achieving high-quality urban development.

According to the tasks and requirements of the spatial planning reform, it is the process of the overall reconstruction of the planning system. The planning work must reflect the supporting role for high-quality development, the promoting role for multi-planning integration and the basic role for resource protection, which also constitutes the complexity and comprehensiveness of the planning work. It can be further summarized into three topics: First, strategic issues, establishing a spatial system that supports the development of high-quality cities; Second, institutional issues, establishing a mechanism for integrated planning and implementation; Third, technical issues, establishing a technical system that meets the requirements of planning reform.

2 Topics of Spatial Planning System Reform

2.1 Strategic Issues: Establish a Spatial System to Support the High-Quality Development of Cities

As China's urbanization enters its later stage, the establishment of a spatial system that supports high-quality urban development and the realization of a more efficient, fairer, and sustainable pattern of land development and protection is at the core of planning work.

(1) Grasp the current focus of turning to high-quality urban development

Embody the transformation of urban development concept and development mode, adapt to the transformation requirements from scale growth to quality growth mode, and promote changes in quality, efficiency, and drivers of urban development on the premise of protection.

(1) High-quality development requires the transformation of urban development mode. Establish the development mode of ecological priority, connotation development, bottom line thinking and dynamic adaptation.

(2) Livability and quality of life are the basic connotations of high-quality urban development. The quality of urban development ultimately reflects the satisfaction of people's aspirations for a better life, and the foundation of high-quality development is to achieve a high quality of life.
(3) Regional coordination and urban–rural integration are the thresholds that must be crossed for high-quality urban development.
(4) Urban renewal has gradually become the focus of urban planning and construction management, which needs to change from urban management to urban fine management and urban refined management and governance.
(2) Grasp the requirements of high-quality development of cities facing the future.

Looking toward 2035, high-quality development will face increasing challenges and contradictions due to uncertainties arising from the scientific and technological revolution, economic globalization, social transformation and environmental change.

(1) Ecology, culture, livability and innovation are the foundation of urban competitiveness and sustainable development. Guided by ecological civilization, cities must adopt people-oriented strategies centered on human resource development as their basic pathway in the knowledge economy era.
(2) High-quality development is a process of shifting from old to new growth drivers, and it is also a process of comprehensive urban transformation. It involves four levels: improving governance capacity, creating quality of life, upgrading urban functions and cultivating industrial kinetic energy. These four levels have a progressive logic, and improving governance capacity is the basic level, with the ultimate goal of realizing a new economic development model.
(3) The future competition is open competition. Urban development should have a more open vision, and the future status of cities depends on the overall influence and competitiveness of the region. Regional coordination is the basic idea of urban development.
(4) Resolving the contradiction between urban and rural areas is a national strategy and a task of local high-quality development. Promoting rural revitalization through urban–rural integration is a long-term challenge of national modernization.
(3) Adjustment and optimization of multi-level spatial structure

Establishing a spatial system to support the high-quality development of cities involves the adjustment and optimization of spatial structure at macro, meso and micro spatial scales, as well as the control and guidance of the organizational relationship of various spatial elements.

The spatial strategic framework proposed in the *Shanghai Overall Urban Planning (2017–2035)* (referred to as "Shanghai 2035") defines the key points of spatial strategic adjustment at three levels, forming a "3 × 3" analytical framework[7,8]. At the macro level, it focuses on the strategic adjustment of spatial layout of global urban goals, including the relationship between global urban functions and spatial

[7] Zhuang Shaoqin et al. (2017).
[8] Zhang Shangwu et al. (2017).

layout adjustment, the relationship between Shanghai and the Yangtze River Delta, and the development relationship between Shanghai's central city and suburbs; At the meso level, it focuses on improving the structural efficiency of high-density and super-large-scale cities, including the relationship between ecospace protection and urban development, the relationship between transportation and multi-center spatial system, and the relationship between differentiated guiding strategies in different divisions; At the micro level, it focuses on the improvement of built environment quality and quality of life, including the relationship between urban characteristics and cultural connotation, the relationship between life quality and community life circle, and the relationship between urban life and public space quality.

2.2 Institutional Issues: Establish Multi-planning Integration for Planning Implementation

In terms of implementation, establishing an operational mechanism to integrate various plans into a single framework and coordinating planning, construction, and management are key aspects of planning reform and an important direction for planning preparation.

(1) A blueprint: reflecting the role of spatial planning in the new era

Spatial planning reform is a two-way reform process of the internal technical system and the external operating environment, and it is the design of planning and operating mechanism to enhance urban governance capacity and promote multi-planning integration. *Several Opinions* pointed out that "territorial space development is the guide of national spatial development, the spatial blueprint of sustainable development, and the basic basis for various development, protection and construction activities". This is the basic orientation of spatial planning, which includes four connotations: "strategic blueprint, legal basis, policy platform and programme of action". First, the strategic blueprint. As a strategy, spatial planning is a blueprint to lead the future development; The second is the legal basis. As the basis of natural resources protection and space use control, spatial planning is the blueprint for implementing bottom line management; The third is the policy platform. As a spatial platform for system coordination, spatial planning is a blueprint for coordinating various spatial development policies and implementing planning management; Fourth, the programme of action. As a public policy, spatial planning must achieve social consensus and is a blueprint for guiding the actions of the whole society.

The status of planning is determined by the external institutional environment. It is necessary not only to face the implementation of planning, but also to improve the forward-looking, scientific, coordinated and operational planning. More importantly, it is necessary to change the organization and working methods of planning and the implementation methods of planning. Planning should strengthen the overall leadership of organizational work, get rid of the limitations of "departmental planning", make planning a social consensus by "adopting an open and participatory approach to plan-making", and establish a series of mechanisms to ensure the effective implementation of planning, so that planning can become a common program for social action.

(2) Spatial planning reform is a process of systematic reconstruction of the planning system

Spatial planning reform is a process of systematic reconstruction of the planning system, and the specific planning work is only a link of the whole system. According to the requirements put forward in *Several Opinions*, the connotation of planning system includes three aspects: planning preparation system, planning management and control system and planning operation system.

(1) Establishing a hierarchical and classified Compilation system of territorial spatial planning. It includes five levels of national, provincial, municipal, county and township planning, and a spatial planning system consisting of three types of planning: overall planning, sectoral planning and detailed planning.
(2) Establishing a spatial planning control system that covers all areas, elements, and processes: spatially covering ecological, agricultural, and urban spaces; managing all types of protective and developmental elements; and ensuring whole-process control, including planning, construction, and management.
(3) Establishing a planning operation system for planning implementation involves four aspects of system reconstruction. First, prepare the examination and approval system, and establish a five-level and three-category preparation system and examination and approval system corresponding to administrative responsibilities; The second is to implement the supervision system, establish monitoring, evaluation, early warning and feedback mechanisms, and support the basic information platform for the implementation of supervision; The third is the system of laws and regulations and policies, and the establishment of laws, regulations and policy systems; The fourth is the technical standard system, which reconstructs a unified technical standard system integrating various plans.

(3) Strengthening the implementation dimension is the core link to establish multi-planning integration mechanism

Planning implementation is the goal of planning preparation, which needs to face planning implementation and operation, and at the same time needs to establish mechanism design of the action dimension. On the one hand, from the perspective of project planning, it is necessary to strengthen the research on planning system, planning transmission and feedback mechanism, pay more attention to planning management and planning implementation administrative responsibilities, and strengthen the research on planning implementation policies and guarantee mechanisms.

On the other hand, from the perspective of planning and implementation, strengthen the role of short-term planning and annual planning in the planning implementation framework, and establish a rolling implementation mechanism of planning; strengthen the construction of laws and regulations system and policy system, including legislative work at national and local levels; strengthen the dynamic monitoring, regular evaluation, supervision and feedback of planning implementation, and establish a planning implementation supervision and dynamic adaptation mechanism; improve the planning committee system, social participation and other organizational systems for planning implementation.

2.3 *Technical Issues: Establish a Technical System to Meet the Requirements of Planning Reform*

Planning itself is a technical work, which should be placed under the overall framework of planning reform, oriented towards planning implementation, and play the role of strategic guidance and rigid control of planning.

(1) The overall operation of the planning system

The five-level and three-category planning system defined in *Several Opinions* constitutes the overall operational relationship, which focuses on strategy, coordination and implementation at the national, provincial and city and county levels vertically, and the overall planning has a binding and guiding role for sectoral planning horizontally. Corresponding to the different powers, different levels and types of planning have different functions in the operation of planning system, which is the basis of establishing an overall planning technical system. For example, the overall planning at the city and county level involves two levels of government, namely, the city (prefecture) level and the county (city) level, but the planning functions are different. The city (prefecture) level planning has the functions of connecting up and down, overall coordination and supervision of implementation, while the county (city) level planning is an important implementation unit of "vertical to the end and horizontal to the edge", and there are differences in the content and depth of planning preparation. In terms of the relationship between the master plan and the detailed plan, the master plan emphasizes the structure and framework, while the detailed plan, as the basis for protection and development, is the basic layer for implementing various management and control requirements and realizing multi-planning integration. The establishment of the overall technical logic requires the consistency of the technical system, and the key technical links are the partition and classification system of uses and the establishment of the overall transimission mechanism. Planning needs linkage from top to bottom, and the construction of a basic information platform can ensure the coordination between planning objectives and system operation.

(2) Bottom line constraint and planning transmission system

Strengthening the bottom line constraint is the key to embodying the basic role of planning, and it is also the focus of planning system reconstruction. At the technical level, it is necessary to strengthen the research work in two aspects. On the one hand, strengthen the identification, evaluation and assessment of major issues, effectively diagnose major risks in cities and key factors affecting the quality and efficiency of urban operation, and make scientific and accurate judgments on binding contents and corresponding planning strategies involving resources, ecology, culture, people's livelihood, safety and major facilities;on the other hand, according to different management and control elements, corresponding to the planning system, an overall management and control system with hierarchical implementation and classified transmission is established, and the implementation can be effectively monitored and evaluated.

(3) Territorial planning and structural control methods

The establishment of a spatial planning and control system for the entire region and all elements involves three levels of technical content.Establishing a global and all-factor spatial planning management and control system involves three levels of technical content. First, according to the requirements of global coverage, establish a reasonable protection and development zoning method, and formulate appropriate use control strategies according to the control requirements of various protection and development elements; second, according to the operation quality of the overall spatial system, strengthen the research on the structural control of the overall layout, and on the basis of the distinction of construction space from non-construction space, clarify the research on the ecospace corridor, the urban functional elements and the overall organizational relationship of the transportation system, and the living circle system; third, aiming at the optimization goal of functional layout and spatial structure, strengthen the overall design of various spatial policy systems, including ecological protection policy, population distribution policy, industrial layout policy, spatial development policy, etc., and guide the formation of a high-quality territorial space development and protection pattern through the integration of spatial layout and differentiated spatial policies.

(4) Management by objectives and process control methods

To ensure effective process management and control from planning to implementation, it is imperative to establish a robust target management framework and process control mechanisms. First, strengthen the research on the overall relationship between planning objectives and supporting means, implementation timing and project arrangement, and establish a strict logic between planning objectives and implementation strategies; the second is to pay attention to the effective monitoring and regulation of the implementation process. The role of the index system is to establish the link between goal-oriented and implementation process control, and it is an important tool to strengthen the monitoring, evaluation and management of the implementation process. At the same time, through the establishment of monitoring, evaluation and maintenance mechanism, the dynamic adaptability of planning is enhanced; the third is to strengthen the research on planning flexibility and blank space mechanism. Shanghai 2035 Overall Planning puts forward three blank space methods: mobile index reservation, strategic space blank space and time sequence rhythm control, which are beneficial explorations[9].

(5) Establishing multi-planning integration planning content system

[9] Zhang Shangwu et al. (2017).

Multi-planning integration is multidimensional. In the spatial dimension, strengthen the transmission of planning control content in vertical planning at different levels; in the system dimension, strengthen the horizontal connection and integrate the arrangement of various special spatial elements in the overall layout; in the time dimension, the role of spatial planning as a platform of "multi-planning integration" is clearly defined, and a policy mechanism of multi-sectoral collaborative planning and collaborative implementation is formed through the rolling preparation and implementation of recent planning.

The "1 + 3" achievement system proposed in the Shanghai 2035 Overall Planning is the first attempt to integrate the content system of multi-planning integration. It is the basic starting point to pay attention to administrative responsibilities and strategic, fundamental and implementable spatial planning[10]. "1" is the "Overall Planning Report", which corresponds to the powers of the municipal government and reflects the strategy and structure; "3" refers to" Zoning Guidelines ", "Sectoral Planning Outline" and "Action Planning Outline" respectively, reflecting the implementation and operability. Among them, "Zoning Guidelines" and "Sectoral Planning Outline" correspond to strategic guidance and rigid control: district and county governments and departments; The "Action Planning Outline" corresponds to the right of administrative responsibilities of county and district governments and departments; the Outline of Action Plan corresponds to the common responsibilities. Through the construction of the "1 + 3" achievement system, the overall, strategic and multi-planning integration platform function of the master plan is reflected, and the content system with clear objectives, clear responsibilities and rights, and process control is formed.

(6) The form of the results reflecting the requirements of the planning reform

The transformation of spatial planning from a technical logic to a spatial governance logic needs to reflect the characteristics of public policy as the attribute of public policy. In the form of achievements, the transformation from technical achievements to policy achievements and the clear communication of policy intentions to the public should be a plan that the public can "understand". To successfully manage the interplay between technical content and public policy intentions, it is imperative to maintain a balance between the fundamental principles of "strategic orientation" and "policy planning" while avoiding excessive focus on legalistic and binding material. In the expression of results, we should pay attention to the differences in the expression of planning results at different levels, especially at the overall planning level, which needs to reflect strategic, policy-oriented, structural and framework, and avoid the "excessive" regulations on land use functions at the overall planning level, resulting in contradictions in implementation and management. To get rid of static thinking, we can't try to be precise and perfect in control elements, planning and implementation contents, etc., but lack the expression and guidance of "space policy orientation". In terms of results content, we should pay attention to operation

[10] Zhang Shangwu et al. (2017).

orientation, deal with the relationship between content expansion and results burden reduction, and embody the compilation principle of "useful, useful and easy to use".

(7) Differentiated technical requirements according to local conditions

It is imperative to prioritize the provision of differentiated guidance and to explore a planning mode that is aligned with the specific realities of various regions throughout China, given the pronounced disparities in development conditions, administrative management modes, and planning management levels.Especially at the city and county level, there are both megacities with a population of over 10 million and county towns with a population of only 10,000 to 20,000. There are great differences in natural conditions, development scale and development stage. In the administrative management system, there are both the model of cities leading counties and the model of provinces governing counties. There are great differences in the planning preparation mode, content depth, approval and planning management mode, so it is impossible to require planning according to one mode and one standard across the board. In this regard, the *Several Opinions* pointed out that "planning should be carried out according to local conditions", "cities, counties and towns of the territorial spatial planning by the provincial government according to local reality, clear planning approval content and procedural requirements."

3 Development Trend in Spatial Planning and Discipline Outlook

The development trend of spatial planning and planning discipline should be understood in the law of urban development. The function of planning and the development of planning theory follow certain rules, but the planning culture is diverse. The planning system in China originates from the planned-economy context, which is different from the development of planning systems in the West. The current spatial planning reform should be placed in the macro background of 40 years of reform and opening up, and its trend is also inseparable from the understanding of the context of Chinese planning culture and China's development issues.

3.1 The Chinese Context of Spatial Planning and the Essence of Planning

From the perspective of spatial planning systems in various countries worldwide, "spatial planning" refers to the sum of all types of spatial planning, and generally does not use spatial planning to define the names of specific planning levels[11]. Generally speaking, what is generally called "territorial spatial planning" originated from

[11] Cheng Yao and Zhao Min (2019).

modern urban planning in the West. Western urban planning is not only the planning of “cities”, but also pays attention to rural areas and the natural environment. The spatial category gradually expands from the built environment to the overall interaction between the built and natural environments, and promotes the cross-integration of multiple disciplines. Just like the difference between “city-oriented urbanization” and “town-oriented urbanization” in Chinese context, using the word “territorial spatial planning” also has “Chinese characteristics”. It is generally believed that urban and rural planning corresponds to the housing and construction department, land use planning corresponds to the land department, and development planning corresponds to the development and reform department. A new concept is needed under the background of “the multi-planning integration”. However, no matter what word is used, the expansion of the category of “space” is a trend, while the essence of “planning” will not change.

Planning is the product of market economy environment. The inception of contemporary urban planning in western countries is rooted in the resolution of urban quandaries occasioned by the market failure that arose in the aftermath of the industrial revolution in those countries. Uncertainty constitutes the most formidable challenge posed by the market economy milieu, and planning represents a process whereby actions are structured based on future expectations and executed in a continuous fashion. Therefore, planning inherently has the attributes of public policy; it is a fundamental and strategic spatial governance tool aimed at addressing urban development challenges and future uncertainties to enable better and more sustainable development.

In terms of the general role of spatial planning, it mainly includes three aspects. First, as a tool of national macro-control, planning formulates the rules governing market operations and safeguards the public interest. Second, as a policy-making and implementation instrument, it coordinates the unified actions of various municipal departments in managing a complex synthesis of elements. Third, as a framework for future urban spatial development, planning provides both guidance and control for spatial growth. Spatial planning, as a social practice for dealing with urban development problems, is constantly developing. Zhang Tingwei divided the intergenerational changes of modern western planning theory into three stages[12]. In the first half of the twentieth century, represented by modern functionalism, it paid attention to material space and focused on the relationship between people and cities; after 1960s, it paid attention to the sociality of cities, and shifted the focus to the relationship between people. At the end of the twentieth century, it paid attention to the ecology of cities and the relationship between man and nature (Table 1). The dimension of urban spatial development is comprehensive, meeting people’s development and needs is the core issue of planning, and highlighting ecological and natural resources protection is the common trends in the context of global urbanization.

[12] Zhang Tingwei (2011).

Table 1 Inter-generational changes of modern western planning theory

The focus of the theory	The first generation theory (rational model, Athens Charter) (%)	The second generation theory (advocacy planning, post-modernist design trend of thought) (%)	The third generation theory (collaborative planning, new urbanism, smart growth, etc.) (%)
Man-city relationship	90	30	35
Person-to-person relationship	5	60	25
Man-nature relationship	5	10	40

Source Zhang Tingwei.(2011). *Urban construction in the 21st Century under the guidance of the twentieth century planning Theory: A discussion on the "third generation Planning Theory"*. Urban Planning Forum (03), 1–7

3.2 Development Trend and Challenge of Spatial Planning

China's urbanization has changed from high speed to high quality development. As an indispensable governance tool, the reform of spatial planning is not only a strategic issue, but also a transformative issue.

First, improving spatial governance capacity will be the fundamental way for Chinese cities to achieve higher quality, more efficient, fairer and more sustainable development, and spatial planning is an important governance tool for the sustainable development of urbanization in China.

Second, it is an integral part of the governance culture with Chinese characteristics and China's modernization road to establish a spatial governance model that meets the needs of social economy development and is guided by planning. It is necessary to change the concept, system and social resource allocation mode, and to establish a corresponding spatial planning system and a series of implementation supervision mechanisms. It requires the overall construction and improvement of laws and regulations system, administrative management system, planning operation system and planning preparation system.

Third, like the successful experience of China's reform and opening up, the reform of spatial planning cannot be accomplished overnight, and it will also be a process of facing problems directly, keeping pace with the times, facing the future and gradually innovating.

The challenges facing spatial planning in the future still come from three levels: strategy, system and technology:

At the strategic level, it is mainly the strategic choice of China's urbanization path. Urbanization is not only a natural process, but also a process of social choice. Great achievements have been made in China's urbanization. The specific national conditions, time and space background and institutional environment have created China's urbanization characteristics, and it will also face new development problems and long-term challenges brought about by scientific and technological changes, environmental changes, social transformation, regional coordination and rural development.

At the institutional level, it is mainly the construction of spatial governance capacity and governance system. The sustainable development of land space is a process of improving spatial governance capacity and building governance system. It is necessary to effectively use the policy tools of spatial planning and promote the participation of government, society and citizens through multiple means such as administration, market and law.

At the technical level, the needs of social development constantly promote the innovation of planning techniques and methods, which need the support from related disciplines fostered by interdisciplinary integration, multi-level specialized personnel training, and intelligent information technology supporting the operation processes of planning, construction, and management.

3.3 Development and Prospect of Urban and Rural Planning Discipline

Urban and rural planning is an applied discipline that actively engages with social practices and evolves accordingly, displaying the hallmark of discipline development. The advancement of this discipline is contingent upon the dual process of innovation, whereby theoretical and practical considerations are theoretical and practical considerations interact continuously and reciprocally, functioning as the fundamental basis for discipline development. Social needs and continuous innovation of discipline system constitute two major driving forces to promote discipline development, but there are still many shortcomings in the face of social development and practice. For example, pay attention to the built-up areas, but not to non-built spaces; theoretical research on urbanization in China is insufficient; planning research is out of touch with planning practice, and more attention is paid to planning itself, but not enough attention is paid to planning function, planning implementation and planning operation system, and so on. It is undeniable that the development of disciplines will also be constrained by the external institutional environment. Typically, it is in the field of rural planning and regional planning, and the attention to rural issues mainly begins after the promotion of policies at the national level. In the area of regional planning, due to the lag of regional planning practice, the internal development of the discipline is not sufficient.

For the development of urban and rural planning discipline, the greatest significance of the reform of the spatial planning system lies in the expansion of "space" and the reconstruction of "system", and the greatest value lies in giving full play to the role of "planning".

It is necessary to actively respond to the needs of society, respect the laws of disciplines, enhance the forward-looking nature of disciplines, and strive to break external constraints. It is necessary to strengthen the integration of multiple disciplines, improve the discipline knowledge system and the construction of sub-disciplines[13], including the disciplines related to the protection and use of natural resources, as well as the cross-integration with the humanities related to society and economy, but there is a process of transformation oriented to social practice. It is necessary to pay more attention to social needs, combine planning practice with discipline development more closely, to find problems in practice, and to improve the ability of discipline development in summing up laws and feeding back practice. It is necessary to have a broader international and historical perspective, and explore the discipline development path in line with China's reality in the process of summing up international and historical experiences.

According to the practical needs of spatial planning, three levels of research need to be strengthened:

First, theoretical research on urbanization and urban and rural development in China. Foresight is the greatest value of the planning discipline. Summarizing the development law of urbanization in China and exploring urbanization with Chinese characteristics and sustainable development of urban and rural areas under the background of ecological civilization are the important theoretical basis and support of territorial space development.

Second, research on the theory and method of spatial planning in China. The research of planning theory and method is the weak link of the current planning work, including planning implementation evaluation, space policy system, planning preparation system, planning management system, planning laws and regulations system, planning implementation and intervention mechanism research, etc.

Third, the innovation of spatial planning technology system, including intelligent planning, big data research, spatial evaluation and spatial diagnosis methods, as well as technology integration and application innovation in multidisciplinary fields. New technologies and methods will be an important force to promote the construction of spatial planning theory and the innovation of planning methods.

4 Conclusion

Establishing and supervising the implementation of the territorial spatial planning system is a major, top-down reform at the national level, which will have a great impact on the planning work and discipline development. It is necessary to clarify the basic requirements, tasks and intrinsic logic of the reform.

The ongoing reform of the spatial planning system is rooted in a profound socio-economic and institutional context. Three key drivers shape this reform: the spatial governance challenges arising from urbanization transformation; the conflicts among existing plans exposed during the shift from a planned to a market-oriented system; and the need to reconcile development with ecological protection under the ecological

[13] Sun Shiwen (2017).

civilization reform agenda. Strengthening the construction of ecological civilization, realizing multi-planning integration and leading high-quality development constitute the basic requirements and tasks of spatial planning reform.

The corresponding reform tasks include three issues: strategy, system and technology, namely, building a spatial system to support high-quality development, multi-planning integration mechanism for planning implementation, and a technical system to meet the requirements of planning reform. Planning itself is a technical issue, so we should understand the change of planning content and method in the overall framework of spatial planning system reform.

The discipline of urban and rural planning, dedicated to improving human settlements, needs to actively respond to social development demands, align its disciplinary evolution with China's practical challenges, and strengthen research on urbanization, spatial planning theory and methods, as well as innovations in technical systems. To embody the discipline value facing the new era, we should not only stick to the core of the discipline, but also play a greater role in realizing the sustainable development and modernization of the country with an open and innovative thinking.

References

Cheng Yao & Zhao Min. (2019). From "Land Use Planning" to "Spatial Planning Orientation" : The British spatial planning reform and its implications for the construction of China's spatial planning system. *Beijing Planning Review* (01),69-73.

Editorial Department of this Journal. (2019). "*Discipline Development under the Background of Spatial Planning System Reform" Academic Pen talk.* Urban Planning Forum (01), 1-11.

Sun Anjun.(2018). *Thoughts on the reform of spatial planning.* Journal of Urban Planning (01), 10-17.

Sun Shiwen.(2017). *Reflections on urban and rural planning education.* Urbanism and Architecture (30), 14-16.

Wu Weijia, Wu Liangyong, Shi Nan, Yuan Qifeng, Mei Yaolin, Ye Yumin... & Wang Kai.(2019). *Spatial planning system reform and discipline development.* City Planning Review (01),17–24+74.

Wu Zhiqiang, Yang Xiu & Liu Wei.(2015). *Intellectual Urbanization or Physical urbanization: Strategic thinking on China's urbanization.* Urban Planning Forum (01),15-23.

Zhang Tingwei.(2011). Urban construction in the 21st Century under the guidance of the 20th century planning Theory: A discussion on the "third generation Planning Theory". *Urban Planning Forum* (03),1-7.

Zhang Tingwei.(2019). *Urban Planning in China: Reconstruction? Rebuild it? Reform?*Urban Planning Forum (03),20–23.

Zhang Shangwu, Jin Zhongmin, Wang Xinzhe, Zheng Degao, Hu Guojun, Fang LAN... & Sun Ye.(2017). *Strategic Guidance and rigid Control: Innovation of Urban Overall Planning Outcome System in the New Era -- The basic idea of constructing the overall planning outcome System of Shanghai 2040.* Urban Planning Forum (S1),52-60.

Zhuang Shaoqin, Xu Yisong, Xiong Jian & Liu Sheng.(2017). *Transformation and reform of megacities' master planning: A new round of urban master planning in Shanghai.* Urban Planning Forum (02),10-19.

Zhuang Shaoqin. (2018). *New Era Reflections on Optimizing National Spatial Planning*. In Proceedings of the 1st National Conference on Theoretical Methods and Practices in National Spatial Optimization.

Strategic Mind of Territorial Spatial Planning

Li Jijun

1 Introduction

The word strategy originated from the military field. In *Sun Tzu's Art of War*, it is commonly known as "stratagem", which is not only used in war, but also in political and diplomatic field. In the West, the English word "strategy" is derived from the Greek word "strategia", meaning military generals and local administrators. The book *Strategia*, written in 580 AD by an anonymous author of the Byzantine Empire under the name of Emperor Maurice, means "the way between leaders and subordinates" or the art of command. It is therefore the strategy of war under the will of the commander.

Since the eighteenth century, the concept of strategy has been continuously completed, highlighting three main characteristics. Firstly, the strategy has a programmatic nature that can influence overall success or failure. In 1799, A.H.D.von Biro of Prussia first distinguished between strategy and tactics in the *Spiritual Essence of the New War System*: "Strategy is the science of conducting military operations beyond visual and artillery range, while tactics is the science of conducting military operations within those ranges. "Secondly, strategy is purposeful, which may be used in war aims, national security or policy objectives, etc. In 1832, C. von Clausewitz of Prussia regarded strategy as "the use of battle for the attainment of the ends of war" in his *Theory on War*. Finally, the main body in the implementation of the strategy has the attributes of a security community, where security is the bottom line and the community is a geographical unit, such as a state, with common interests, emphasizing the use of instruments of power and the overall deployment of resources. According to the *Encyclopedia of Soviet Military* published in 1980s, strategy is "the component and the highest field of military art, which includes the theory and practice of preparation for war, planning and conduct of war and strategic

L. Jijun (✉)
Shanghai Tongji Urban Planning and Design Institute Co., Ltd., Shanghai 200092, China
e-mail: lijijun68@163.com

Z. Peng (ed.), *Theories and Frontiers in Territorial Spatial Planning*, Urban Sustainability, https://doi.org/10.1007/978-981-96-9825-7_6

campaign by the state and the armed forces," that is, taking them as the community of security.

At present, the concept of "strategy" has been extended and widely used in other fields, such as political strategy, diplomatic strategy, economic development strategy, etc., but its programmatic, purposeful and subjective connotations remain unchanged. Therefore, the strategic mind of territorial spatial planning is in the field of spatial planning, which aims to build a community of security in the administrative areas of the country and other levels, to use government power to allocate territorial space resources from the supply side as a whole, and to achieve certain spatial strategic goals with programmatic overall policies.

In the following, the strategic perspective of territorial spatial planning will be interpreted from three dimensions, namely, the national strategic perspective, the local strategic perspective, and the people's city perspective, and finally, reflections on territorial spatial planning will be presented.

2 National Strategic Perspective

From a national strategic perspective, first, the current territorial spatial planning is an important tool for deepening reform. It has both the meaning of "reform" and the content of "revolution", so it has a revolutionary significance, and is a solution to the problems of too many types of planning, overlapping and conflicting contents, complicated approval processes, excessively long cycles, and the unpredictable changes in local planning. It is a reform lever for the construction of an ecological civilization and the planning of territorial space protection and development patterns in the new era.

Second, to establish an international perspective. Based on the current global international situation, countries are more interconnected and interdependent than ever before, and humanity is increasingly becoming a community with a shared future, sharing global responsibilities such as global climate change and global warming. Therefore, it is also necessary for planning to have an international perspective. Therefore, in fact, in addition to geological disasters and extreme weather in the traditional sense, there are heat islands, global warming, sea level rise and other issues in the risk assessment in the double assessment of territorial spatial planning, which need to be considered in the planning. At the 2020 Central Economic Working Conference, China proposed the new goal of "striving to achieve peak carbon dioxide emissions by 2030 and carbon neutrality by 2060" in response to global climate change, which is in fact China's response to its international obligations, responsibilities and perspectives.

Third, it has moved from management to governance. In the future, the government's governing process will no longer be a simple administration based on departmental rules or regulations, i.e., management of affairs by administrative means only, but will become a systematic process of governmental implementation, in which the government will effectively change its functions, strengthen its service functions, and promote the participation of the whole society in urban governance.

Fourth, it embodies ecological civilization. It is necessary to be vigilant in times of peace for reflecting the protection of resources and the environment, and for embodying the concept of sustainable development. The recent communiqué of the Fifth Plenary Session of the 19th Central Committee of the CPC and the Central Economic Working Conference (CEWC) have both highlighted the new development pattern of ecological civilization, and the systematic consideration of how the ecological pattern can be carried in the planning patches and space, and how the new development pattern can be reflected, are all required for the current round of territorial spatial planning.

Fifth, the transformation of kinetic energy. Traditionally, the coast-to-inland development gradient line and the Heihe-Tengchong Line of China's urbanization geographic pattern have divided the country into east and west. In fact, such a division of China's economic trend is quietly changing. During the 13th Five-Year Plan period, the gap between the north and the south of China grew wider as measured by GDP per capita and bounded by the Yellow River. According to the GDP ranking of China's cities in the first three quarters of December 2020, only Beijing ranked among the top 10 cities north of the Yellow River, and Tianjin has been squeezed out of the top 10 list. Therefore, in the compilation of territorial spatial planning, it is necessary to reconsider the positioning of cities, spatial needs, and strategic control in light of the north–south transformation of the country's economic and geographic pattern.

2.1 *Understanding of Policy Space–Time Background*

The strategic requirement of the country in the new period is to consider territorial space as the most important natural resources and assets, and to build a territorial spatial planning and protection system based on spatial planning and taking use control as the main means. It can be seen that the future territorial spatial planning will become a system that is quite different from the original departmental planning.

(1) The context of territorial spatial planning reform.

From the policy line, the mind of spatial planning reform is becoming clearer (Fig. 6.1).The construction of ecological civilization was proposed in the Eighteenth National Congress of the Communist Party of China, and the Communiqué of the Fifth Plenary Session of the 19th Central Committee of the CPC proposed to optimize the layout of territorial space and promote regional coordinated development and new urbanization. It is worth mentioning here that new urbanization has been proposed again, which does not mean that the previous urbanization was not effective, but at the present stage, with the increasing needs of the people, higher demands are being made on urbanization, especially on indicators, spatial performance and public resource allocation. In addition,

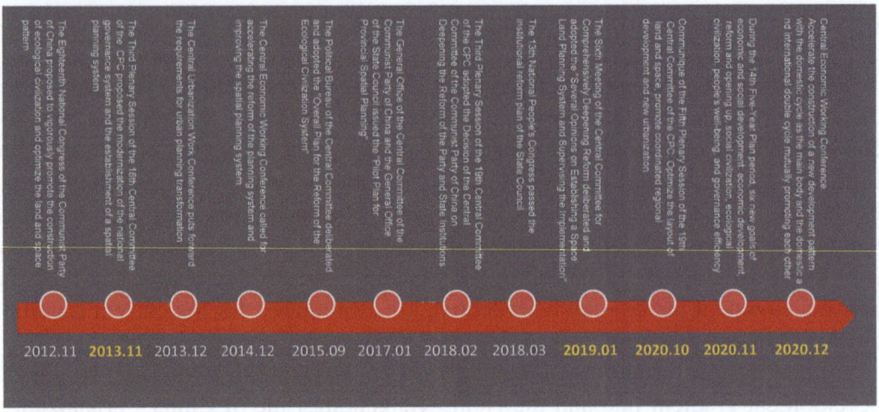

Fig. 6.1 The context of territorial spatial planning reform. Source: Self-painted by author

there are six "new goals" put forward in the "14th Five-Year Plan" and the new development pattern put forward by the Central Economic Working Conference. These national strategic keywords are the problems that need to be dealt with, responded to and considered in the technical work of the current round of territorial spatial planning.

(2) Framework for promoting the reform of territorial spatial planning

The reform of territorial spatial planning is promoted in three stages.

The first stage is system construction, and the territorial spatial planning system is established. Territorial spatial planning is summarized in a broad framework of "five levels, three categories and four systems", where "five levels" corresponds to China's administrative management system, namely the national level, the provincial level, the municipal level, the county level and the township level, in terms of the content of the hierarchy. National plans prioritize strategic planning, while provincial plans focus on coordination, and municipal, county, and township plans concentrate on implementation. The term "three categories" are the types of plans, which are divided into master plans, detailed plans, and related special plans. The overall plan emphasizes its comprehensive nature as a global arrangement for protecting, developing, utilizing, and restoring the territorial space within a specific area, for instance, the entire administrative district. Detailed planning emphasizes implementation, which is generally organized below cities and counties, and is an implementation arrangement for specific land use and development intensity. Outside the urban development boundary, village planning is taken as detailed planning. Special planning with a focus on specialization is usually organized and developed by the Ministry of Natural Resources or other relevant departments. It can be prepared at national, provincial, city, and county levels, especially for specific district areas or river basins. These plans include specific arrangements for the development, protection, and utilization of space to reflect particular functions. The "four systems" are the establishment of planning and compilation approval system, implementation and supervision system, regulation and policy system, and technical standard system for the multiple compliance.

The second stage is institutional development, including regulation and policy system, technical standard system, monitoring and early warning mechanism, performance assessment mechanism, and a system for developing and protecting territorial space through unified use control. In fact, the compilation of the current round of territorial spatial planning takes 15 years as a cycle, which includes not only the preparation of "1 + 1", namely a general report and an information platform, but also a series of institutional documents, such as laws, regulations, technical standards, implementation rules, guidelines, etc. In the current round of territorial spatial planning, we must not just wait passively for the State and the Ministry of Natural Resources to issue documents. The documents that have already been issued and implemented on a trial basis mainly provide more guidance, including the categorization of land and sea use, the general urban design guidelines, and the guidelines for municipal territorial spatial planning. However, on a large scale of the country, there are huge differences in industrialization level and urbanization stage in different regions of the country, so it is difficult to fully cover and use only one technical specification. Therefore, all parts of the country should respond to the national strategy and formulate differentiated and local policy documents on the basis of absorbing and understanding the value orientation of the published documents. For example, in the process of territorial spatial planning, the question of how to build in resilient development zones often arises, which is not solved in the existing national policies and documents, so it is necessary for municipalities to formulate detailed regulations to determine the procedures for initiating the construction of flexible development zones in cities and towns, including the requirements for the compilation of the plan, the approval process, record keeping procedures and land use permits.

The third stage is capacity building, and the highest requirement for territorial spatial planning is to complete it, not only for the work of planning, but also to form the standpoint of planning patterns, with the main purpose of promoting modernization. After the "four modernizations" of industrial modernization, agricultural modernization, national defense modernization, and science and technology modernization, General Secretary Xi Jinping proposed the "fifth modernization", which is to promote the modernization of the country's governance system and governance capabilities. Before the territorial spatial planning was rolled out, there were about over eighty kinds of plans in all departments. In fact, all the elements of spatial resources elements were controlled, but the effect was not satisfactory, and the methods, concepts of control and governance systems could no longer adapt to the development needs of the new era and the new changes in people's needs. Therefore, China has proposed the compilation of a territorial spatial planning system, in order to re-compile a plan that is useful, functional and good, and to realize a higher quality, higher efficiency, more equitable and more sustainable development and protection of territorial space.

(3) Overall requirements of territorial spatial planning

The great significance of the current round of territorial spatial planning is the construction of national spatial governance capabilities building. It requires implement the new development concept, establish a nationally unified, clearly defined, scientific and efficient territorial spatial planning system, make overall plans for the development and protection of territorial space in the new era, and promote the modernization of the country's governance system and capabilities. The key words are "the establishment of system, the planning of pattern and the promoting of modernization". The overall requirements are to implement the new development concept, improve the system of territorial space development and protection, and reflect the strategic, scientific, authoritative, coordinated and operational nature of the system.

2.2 *Strategic Rationale for Reform*

The report of the Nineteenth National Congress of the Communist Party of China clearly stated that it is necessary to change from high-speed growth to high-quality growth, that is, double high goals. The communiqué of the Fifth Plenary Session of the 19th Central Committee of the CPC proposed to take two centenary goals as the main goal. It conveys some new signals. First, the new expression of the national macro-strategic layout, China has basically achieved the goal of building a moderately prosperous society in an all-round way, achieved the goal of the first centenary struggle, and started a new journey of building a socialist modern country in an all-round way, reflecting the layout of the system and governance system at the national level. In fact, the current round of territorial apace planning is also a part of national reform and system building. Second, the word "powerful country" appears frequently, such as cultural power, educational power, talent power and sports power, which shows that China has changed from emphasizing "big" to "strong" and emphasizing the quality of development. Third, a new pattern of development should be built, with economic and social development centered on the theme of promoting high-quality development, deepening supply-side structural reform as the main line, reform and innovation as the fundamental driving force, and meeting the people's ever-increasing needs for a better life as the fundamental goal. Fourth, we should insist on the core position of innovation, and the most basic element of innovation is high-quality scientific and technological innovation, and high-quality scientific and technological innovation should be used to support and lead high-quality development. Another thing that needs to be emphasized is conceptual innovation. If there is no conceptual innovation, it will be difficult to promote the development and protection of the entire territorial space with high quality and high performance. Fifth, the rural revitalization should be comprehensively promoted and the development of agriculture and rural areas should be given priority. The urbanization of rural population is an irreversible trend, but it does not mean that the rural areas will die out. In addition to traditional village planning, rural areas should also have differentiated control requirements and development goals, including the development goals of industries, land use and population, as well as support policies for talents returning to their hometowns to start their own businesses and the allocation of relevant public service facilities. For example, the planning of Moganshan Township in Deqing County, Huzhou City should be different from the traditional simple supporting commercial and medical facilities, and some rural areas can even be combined with scientific research, the display of agricultural civilization and the development of agricultural science and technology. Sixth, we should form a new pattern of opening up, and open up at a higher level in the new round. We will build a domestic and international "double cycle", with the domestic cycle as the mainstream, and through the internal cycle, we will open up the whole industrial chain and supply chain, so as to promote a more proactive position in the external cycle, and cultivate a new competitive advantage in the international arena. Seventh, more tangible and substantial progress will be made towards common prosperity, with per capita gross domestic product reaching the level of that of middle-developed countries by 2035 and the middle-income group expanding significantly. The term common prosperity is not average prosperity, but differentiated prosperity. It is found that with the emergence of Taobao Village, the economic development in many remote areas has been greatly promoted, and the income of farmers has been increased. However, the relatively more economically backward the area, the better the development of Taobao Village. Inversely, the development is not good in Taobao Village around big cities. In addition, in areas with relatively low income, the happiness of ordinary people is much higher than that of big cities, so happiness is not completely positively correlated with income. Common prosperity in the new era should be endowed with new connotations, not limited to material prosperity, but also spiritual and cultural prosperity. Eighth, it is necessary to implement the national strategy for actively coping with the aging of the population and focus on the development of a multi-level, personalized, high-quality and accurate pension system and service provision.

There are some hot words in the communiqué of the Fifth Plenary Session of the 19th Central Committee of the CPC, including "development", "newness", "comprehensive", "perseverance", "strategy", "system", "people", "modernization" and "innovation" (Fig. 6.2), among which the core is "development" and "newness", which appeared 72 times and 66 times respectively, actually indicating the need to switch to a new way of mind about development. In the new period, development is still the main goal, but it has been transformed into development with bottom line and goals, emphasizing development quality and efficiency, and solving various problems in the process of development.

The Central Economic Working Conference in 2020 put forward seven key words: new development pattern, demand-side management, dynamic equilibrium, no predictable changes in policy, scientific and technological strategic support, strong domestic market and strengthening anti-monopoly. These seven aspects reflect China's institutional self-confidence, scientific and technological strength and positive expectations for future governance. Therefore, in terms of methods and technologies, the current round of territorial spatial planning should be able to respond to

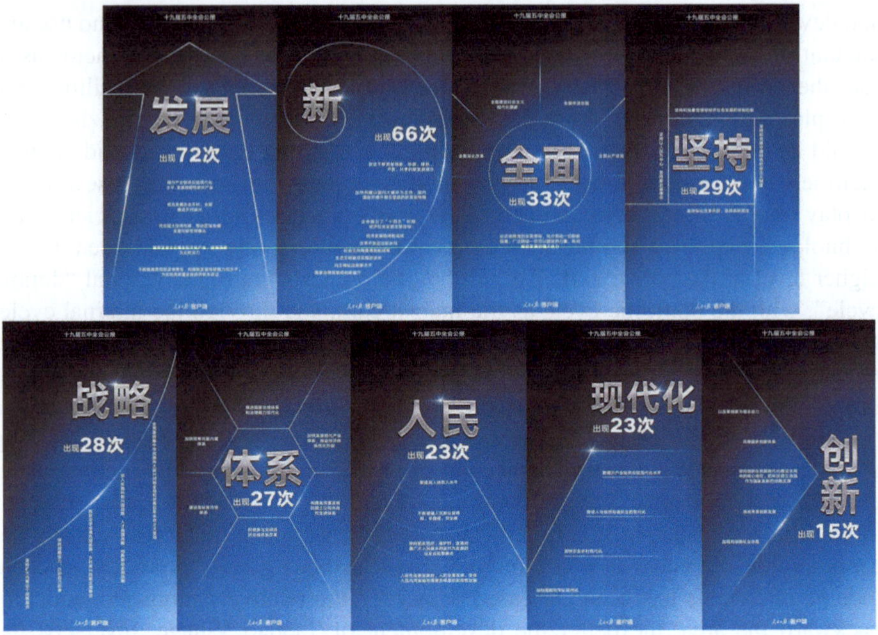

Fig. 6.2 Hot words in the communiqué of the Fifth Plenary Session of the 19th CPC Central Committee. *Source*: people.com.cn

the modernization trend of national governance and undertake the national position, attitude and strategy.

The current round of territorial spatial planning has changed from technical documents and departmental planning to government planning, which embodies the policy agenda of the whole government. This policy agenda does not only rely on the unilateral power of the government, but should also be a national consensus, a stable blueprint and goal for its formation, which will not be given up until the goal is reached.

3 Perspective of Local Strategy

For local governments, the core work of compiling territorial spatial planning from the national strategic perspective is to undertake five levels and three categories, which are decomposed and transmitted in the process of undertaking. In the process of transmission, on the one hand, it is necessary to respond to the national strategic intention, on the other hand, it is necessary to combine the actual local development, that is, to carry out the translation.

The core of local area is development, but the path of it is different. China has entered a period of shrinking in some cities. According to statistics, nearly 30% of municipal units are currently shrinking, and planning practitioners are still accustomed to incremental planning, which will inevitably lead to certain problems. The common forecasting method in planning is the trend extrapolation method of scale, including the use of popular cellular automaton model to calculate the annual land supply and growth in the future. The problems with this approach are manifold: on the one hand, city size does not grow linearly at a constant rate, but rather curvilinearly non-linear growth, with different growth rates in the short, medium and long term; and on the other hand, it reflects only size, but not proportionality, structure and performance, which are at the core of the current round's measurement of high-quality development. In fact, multiple scenario schemes of urban growth should be proposed, with the junior scheme being the one under maximum constraints and the senior scheme being the one under strategic leadership, reflecting different strategic intentions.

The combination in development of individual and planning and collaboration is also an innovation in the compilation of the current round of territorial spatial planning. City, county and township-level territorial spatial planning can be adapted to local conditions. They can be compiled with combination, and multiple townships with adjacent spatial locations and similar industries can also be compiled as a unit. At the village planning level, villages in the agglomeration and upgrading category outside the township government area may compile village plans independently or in conjunction with several surrounding villages, while villages in the suburban integration category may compile detailed plans in conjunction with the construction areas within the urban development boundary, or may compile village plans independently or in conjunction with each other.

There is also planning and collaboration between urban and rural areas. Under the background of ecological civilization construction, the land use pattern in developed areas has a new trend from incremental expansion to the reduction of existing urban and rural development land with potential, and a new concept of " total amount of urban and rural development land framing" has emerged. The expansion of town site needs to rely on the increase and decrease allocation target of village development land reduction, that is to say, without the decrease of village development land, urban development land cannot be increased. However, at present, the provincial capitals control the increase and decrease allocation target, so the flow indicators that can be used for urban development land are also limited.

Local innovation policies need to be developed at the local level, such as which sites can and cannot be located in resilient development zones and what the criteria for access are. There are two types of non-buildable land in Shanghai overall planning, one is functional non-buildable land, that is, the nature of land use is not determined for the time being, and the type of land use is not given, but there are indicators of development land use; The other is non-buildable land for strategy to deal with major events, decisions and national strategies in the future. For example, if Shanghai plans to host major events such as the Olympics or the World Cup, the bid may not be successful in the short term, but there will be some reservation in terms of land indicators and expectations. For local areas, the scope of peripheral expansion is not necessarily limited to 8%-15% stipulated by the state, but it can be relaxed to

26%-30% in strategic areas, which will be vary from city to city. In addition, local governments also need to formulate policies and norms for starting the construction of resilient development zones, and the state has not yet issued clear requirements.

Emphasis should be placed on guarantee and innovation of institution and mechanism. In the traditional text of the urban and rural overall planning, although this part is included, it actually plays a limited role. In the current round of territorial spatial planning, the chapter on guarantee of institution and mechanism needs to be considered and discussed. For example, the institution and mechanism can be started by adjusting the planning and approving the People's Congress and the higher authorities. Therefore, the importance of this part has been greatly enhanced, and planning practitioners need to pay more attention to it.

3.1 Scale and Quality

(1) The challenges in development patterns of urbanization and spatial governance

The urbanization trajectory of a country or region is recognized as a slightly flattened "S-shaped curve" (Northam Curve). The urbanization develops slowly in the initial stage, accelerates in the middle stage, and slows down or even stagnates after the second half of urbanization in the mature stage. China's urbanization level varies greatly. The urbanization rate in a large number of inland areas is only about 30%, which is still in the initial stage. And there is much room for future growth, which needs to increase by 2% or more every year. The urbanization in coastal areas has entered a mature stage, and the urbanization rate has increased by about 1% every year.

When the planned urbanization rate increases by more than 2% and 3% every year, it is particularly important for the local government to have corresponding planning policies as support and put forward more active population policies, otherwise it is difficult to continue to achieve the rapid growth of urbanization. As China's urbanization rate enters the second half, and there has been shrinking cities of around 30%. The report of the Nineteenth National Congress of the Communist Party of China puts forward that it is necessary to shift from high-speed growth to high-quality growth. The Central Urbanization Working Conference also pointed out that it is necessary to follow the law and strengthen the macro management of urbanization. However, planning practitioners are still accustomed to incremental planning, which brings wrong guidance to government departments, leads to equally difficult to advance or retreat in the exercise of government affairs and the formulation of local policies, and cannot better guide urban construction.

The country is now revisiting the new urbanization, and the scale and quality of cities becomes important spatial governance challenges. Cities grow at different rates at different stages of development, with different levels of efficiency in economic output and population concentration. Therefore, it is

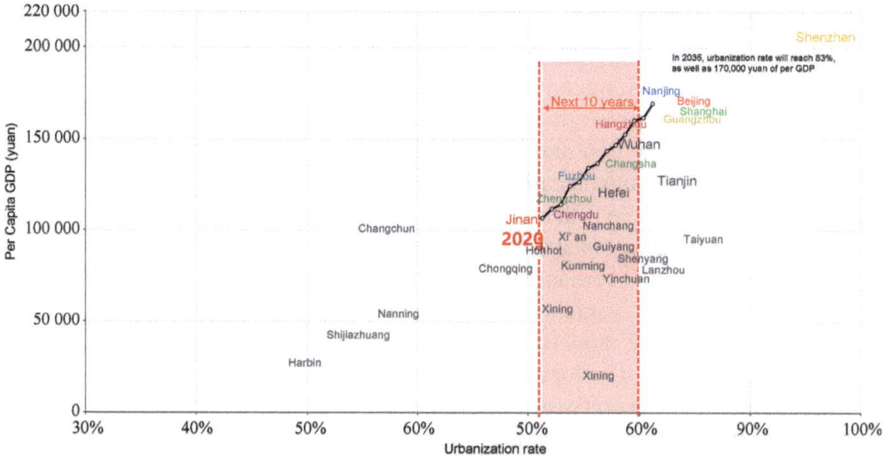

Fig. 6.3 Pre-judgment of urbanization rate and per capita GDP planning in Jinan from 2020 to 2035. *Source*: Self-painted by author

necessary to be in accordance with local conditions and explore the development mode of urbanization according to local conditions. Taking Jinan as an example, it is planned that the urbanization rate of Jinan will reach 80% from 70% in the next decade (Fig. 6.3). This process of urbanization is also the most important process of urban transformation in Jinan, and even at the risk of falling into the middle-income trap. Therefore, the country regards Jinan as a pilot demonstration area for the transformation of old and new kinetic energy, which is actually a measure taken according to the pattern of urban development.

(2) Demand response of new urbanized population

From the perspective of people, attentions need to be paid in planning to the demand of public service facilities for the population moving into the cities after they have changed from an agricultural to a non-agricultural population. After urbanization enters the second half, what the newly urbanized population needs most is not the allocation of facilities according to the 1,000-person target or the living area, but vocational training for workforce transfer. Without employment training, it is difficult to make a living in cities, and is impossible to really enjoy the dividends of urbanization. Thus, there will be unstable factors in cities. In fact, many areas with more than half of urbanization are facing such a problem. The population entering the city is no longer an effective workforce, but most of them are children, women and the elderly. At this time, the public goods that cities need to provide need to be targeted and dynamically adjusted according to the structure of the population entering the city.

Housing policy is an important policy guarantee for new urbanization. The government and the market should jointly build a fairer social environment. The government should provide more active housing policies, regulate the housing structure, and provide more low-rent housing, public rental housing, talent apartments, etc. There is also the concept of affordable housing in foreign countries. In November, 2020, Hongkou District in Shanghai provided public rental housing for 60 couriers to recognize their contribution to supporting the circulation of epidemic prevention and living materials in the city during the epidemic, reflecting the humanistic care of the city.

(3) The foundation of the fifth modernization is the modernization of human beings

The ultimate goal of territorial spatial planning is the modernization of governance system and governance capabilities, and the foundation is the modernization of people. Government officials, planners, teachers and designers of design institutes should be modernized, and the whole nation should be modernized, while modernized development concepts and planning and compilation techniques must also be updated. Only when people achieve modernization can they practice the modern governance system.

3.2 Programme and Planning

(1) Contradiction between planning function and the multiple compliance in market transition

In the past, there were serious conflicts in various spatial planning, which led to out-of-control management and unclear responsibilities due to non-interlaced planning and uncoordinated management. The planning of the three major departments, namely the land department, the development and reform department, and the construction department, and the various types of spatial planning, such as urban and rural planning, land use planning, development planning, and environmental protection planning, overlap in space and clash in content, resulting in problems such as fragmentation of the management system, self-expansion of departmental interests, ambiguity in the boundaries of management authorities, and decentralization and inefficiency in management decision-making (Fig. 6.4).

The difference between urban and rural planning and land use planning is the most prominent, which leads to the decline of planning space control ability, the inability to use a large number of development land indicators directly, and the difficulty in revitalizing the existing urban and rural development land with potential. Taking Xiamen as an example, there are 120,000 difference spots between urban and rural planning and overall planning for land use, with a total area of 306 km^2, of which about 55 km^2 of development land indicators cannot be used directly, making it difficult for annual key construction projects to be implemented. When a project needs to be implemented, due to the lack of detailed control planning, the government will usually come up with a set of control indicators specifically for the project, but in fact, these indicators lack systematic consideration at the urban level. Although in the short term, it solves the problem of the project's pre-implementation at the moment, over time, it

Fig. 6.4 Overview of the main plans of the original departments. *Source*: Self-painted by author

	Planning Name	Laws and Regulations
1	National Economic and Social Development Plan	*Constitution of People's Republic of China*
2	Main function Area Planning	Strategy and System
3	New Type Urbanization Planning	----
4	Land Planning	----
5	Land Use Planning	*Law of the People's Republic of China on Land administration* *Implementation Regulations of the land administration Law of the People's Republic of China* *Property Law of the People's Republic of China*
6	Urban and Rural Planning	*Urban and Rural Planning Law of the People's Republic of China* *Measures for the implementation of the regulations on the administration of the planning and construction of villages and towns* *Regulations on the Protection of Famous Historical and Cultural Cities, Towns and Villages* *Measures for the Preparation, Examination and Approval of Detailed Control Plans on Cities and Towns*
7	Ecological Function Zoning	---
8	Environmental Protection Planning	*Environmental Protection Law of the People's republic of China*
9	Mineral Resource Planning	*Mineral Resources Law of the People's Republic of China* *Interim Measures for Planning and Management of Mineral Resources*
10	Agricultural Regionalization	*Agriculture Law of the People's Republic of China*
11	Marine Functional Zoning	*Law of the People's Republic of China on the Administration of the Use of the Sea Areas*
12	Island Conservation Planning	*The People's Republic of China Island Protection Law*
13	The Protection, Construction and Utilization of Grassland Planning	*Grassland Law of the People's Republic of China*
14	Water Function Zoning (Integrated Water Resources Planning)	*Water Law of the People's Republic of China*
15	Nature Reserve Planning	*Regulations of the People's Republic of China on Nature Reserves* *Law of the People's Republic of China on the Protection of Wildlife*
16	Soil and Water Conservation Planning	*Law of the People's Republic of China on Water and Soil Conservation*
17	Planning for Prevention and Control of Desertification	*Agriculture Law of the Peoples Republic of China* *Law of the People's Republic of China on Desert*

will be easy to cause confusion in the urban layout, and will become the wrong guidance of urban development, and it will be difficult for the planning to play the role of guiding the development of the city.

(2) The role of planning in supply-side reform

With regard to the transmission of the focus on indicators and the scale of territorial spatial planning, there is a view that the state has begun to emphasize "planning" again. This view is incorrect. "Programme" and "planning" are essentially different. The programme emphasizes decomposition and top-down transmission, while the planning emphasizes local differentiation strategy and response. For example, in some regions, before 2035, population and land use may go through a process of increase and then decrease, and industrial land may go through a process of preservation and then retreat, which requires local governments to formulate differentiated and innovative policies during periods of the 14th, 15th or even 16th Five-Year Plan.

The current round of territorial spatial planning compiling needs a span of 15 years, which includes the establishment of four systems: planning and compilation approval system, implementation and supervision system, regulation and policy system, and technical standard system to ensure the implementation of the planning.

(3) Global resource element management

At the end of 2020, *The Guide to Classification of Land Use and Sea Use for Territorial Space Survey, Planning and Use Control (Trial)* was issued. On the basis of integrating the original classifications such as *Land Utilization Status Classification*, *Urban Land Use Classification and Planning and Development Land Use Standard*, and *Sea Area Use Classification*, it established a unified national guide to classification of land use and sea use, and carried out full coverage of land use classification on the ground and underground, sea area and land, and inside and outside the development boundary. It makes up for the defects that traditional urban and rural planning only pays attention to the surface, land and development land. China will carry out overall planning for mountains, water, forests, fields, grasses, lakes and oceans, and integrate them into territorial spatial planning, so as to carry out control over the whole area, type and resources, and transform the original vertical line control into horizontal network and systematic control.

Within the urban development boundary, the management and control mode of "detailed planning + planning permission" is adopted. The current round of detailed planning map should not only include the traditional 9 or 12 indicators, such as entrance direction, floor area ratio, height limit and green area ratio, but should also include the response to the aging population, such as the accessibility indicator, the allocation of vocational education facilities for the new urbanized population, the employment indicator, the output and performance of industrial land, the indicator of carbon emission, the per capita production value and the tax revenue. There are great differences in industrial performance in different regions of China. For example, Liaodong Bay New District in Panjin City, Liaoning Province has low industrial performance, with an output value of only about 300 million(yuan) per square kilometer, which is far from advanced regions such as Shenzhen and Shanghai (the output value per square kilometer of Hongqiao in Shanghai is as high as 8 billion(yuan)). Therefore, the map indicators of detailed territorial spatial planning should reflect the carrying capacity of employment, the average output tax revenue, and the requirements of energy conservation and emission reduction, the capacity of carbon sinks, and respond to the national strategy and layout.

Outside the urban development boundary, the management and control mode of "constraint indicator + district access" is adopted. In fact, the constraint indicator means that the local government decomposes and undertakes the national strategy, formulates the local strategy, transforms the strategy into a goal, and then transforms the goal into indicator.

3.3 Development and Protection

(1) Transformation of spatial governance concept and reform for promoting ecological system

China is a country with fragile ecological environment, shortage and uneven distribution of natural resources. It is necessary to take an intensive and ecological development road on the premise of ecological resources and environmental carrying capacity. In the past few decades, the local development in leaps and bounds has resulted in an increasingly prominent contradiction between development and protection. Under the background that the concept of spatial governance needs to be changed urgently, the state issued the *National Main Functional Area Planning*, which makes overall planning for the future population distribution, economic layout, land use and urbanization pattern according to the carrying capacity of resources and environment, existing development density and development potential of different regions, and gradually forms a spatial development pattern in which population, economy, resources and environment are coordinated.

According to the development content, the main functional areas are divided into: urbanized areas, major agricultural production areas and key ecological function areas are actually the predecessors of the three areas and three lines proposed in territorial spatial planning. Zoning is the concept of main function, and there are still various types of land under zoning. For example, in the major agricultural production areas, there will also be development land and ecological land; In key ecological functional areas, there will also be farm land. The main functional area only represents the dominant function.

The unit where the main functional area descends to the district and county level is called the leading functional area, that is, the first-level planning zoning. According to the strategic positioning of the main functional areas of cities and counties, combined with the development strategy of territorial spatial planning, the whole territorial space of cities and counties is divided into six basic divisions: ecological protection area, ecological control area, farmland protection area, urban development area, rural development area and marine development area. Urban development area, rural development area and marine development area are subdivided into second-level planning areas.

The detailed planning level of territorial space will be divided into several control units. These units within the urban development boundary of the central city are regulatory detailed planning units, suburban units are outside the urban development boundary of the central city, and township administrative boundaries are taken as units around the central city. Different control units need to refine the corresponding control requirements: some control units need development and construction, some need reclamation in farm land, and some need comprehensive land improvement and ecological restoration. Therefore, "transmission" is the key word in territorial spatial planning, including five-level downward transmission and transmission from overall planning to detailed planning.

In the *Overall Plan for the Reform of the Ecological Civilization System* issued by the CPC Central Committee and the State Council in 2015, it proposed that the reform for promoting the ecological system consists of eight parts, namely, the system of property rights of natural resources, the system of territorial space development and protection, the system of spatial planning, the system of overall resource management and comprehensive saving, the system of paid resource use and ecological compensation, the system of environmental governance, the system of market economy of environmental governance and ecological protection, and the system of performance evaluation and accountability of ecological civilization. The eight systems constitute an ecological civilization system with clear property rights, multiple participation, equal emphasis on incentives and constraints, and complete system, which promotes the modernization of national governance system and capabilities in the field of ecological civilization.

(2) The internal rationale of spatial planning reform

Spatial planning reform has evolved from guiding construction to guiding development to spatial governance, with the goal of establishing an ecological civilization-oriented spatial governance model for high-quality development, emphasizing the strengthening of bottom-line management, promoting multiple compliance, and guiding high-quality development, with the core of grasping the relationship between planning reform and high-quality development.There are three major issues in spatial planning reform. The strategic issue is to establish a spatial system to support the high-quality development of cities, the institutional issue is to establish the multiple compliance mechanism for planning implementation, and the technical issue is to establish a technical system to meet the requirements of planning reform.

3.4 Reform

With the deepening of reform, the management system and mechanism of natural resources in China are undergoing important changes. According to the institutional reform plan of the State Council reviewed and approved at the the First Session of the 13th National People's Congress of the People's Republic of China, the Ministry of Natural Resources was established. Therefore, the current round of territorial spatial planning is not only the wide implementation of national strategy, but also the shadow of political system reform. The state and local governments have merged departments and set up Ministry of Natural Resources, merging the former Ministry of Land and Resources, National Development and Reform Commission, Ministry of Housing and Urban–Rural Development, Ministry of Water Resources, Ministry of Agriculture, State Forestry Administration, State Oceanic Administration, State Oceanic Administration and State Bureau of Surveying, Mapping and Geographic

Information into the Ministry of Natural Resources, exercising the functions and powers of spatial planning, development control and property rights management in a unified manner. Spatial planning refers to compiling a unified spatial planning, promoting the multiple compliance, and realizing the organic integration of land use planning and urban and rural planning. Development control refers to the unified exercise of all territorial space use control duties, and the control of territorial space use based on spatial planning. Property rights management refers to the unified exercise of ownership over all types of natural resources, including mountains, water, forests, fields, lakes and grasslands, and is responsible for the management of all natural resources.

The main body of territorial spatial planning is the Ministry of Natural Resources, which needs to organize the compilation and supervise the implementation, and other departments also need to participate and cooperate in the process of it, including National Forestry and Grassland Administration, Ministry of Agriculture and Rural Affairs, Ministry of Water Resources and so forth. Therefore, if there is only a "1 + 1" (general report + information platform) result system, it cannot meet the needs of local government planning and implementation. Referring to the achievement system of Shanghai overall planning, the problem of horizontal departmental coordination is well solved. Shanghai overall planning 2035 adopts the achievement system of "1 + 3", "1" is the report of Shanghai overall planning (2017–2035), and "3" is the zoning guide, special planning outline and action planning outline respectively. Among them, the special planning outline includes 7 sections and 19 special fields, corresponding to the authorities of departments, and the strategic orientation, system layout and key indicators are clearly defined in sub-systems to ensure the deepening and implementation in each special system.

4 Perspective of People and City

From a personal perspective, the core of a city is "people", and urban planning and development should always put people at the first place. Through the comprehensive upgrading of urban planning, development, management and service, the city will be warmer, and the "people" living in the city will have a sense of well-being and belonging. The plan seeks to establish a performance appraisal system, such as annual check-ups and five-yearly evaluations, with the ultimate goal of measuring whether people have a sense of well-being and belonging. Even if a city is built to be as prosperous as Shanghai, Guangzhou or Shenzhen, if the people do not have a sense of well-being, the plan will not be successful. From a regional perspective, the perspective of people and city includes new urbanization, urban–rural coordination and other important contents. Urban planning should adhere to the all-round development of people and promote new urbanization with people as the core. For example, the special importance of the urban experience is reflected in Shanghai's proposal to build a city that is "readable, walkable, and warm."

For another example, Taiyuan has established a "street" that will never be widened, which is called "street" instead of "road" here. What is the definition of street? It is people-oriented, mainly based on pedestrians priority, while roads may be mainly based on cars priority. It is impossible to simply evaluate whether it should be guided by "pedestrians priority" or "cars priority", because these are two different stages of development. However, if the strategic perspective and value orientation of planning changes, the control techniques need to change as well.

In addition, there should be opportunities to participate in urban development from an individual perspective. Apart from the participation in planning research and the planning public announcements, public participation should be carried out throughout the entire planning process. At the same time, it is necessary to adopt different participation methods and contents according to the age structure, income structure and location differences of local citizens. The "people" living in the city are part of the urban system, and only by giving them a sense of well-being can they co-prosper with the city, so that the city will show its vitality, and the urban community of life formed by the people and the city will be more attractive.

4.1 Demand Change

The Fifth Plenary Session of the 19th Central Committee of the CPC proposed that the common prosperity should be promoted in a down-to-earth manner, and the people's sense of well-being, happiness and security should be continuously enhanced. By 2035, people's lives will be better, and more obvious substantive progress will be made in the all-round development of people and the common prosperity of all people. And what exactly is common prosperity? Taking Shanghai as an example, now that there are fewer and fewer young people in the city and housing prices are getting higher and higher, the government needs to change its housing policy to meet the housing needs of young people in accordance with the structural contradiction between supply and demand.

Another example is that under the traditional development model, urban parking is an important issue that needs to be urgently resolved in the process of urban expansion due to economic factors and capital investment. Many city governments have found this problem and actively taken control measures. For example, in actual projects, a city constantly requires to increase the floor area ratio of parking spaces, and its ratio has increased from the initial 0.8 to 1.0, then to 1.5, and even to 1.8.

For this problem, different areas have different parking space allocation standards, so it is necessary to allocate corresponding parking spaces according to the development needs of the region. However, if we want to solve the parking problem, we don't have to increase the floor area ratio of parking lots. Taking *Jinan Overall Planning of Territorial Space (2021–2035)* (process draft) as an example, the plan proposes to build localized control units, i.e., slow-moving units or bicycle units, in the central city. Slow-moving traffic will be used within the units, and TOD or medium-capacity transit will be used to connect the units, rather than simply increasing floor area ratio to solve the problem.

Countermeasures of "The Worst Traffic Jams" Jinan

Jinan once became the "The Worst Traffic Jams", that is, the most blocked city in China in 2017.Jinan is about 108 km long from east to west, and is a typical belt-shaped city, separated from north to south by the Yellow River and Mount Tai, which makes the congestion phenomenon very obvious. Planners need to change their mind and use new urbanization concepts to deal with it, so as to enhance people's sense of well-being.

As early as 2013, Long Ying et al. took the national population and land area data in 2000 and 2010, and measured the change in population density in each township and street across the country by dividing the population density in 2010 by the population density in 2000.The results showed that the population density of 4 provinces, 105 regions/city areas, 897 counties and 19,822 township sub-district offices in China has decreased, with 1/3 of the territorial area experiencing a decline in population density at the township sub-district office scale. In fact, this reflects the differentiation of urban development in China, which can be divided into agglomeration development and contraction development.[1]

We need to understand that in the process of wide implementation of the national strategy, local strategies are needed to implement with it, so as to enhance the people's sense of well-being. The author believes that the master of the city in the future is "the people", if they can not have a real sense of experience, or we can not let them understand the process of national strategic transmission, only rely on restriction and control, then the planner is just doing some digital games, planning will be meaningless.

4.2 *Governance Change*

Territorial spatial planning is the wide implementation process of the top-level strategy of the CPC Central Committee, as well as the policy agenda of local governments, the national consensus, and the full expression of local differentiated development demands. Shanghai has earlier explored and practiced in this regard, for example, the *Opinions of the Shanghai Municipal Committee of the Communist Party of China on Deeply Implementing the Important Concept of "People's Cities are Built by the People, and People's Cities are for the People", and Writing New*

[1] Long et al. (2015).

Chapters of the People's Cities in the New Era puts forward the following five "everyone": A city where everyone has a chance to excel in life, a city where everyone can participate in governance in an orderly manner, a city where everyone can enjoy quality of life, a city where everyone can feel the warmth of the city, and a city where everyone can have a sense of belonging and identity. In practice, the banks of the Huangpu River in Shanghai run through 45 km, and the public service facilities are updated appropriately, the public service is provided accurately, the response to the aging society, the corresponding housing demand for young people, the construction of international communities, etc.. The author believes that this is a model of the local government's own work.

Cities are people's cities, and the related planning of cities must be closely related to people. In the process of compiling Shanghai's overall planning, its public participation is also worth learning from. Throughout the planning process, public participation is organized: There are public documents specially compiled for people, and they will participate in discussions during the preparation process. Therefore, due to the high participation of them, most citizens know that "an excellent global city" is Shanghai's long-term goal after compiling the overall planning.

(1) Rationale for the transformation of governance foundation: changes from "quantity" to "flow"

In the process of unbalanced development of cities, the population agglomeration is uneven in space. Planners used to know the phrase "10,000 people per square kilometer" is actually quite problematic. Planners need to change the mind, from the original "quantity control" to "flow control", that is, planning not only to control the total amount of scale, but also need to pay attention to the agglomeration of people and the implementation of future strategic intent in space. Therefore, planners should pay attention to the "bottom line", face control, support control, innovate control, carry out bottom line constraint and effective transmission, implement structural element identification and zoning guidance strategy, and carry out target management and process control.

(2) Emphasis on the compilation process and establishment of four major planning systems

The compilation of territorial spatial planning should be strategic, scientific, coordinated and operational, with emphasis on planning implementation and management, and four systems should be established: the system of examination and approval for compilation, implementation and supervision system, regulation and policy system, and technical standard system, so as to ensure that the planing is usable, effective, and easy-to-use. The examination and approval system for compilation should correspond to the authorities and form a hierarchical compilation system and examination and approval system; To implement the supervision system, it is necessary to establish monitoring, evaluation, early warning, feedback mechanism and information platform; The regulation and policy system should be guaranteed by laws, regulations and policies; The technical standard system should reconstruct the technical standard system of spatial planning for multiple compliance.

4.3 Main Goals of Reform

The goal of the current round of territorial spatial planning reform is to establish a spatial governance model oriented by ecological civilization and with high quality development; Its task is to strengthen bottom line control, promote the multiple compliance and lead high-quality development; The core of its work is to grasp the relationship between planning reform and high-quality development. Promoting high-quality urban development is the purpose of reforming the spatial planning system, which is the guarantee for leading high-quality urban development.

In the process of planning and compilation, planners should firmly grasp the three main goals of reform: spatial pattern, compilation system and the multiple compliance, and establish a response mechanism of action dimension.

(1) Spatial pattern under the guidance of high-quality development

It is necessary to do a good job of paying attention to two "concerns", that is, to build a new pattern of territorial space development and protection under the guidance of high-quality development, to optimize the pattern of the space in urban, agricultural and ecological areas, and do not give up until the goal is reached. First, it is concerned with the change in development orientation. China's urbanization has entered the second half of the period, which requires changing the urban development concept and development pattern from upgrading the scale to upgrading the quality, reflecting the requirements of high-quality urban development and high-quality life. Take the road of ecological priority and green development while adhering to connotation development. Second, it is concerned with the requirements of high-quality urban development. The essence of it is a people-oriented city that can be felt by the people, so the livability of the city and the improvement of the quality of life are the ultimate goals of high-quality urban development. High-quality development needs to be supported by regional coordination and urban–rural planning, and the level of urban spatial governance needs to be upgraded to sophisticated governance.

(2) Planning system adapted to the requirements of control.

According to Ministry of Natural Resources, in accordance with the principle of "approving what is managed", the territorial spatial planning should be compiled in a way that is implementation-oriented, control-oriented, supportive of control, innovative in control, and gives full play to the dual roles of planning as a target leader and a bottom-line controller.

Bottom line constraint and effective transmission are formed in transmission, structural element identification and zoning guidance strategy are carried out in space, and target management and process control are carried out in time. In terms of innovation in planning achievements, the experience of *Shanghai Urban Overall Planning (2017–2035)* is worth learning from. Its planning achievements are divided into three versions, which are for the higher-level government, for the local control and for the public. The first is the submission documents, namely, the achievements submitted for approval to higher level governments, which need to be determined by higher level authorities. The second is the policy documents, namely the technical documents for local control, which can effectively achieve the transmission of control requirements to the lower level planning. The third is public documents, that is, transforming the planning achievements into public language that people can understand, and illustrating the main contents of the whole territorial spatial planning.

(3) Strengthen the planning and operation mechanism of the implementation dimension

The ultimate goal of spatial planning reform is the implementation of planning. It is necessary to establish a planning operation mechanism in implementation dimension, and improve the examination and approval system, implement supervision system, regulation and policy system and technical standard system, which reflect the overall requirements of planning, construction and management, and do not give up until the goal is achieved. Territorial spatial planning is not only a strategic blueprint with forward-looking leadership, but also a legal rationale for the protection of natural resources and the control of spatial use; as well as a platform for collaboration on spatial systems and an action program, which is an important tool for the government to implement government affairs.

(4) Establish an action response mechanism.

A response mechanism for the current round of territorial spatial planning will be established in six aspects. First, improve the compilation system, including five levels and three types of planning systems; Second, establish a rolling implementation mechanism with short-term planning as the carrier; Third, strengthen the design of regulation and policy system, and formulate planning implementation regulations; Fourth, establish a planning evaluation and adjustment mechanism, and the planning evaluation process is also a process of evaluation and supervision; Fifth, improve monitoring and early warning and implement supervision mechanism; Sixth, improve the planning committee system.

With regard to the short-term planning, it is necessary to establish a response mechanism for integrated evaluation of the status of urban development. Short-term planning is an important carrier for the rolling implementation of the overall planning of territorial space, but also a carrier for the dynamic maintenance of it. The status of the short-term planning should be clarified, a rolling implementation mechanism should be established with the short-term planning as the carrier, annual check-ups and five-year assessments should be combined with the compilation of the short-term planning, and a timely feedback and dynamic adjustment mechanism should be formed.

4.4 *Perceivable Space of People's City: A Case Study of Shanghai University Road*

(1) Overview of University Road

The Road is located in Wujiaochang, Yangpu District, connecting Wujiaochang Business Circle in the west and Fudan University in the east, and the reconstruction project was completed in 2006.Its planning is positioned as a mixed area of residence and entrepreneurship, which was originally an ordinary road. Because the whole road space is controlled by pedestrians priority in the planning, it is an ordinary road during the day, but a large number of outsiders gather at night, which makes it extremely lively. The location is close to Fudan University and Tongji University, attracting many young people to start their own businesses and activities, and has now developed into a internet celebrity street that brings together artistic youth and student communities. University Road has a total length of 700 m. There are two normalized open-air markets, namely Tiandichuang Market and Trunk Party. The open-air market is open to the public every Friday to Sunday and holidays from 11:00 to 22:00. The theme of the market changes every month with dazzling products, which attracts young people.

For example, in Cambridge, Boston metropolitan area, which is home to two schools, Harvard University and Massachusetts Institute of Technology. And there are more young people including college students. On special festivals, the mayor has to sign a decree on closing roads for 8–10 h, and then various activities will appear on the roads one after another. At this time, land use management becomes spatial management, and indicator management becomes behavior management. Markets and flea markets have become places for young people to socialize and meet. In fact, operators do not rely on market transactions to make profits, but because of this form, the market becomes a kind of social interaction, which makes an urban scene with happiness knowing no bounds, a more stable society and a more attractive city.

The national voice of street vendor economy

In June 2020, when Chinese Premier Li Keqiang visited Yantai, Shandong province, he said that the street vendor economy and the retail store economy are important sources of employment, which embodies the hustle and bustle, and like "high-end", they are the vitality of China.City market is a form of street vendor economy and an important spatial carrier to stimulate urban vitality.

Tiandichuang Market: Exit 9 of Jiangwan Stadium Station of Metro Line 10.

Trunk Party: The intersection of University Road and Jinjia Road.

(2) The spatial background was laid by people-oriented street space

Shanghai has always emphasized the design of street space, and University Road is a model of street space construction in Shanghai, which is a good street scale and environment formed under the guidance of design. Therefore, from the beginning of its establishment, it attracted the spontaneous entry of college students' market and formed a unique street culture.

(3) Compound utilization of space to enhance the vitality of places

The mutual integration of different spaces is used to create places on University Road. The retreat space of shops along the street is integrated with the market, which brings people to the shops, and the shops provide water, electricity and warehousing support for the market. Parking lots, street corners, squares and markets are integrated to form a symbolic stopover space, and characteristic booths are gathered around them to provide diversified choices for pedestrians.

(4) Continuous renewal of commercial activities to maintain the vitality of the place

Staggered development is achieved with market function, street level commercial and complex. Commercial activities planning is characterized by "small, many and new". "Small" means that all kinds of goods are small in scale, and the goods in a store may be filled in several boxes; "Many" means that there are many kinds of goods; "New" means new ideas, which refers to the theme of University Road Market will change every month.

(5) The unique trunk mode strengthens the characteristics of the place

Trunk Party uses trunk as booth, which not only saves booth space, but also enhances the convenience of setting up stalls. Trunk market has the characteristics of small floor area and novel form, which effectively alleviates the negativity of parking space, combines parking spaces with street and square space to form a vibrant public space, and at the same time creates the "topicality" of University Road Market.

5 Some Thoughts on Territorial Spatial Planning

5.1 Beware of Four Tendencies

Tendency one: The smaller the key ecological function area of dual evaluation is the better. The smaller the authorities reported to the higher-level government is the better, but this does not mean that the key ecological function area of dual evaluation or ecological red lines is the ecological bottom line, and the ecological space outside does not need to be controlled. In fact, each city should set the bottom line requirements according to local characteristics, bring the space with high ecological value outside the ecological red line under control, and formulate the corresponding control requirements.

Tendency two: In many cities, the permanent prime farmland protection line and ecological red line are separately dispatched under the circumstances that the spatial planning has not yet been determined. This working method has certain overall significance, because the "lines" drawn by the previous forestry department, agricultural department, environmental protection department and other departments conflict with each other, which requires coordination and overall planning in the current round of territorial spatial planning. However, the essence of the three-line demarcation is not a technical problem, but a strategic decision-making problem. It's method is closely combined with the strategic orientation of urban development. If the spatial planning is put aside, and it cannot be effectively combined with the urban development strategy and the actual needs of people, the delineated "line" may be unreasonable and untenable.

Tendency three: It's better for more indicator increments. In current round of territorial spatial planning, the indicators of development land issued by higher-level government are showing a tight trend, and all local governments are striving for indicators, which means that "the more the incremental indicators, the better". It is impossible to reasonably determine the truly usable and usable amount according to the attitude of seeking truth from facts, failing to reasonably determine the amount of land that can be truly used and usable in accordance with a realistic attitude, and failing to face the current situation of existing urban and rural development land and ineffective land use squarely. According to the principle of "indicators follow the project and the performance", the land for development projects should be guaranteed and the performance of development land should be controlled. In the future, if the local governments fail to achieve the performance, not only can they no longer be provided with land, but they may also face the risk of administrative accountability.

Tendency four: Waiting for superior guidance to planning. The guidelines for compilation of territorial spatial planning issued by the state focus on principle and guiding requirements, which do not have strong local adaptability, and need local practice and exploration to implement and supplement the content of the guidelines locally. Planning can be explored and innovated appropriately while waiting for the guideline.

5.2 *Clarify Three Major Strategies*

(1) Strictly adhere to the bottom line strategy in planning expansion

In addition to the hierarchical control in the ecological conservation red line, local governments can also establish their own ecological control system according to their needs. Taking Shanghai as an example, an ecological network system is established through wind environment, green corridor, slow-moving system and ecological loop, which is controlled by Shanghai local government. At the same time, this network system needs to be equipped with necessary public service facilities such as toilets, bicycle lanes, baby care rooms, etc. These nodes with international vision reflect the development goal of Shanghai's "global city" and the meticulous care of Shanghai's "people's city", which is the bottom line strategy with Shanghai's local characteristics.

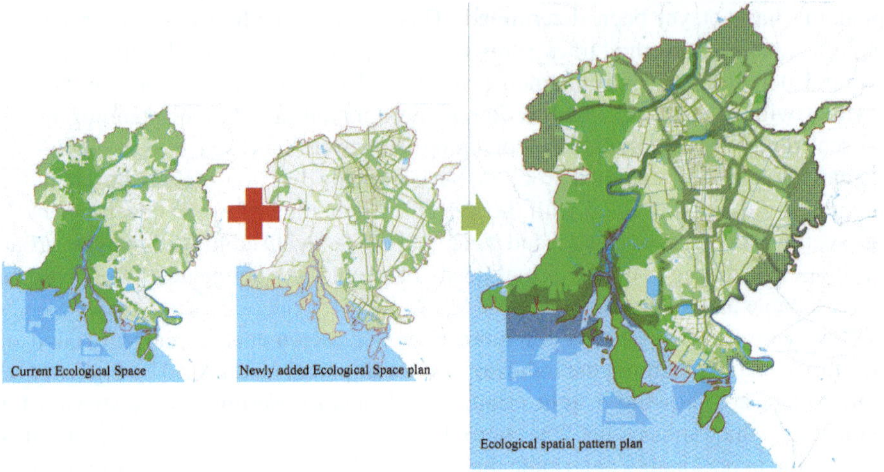

Fig. 6.5 Analysis of ecological spatial pattern in Panjin city. *Source*: Fifth Insitute of Shanghai Tongji Urban Planning and Design Institute Co., Ltd. Project Team of *Overall Plan of Panjin City Territorial Space (2021–2035)*

For example, in the *Overall Plan of Panjin City Territorial Space (2021–2035)* (process draft) of Liaoning Province, it is necessary not only to ensure the total scale of ecological land, but also to establish its ecological network structure to ensure the ecological conservation red line and biodiversity. Through the establishment of ecological space system oriented by wetlands, the urban wetland pattern surrounded by country parks and covered by green nets will evolve.In addition, a 150 km greenway is planned to respond to the development goals of Panjin as an international tourist city and an international wetland city (Fig. 6.5).

(2) Strengthen the quality strategy in land use layout

The layout of territorial spatial planning should deal with the relationship between human and land, highlighting the correspondence between human and land and that between production and land, and consider quality as a major strategy of urban development. In Panjin, Liaoning Province, for example, there are several spatial levels of planning. At the central city level, there is a human-land correspondence of 10,000 people per square kilometer, while the peripheral Liaodong Bay area can be treated differently. Another example is Xiongan New Area, because there are no large-scale factor-intensive industries, it is planned to support 2 million people in 100 km^2.Many cities in China are mainly equipped with traditional living facilities, and it is also possible to support 12,000–15,000 people per square kilometer. Therefore, human-land corresponding planning should be formulated according to local conditions. Panjin, Liaoning Province is still taken as an example to show the corresponding between production and land. At present, the average GDP of industrial land in Liaodong Bay is 359 million yuan/km^2. During the planning, the project team made a horizontal comparison of heavy chemical industrial parks and ports at home and abroad, including Ningbo, Shenzhen, Seoul, New York, etc. After reasonable benchmarking, the average land productivity target is set at 1.5 billion yuan/km^2, which is also an important evaluation indicator for urban check-ups (Table 6.1, Table 6.2).

The correspondence between human and land and that between productive areas in Panjin City Forecast of land use scale in central city and county town in northern Panjin:

The current situation is 180 m^2/person, which will be optimized to 115 m^2/person in 2025 and 100 m^2/person in 2035.The carrying capacity of the central city in Xiongan New Area is 20,000 person/km^2, and the target of Panjin is 10,000 person/km^2.

Northern Central City: In 2025, the land use scale of the central city will be controlled at 94.3 km^2, and in 2035, it will be controlled at 105 km^2.

Dawa City Group: In 2025, the land use scale of Dawa City will be controlled at 10.35 km^2, and in 2035, it will be controlled at 15 km^2.

Panshan County Group: In 2025, the land scale of Panshan County will be controlled at 5.8 km^2, and in 2035, it will be controlled at 8km^2.

Forecast of land use scale in Liaodong Bay: land use determined by production

The current situation is 359 million yuan/km^2, and the planning is optimized to 1.5 billion yuan/km^2. It contributes 1.689 billion-2.6 billion yuan/km2 to the GDP of industrial land in Ningbo + Shenzhen, and Panjin aims at 1.5 billion yuan/km^2.

GDP forecast of secondary and tertiary industries: 62.8 billion in 2025 and 245 billion in 2035.

Average land productivity: At present, the added value of secondary and tertiary industries is 11.87 billion, the urban development land area is 37.3 km2, and the average land GDP is 318 million yuan/km2. The target value of average

Table 6.1 Comparison of GDP per unit of industrial land (100 million yuan/km^2)

Domestic	Liaodong Bay	Hangzhou	Shenzhen	Ningbo	Hongkong
	3.59	20.1	26	16.89	30
Overseas	Seoul	NY	Tokyo	Singapore	London
	53	52	57	75	98

Table 6.2 GDP Scale Forecast of Secondary and Tertiary Industries in Liaodong Bay and Choice of Average GDP

Programs	Year of 2019	Year of 2025	Year of 2035
GDP of Secondary and Tertiary Industries (RMB 100 million)	118.7	628	2450
Average Output of Secondary and Tertiary Industries (100 million yuan /km^2)	3.2	6.5	15
Urban Construction Land (km^2)	37.3	96.6	164

land GDP is set for major domestic cities. By 2025, the urban development land will be about 96.6 km2, and in 2035 it will be 164 km2.

(3) Highlighting performance strategies in spatial goals

Because there are differences in development conditions and development goals between regions, performance must not be done according to a set of standards, but based on the local reality and future vision, targeted performance targets should be put forward, such as Panjin Wetland, whose assessment indicators will include GDP, carbon sink capacity, carrying capacity, etc., and its performance evaluation is more biased towards ecological aspects. For a super-large city like Shanghai, which is building a "global city", its assessment indicators are more inclined to the examination of innovation, openness, internationalization and other capabilities, while its performance evaluation focuses on whether it conforms to the city's development goals and directions. Highlighting performance strategy in spatial goals effectively enriches the goal system of traditional planning, and makes the development more rational, more efficient and more prominent in urban characteristics.

5.3 *Attention to Five Issues*

First, grasp the role of territorial spatial planning as a policy agenda rather than departmental planning, and do not do "director's planning", but do "mayor's planning" or "secretary's planning". As a means of spatial governance, territorial spatial planning is also an important leverage for urban development. Only by reaching the same development direction and mode with local leader and deeply understanding the administration intention can it be more handy in the implementation stage. At the same time, the design of strategic intention at a higher level can also avoid conflicts between departments over ownership of planning.

Second, to understand the connotation of the current round of territorial spatial planning unchanged for fifteen years means that the strategic intention, strategic goal and planning vision of urban development remain unchanged for fifteen years, rather than the specific planning drawings remain unchanged for fifteen years. As a programmatic document of urban development, territorial spatial planning should put forward a certain forward-looking and predictable strategic intention, goal and vision direction of urban development after in-depth study, and try to ensure that the general direction will remain unchanged before the end of the validity period of the current round of territorial spatial planning. In the concrete implementation process, the specific strategies and drawings of action planning can be updated accordingly with the continuous changes of urban development needs and the continuous improvement of urban development conditions.

Third, to understand the "strictly" protected nature reserves are not simply establish the nature reserve system, but divide the spatial levels and system construction such as ecological protection, ecological restoration and moderate production on the basis of ecological spaces such as nature reserves and ecological conservation red line. Obviously, the current round of territorial spatial planning is to make "easy-to-use" planning, and the simple ecological protection system can not meet the purpose of establishing different levels of ecological systems in local areas, nor can it practice the concept of ecological civilization. The "one size fits all" approach is no longer suitable for the goal of spatial governance in the new era. It is necessary to flexibly understand that under the premise of "strict" protection, making better use of ecological resources is also a way to protect ecological resources.

Fourth, it should be specified that public participation is not just publicity, but requires deep public participation from the planning stage, publicity stage and implementation stage of territorial spatial planning. As one of the important means of this territorial spatial planning reform, public participation needs to carefully design its scheme. In the overall planning stage, the public's wishes are deeply understood by means of online questionnaires, field interviews and online participation. In the detailed planning stage, the community planner system is fully utilized to conduct extensive and in-depth communication with stakeholders, actively adopt opinions and suggestions conducive to the better implementation of the scheme, and deeply understand the importance of the public's experience and wishes as "users" of the city.

Fifth, the achievements of "1 + 1" (General Report + Information Platform) are not enough for the new stage of territorial spatial planning, because it involves local differences and further reform of departmental systems and mechanisms. It is necessary to further combine the exercise of authorities by local governments to formulate more local rules, regulations, guidelines, etc., which fully reflect the local response in the process of national strategic transmission, and make spatial planning more suitable for local and "easy-to-use".

6 Summary

All the compilation and control work of territorial spatial planning need to be oriented towards the exercise of governmental affairs, which is reflected in the refined governance. The ultimate goal of all government affairs is to enhance people's sense of experience. Planners must effectively link the current round of territorial planning with the national strategy and national intention, and at the same time combine it with the construction of the "people's city", so as to truly achieve consensus across society. From a strategic point of view, it is only in this way that there may be better technical solutions to the problems related to the three zones, the three lines, the construction targets and so forth.

Reference

Long Ying, Wu Kang, & Wang Jianghao. (2015). *China's shrinking cities and their research framework*. Modern Urban Research, (09): 14–19

Practice

Elastic Mechanism in the Comprehensive Territorial and Spatial Planning of Cities and Counties: Taking the Urban Development Boundary as an Example

Wang Xinzhe and Xue Haoying

The establishment of territorial and spatial planning system is an important measure to improve the spatial governance capacies in China. On the one hand, it solves the problem of overlapping spatial planning, on the other hand, it helps to clarify the ownership of natural resource assets property rights. According to the report delivered at the 19th National Congress of the Communist Party of China, the three lines of ecological protection red lines, permanent basic cropland and urban development boundary are regarded as an significant part for the construction of ecological civilization system, and has also become the core control content of territorial space planning, while the urban development boundary is undoubtedly the most concerned by urban planning practitioners.

The seriousness of planning has been emphasized many times in various documents of territorial space planning, and the goal is to reform the planning system through various means such as institutional reform, legal system construction and mechanism reconstruction, so as to avoid overlapping and conflict of planning contents and constantly changing rules. Planning rigidity is the focus and consensus of this round of spatial planning, which leads to the fact that spatial planning in practice is often more detailed and rigid. However, due to the complexity of reality, adaptive adjustment should be made according to the actual situation. At the same time, problems usually have "multipal solutions," so we have to give more possibilities for the future. How to adopt a more flexible way to ensure the scientific and serious spatial planning is an important issue that needs to be discussed urgently.

W. Xinzhe (✉) · X. Haoying
Shanghai Tongji Urban Planning and Design Institute Co., Ltd, Shanghai, China
e-mail: wxinzhe@163.com

Z. Peng (ed.), *Theories and Frontiers in Territorial Spatial Planning*, Urban Sustainability, https://doi.org/10.1007/978-981-96-9825-7_7

1 Rigidity and Elasticity in Planning

1.1 Rigidity and Elasticity of Urban Master Plan

Urban master plan is an overall, comprehensive and strategic plan that guides urban development and construction from the macro level. Through the transmission of master plan-detailed plan, planning permission system and a large number of technical reviews, in a long period of time, through the "rule of man", the rigidity and elasticity of the comprehensive territorial and spatial planning have been well implemented.

However, with the rapid development of economy, especially the strong intervention of local government, the urban master planning has been updated many times. With No.13[2002] of the State Council "Circular of the State Council on strengthening the supervision and management of urban and rural planning" as a symbol, the related governments strengthened the seriousness of urban and rural planning and gave full play to its guiding and regulating role in urban and rural construction as the orientation of the reform. To implement it, the Ministry of Construction issued the Interim Provisions on Mandatory Contents of Urban Planning, and then the mandatory contents were included in the "Measures for the Formulation of Urban Planning" promulgated and implemented in 2006. The Urban and Rural Planning Law of the People's Republic of China, which was revised and implemented in 2008, further emphasized the mandatory content in urban planning, and the subsequent "integrating various types of plans into a single plan" provided technical support for strengthening the rigidity of the comprehensive territorial and spatial planning.

Planning rigidity ultimately lies in the control of planning administration by laws and regulations. The related governments reduced discretionary authority and avoided power abuse and corruption caused by "multi-regulation" conflicts and repeated revisions. But at the same time, it also brings the phenomenon of "a little breakthrough and total negation of the achievements", which leads to the problem that the compilation time of the comprehensive territorial and spatial planning is too long and the related departments are hesitate to reply..

1.2 Rigidity and Elasticity in General Land Use Plan

The comprehensive land use planning is based on rigid management and control, which bears some planned economy flavor. Usually, various land indicators are used to realize effective management and control of land development, with clear management logic and rigid planning control. On the one hand, the comprehensive land use planning is too rigid and rational. Compared with the comprehensive territorial and spatial planning, it is bound to lack flexibility, and there are a large number of indicators that cannot be implemented, which need to be reflected; On the other hand, because the rules of the comprehensive land use planning are more clear, with the strengthening of the flexibility consciousness of land managers, the setting of various flexibility indicators and flexibility contents, the planning runs orderly under clear rules, and the effectiveness of its flexibility contents is worth learning.

Taking cultivated land as an example, the index of cultivated land quantity is one of the main control indexes of the comprehensive land use planning, but in the initial stage of planning implementation, its control often fails due to lack of flexibility. Therefore, flexible regulatory easing measures have been introduced one after another, from unchangeable cultivated land quantity index to dynamic balance of total cultivated land, and various types such as "replacement index", "deduction index" and "turnover index" have been introduced to permit some flexible demand.[1]

Up to now, the comprehensive land use planning has also formed a fixed and mature practice in setting flexible contents. In terms of agricultural land use, the rigidity of management control is emphasized for prime farmland protection areas, and reserve areas are reserved as coordinated areas for prime farmland adjustment; For general cultivated land, due to the change of farming conditions, the farmer is given appropriate development options in land use. In terms of construction land, binding indicators are adopted for urban and rural construction land and full-scale construction land, while anticipated indicators are adopted for urban construction land. At the same time, indicators and land use (conditional construction areas) are adopted to deal with the uncertainty of urban construction. Although there are not many flexible contents in the comprehensive land use planning, its operating conditions are clearly stipulated in laws and regulations. Flexible indicators are transmitted from top to bottom, which are usually retained by the superior government and distributed according to projects after examination and approval. Its clear management logic should also be used for reference by spatial planning.

1.3 Rigidity and Elasticity in Regulatory Detailed Plan

Regulatory detailed planning is an innovation of China's planning system. In order to solve the incompatibility of constructive detailed planning and compile the content that should control. Regulatory detailed planning is created in combination with foreign zoning systems. Comparatively speaking, the implementation of regulatory detailed planning has a stronger rigid control effect. However, since regulatory detailed planning was created to adapt to the uncertainty of construction and do something different, the research and practice of flexible planning in regulatory detailed planning are better than the comprehensive urban planning, reflecting the reason that rigidity and flexibility complement each other in planning. Flexible strategies of regulatory detailed planning include formulating general rules, adjustment possibility under special circumstances, interval control of indicators, dynamic organizational

[1] Zheng Zhenyuan (2004).

planning, etc.[2] The elastic strategy is developed to improve the applicability of planning and to involve more variety in space and construction. These elastic planning methods have been well practiced in Shanghai, Shenzhen, Chengdu and other cities.

1.4 Rigidity and Elasticity in Law

Even laws have flexible mechanisms. Taking administrative discretion as an example, administrative discretion is the disposal power that the state endows administrative organs with certain choice within the range and scope stipulated by laws and regulations. It is an important part of administrative power to deal with situations that laws and regulations cannot cover or need further discretion. Under the complexity and variability of society, it is an important means to improve administrative efficiency and accuracy of administrative decisions.

No one will question the seriousness of the law because of the existence of flexible mechanism, which is because the flexibility in the law is restricted by rigid rules. Usually, flexible mechanism is restricted by the power of relevant laws on the one hand, and supervised by rigorous procedures on the other hand. In all countries of the world, discretion is restricted by judicial review. In 1990, China issued the Administrative Litigation Law, which provided a legal basis for the judicial review of administrative discretion, and included abuse of power and other issues in judicial review. At the same time, strict administrative law enforcement procedures and open and transparent regulatory environment can effectively guarantee discretionary procedural justice. As jurist Taniguchi Yasuhei said, "Procedural justice is the origin of substantive justice, and whether the trial result is correct or not is not measured by some external objective standard. Enriching and attaching importance to the procedure itself to ensure that the result can be accepted is its common spiritual essence".[3]

The flexible mechanism ensures the accurate and efficient operation of rules, and the rules and procedures frame the reasonable category of flexibility. As a statutory planning in China, the operation logic of spatial planning is the same. Elasticity and rigidity guarantee and optimize each other, and jointly promote the legalization process of planning.

[2] LIU Kun et al. (2012).

[3] TAniguchi Anping (1996).

2 The Formulation of Territorial Space Planning Needs Flexibility

2.1 Flexible Planning is the Embodiment of Scientificity

Society is an extremely complex living body, and no planning can accurately predict the future trend. Retaining flexibility[4] itself is the embodiment of scientificity. Many literatures have discussed the bounded rationality of planning and the deviation between practice and theory under the rapid development of urban and rural areas in China.[5] Urban development boundary or urban growth boundary, its role as a policy tool has been recognized by most scholars. Urban development boundary needs rigid and flexible mechanism, as well as reasonable evaluation and adjustment mechanism. On the basis of rigid principle, a large number of scholars have put forward innovative thinking on flexible application mechanism of control boundary.

Although the current planning technology has made great progress, the social development has entered a relatively stable "new normal", the uncertainty is decreasing, and the management and control mode of spatial planning is also changing. Taking "urban development boundary" as an example, it has experienced the evolution process of "ultimate blueprint" to "elastic boundary" to "compulsory control line," which is essentially a policy design process and a comprehensive policy toolkit.[6] As a policy tool, it should have its corresponding supporting mechanism, where flexibility is important and indispensable.

2.2 Flexible Planning is the Stabilizer of Planning Seriousness

Territorial and spatial planning is different from traditional master planning in terms of contract mode. The control right agreed in advance in territorial and spatial planning becomes stronger, and the uncertainty of urban development still exists, but the cost of adjustment becomes higher. Through the flexible planning method, the planning can adapt to uncertain and changeable situations, greatly reduce unnecessary adjustments, make the planning more resilient and stable, and thus guarantee the credibility and seriousness of the planning.

Flexible rules are often a supplement to rigid content, which can provide regulatory cushion for rigid content. It not only adapts to various situations and local needs, but also ensures the adoption of equal procedures and rules in different situations, thus maintaining the fairness and justice of planning implementation.

---

[4] Cao Kang and Henry Hui Wang (2009).

[5] Yang Baojun (2003).

[6] Zhang Bing et al. (2014).

2.3 Flexible Rules Are the Guarantee for the Effective Implementation of Rigid Contents

The seriousness of planning emphasized by spatial planning should be reflected not only in the coercive force and accuracy of results, but also in the standardization of processes, which complement each other. Because of the complexity and uncertainty of planning results, the latter is often more important. Flexible rules are rules oriented to principles, methods and processes, which make flexible planning have evidence to follow, avoid corruption by stipulating process rules such as formulation, implementation, revision and supervision, and ensure social justice under various unpredictable scenarios. Flexible rules provide a code of conduct for planning, under which procedural justice in planning can be realized.

Rules that serve flexible content often define the tolerance and consistency judgment rules of flexible content to protect procedural justice in gray areas. It not only ensures the diversity and flexibility of planning through "flexibility", but also ensures the effective implementation of management and control through "rules".

3 Establish a Flexible Mechanism for Rigid Control of Territorial Space Planning

3.1 Elastic Mechanisms in the Formulation of Territorial and Spatial Master Plans

(1) Hierarchical and Graded Differentiated Constraints

Being a system of "five levels and three categories", the territorial and spatial planning not only includes the transmission from master plans to detailed plans but also requires transmission between master plans at different levels. Except for a few elements throughout all levels, plans at all levels and types are required to have varying degrees of flexibility to ensure the in-depth transmission and implementation from upper-level to lower-level plans. Plans at each level need to adopt graded differentiated constraints based on the power hierarchy. The so-called "graded differentiated constraints" mean that upper-level plans should leave room for flexibility. The relationship between upper-level and lower-level plans is not repetitive or contradictory but correlative and progressive. The distinction between strategic, coordinating, and implementable roles in the planning system clearly defines this progressive relationship from shallow to deep, thus resulting in different planning flexibility and depth.[7]

Regarding specific schemes for graded differentiated constraints, Dong Ke and Zhang Jing proposed 6 regulatory methods. They are setting rules, quantifying indicators, defining structures, delimiting boundaries, shaping forms,

[7] Zhang Ruixia and Lin Zhiming (2019).

and ordering sequences. Mega city, super large city and large city have more hierarchical plans, so they can set more "flexible" mandatory contents in the master plan stage; while medium and small cities can "achieve clarity in one step" by setting as "clear" mandatory contents as possible. Shanghai Tongji Urban Planning and Design Institute proposed six types of elements: "standards, zoning, lists, relationships, locations, boundaries" and four methods: "implementation, elaboration, optimization, and supplementation". In specific implementation, these six elements and four methods should form an intersecting matrix (Table 1).

(2) Planning Reservations

The Shanghai Master Plan proposed a reservation mechanism, which exerts elastic effects in three aspects: spatial flexibility, indicator flexibility, and timing flexibility of planning. Planning reservations in territorial and spatial master plans can be divided into two categories: functional reservations and strategic reservations. Functional reservations mainly address the issue that land use functions in long-term planning are difficult to predict. They involve arranging comprehensive land-use and "white" land-use in planning without setting functional requirements, which can be further clarified and detailed as urban development changes. Strategic reservations aim to reserve construction space in advance to cope with unforeseen major projects. For example, the Guangzhou Territorial and Spatial Master Plan repeatedly emphasized the need to reserve land for major events and projects and establish a spatial reservation mechanism.

Indicator flexibility is common in land management, known as reserved indicators or "non-landed indicators", which have been practiced in plans at all levels. In the plans of Beijing and Shanghai, such indicators are included in the reservation mechanism, but for other cities and counties, most indicators are under provincial authority, so they cannot be reflected spatially. The Notice of the Ministry of Natural Resources on the Management of Land Use Plans in 2020 proposed reforming the land use planning method in 2020, adhering to the principle that land elements follow projects. Although this is only a reform targeting the annual land plan, it has had a significant impact on various regions. Shandong, Zhejiang, and other provinces have issued management measures for "land elements following projects", which to a certain extent solve the problem where some cities rush for indicators during plan formulation but face shortages

Table 1 Scheme for graded differentiated constraints and transmission mechanisms

Elements/ methods	Standards	Zoning	Lists	Relationships	Locations	Boundaries
Implementation	●	●	●	●	●	●
Elaboration	●	●	●	●	●	●
Optimization	●	●	●	●	●	●
Supplementation	●	●	●	●	●	●

Source by the author

during implementation, while others have unused indicators. It can be predicted that the proportion of provincially coordinated parts in the land use indicators of cities and counties will increase in the future. However, against the background of strictly controlling the scale of construction land, there is still a lack of effective mechanisms for how "non-landed indicators" can be landed and aligned with plans. From the current situation, "indicator reservation" is a relatively effective mechanism carrier to implement this policy.

There are two options for the spatial implementation of indicator reservations: urban concentrated construction areas and urban elastic development areas. In urban concentrated construction areas, land zoning is clear, which conforms to the logic of functional regulation but may actually increase planned construction land; urban elastic development areas are more conducive to the management and control of incremental indicators but are not conducive to the optimal site selection of key projects in cities. The author tends to place indicator reservations in urban concentrated construction areas. At the same time, it is suggested that as a supporting mechanism, on the one hand, the accuracy of the long-term construction land scale should be weakened, and the scale should have a certain elastic range through reservations; on the other hand, the annual monitoring of land use should be strengthened.

(3) Urban Elastic Development Areas

In multiple draft discussions of the Guidelines for Urban Development Boundaries, it is basically clarified that the development boundary includes urban concentrated construction areas, urban elastic development areas, and special-purpose areas. The establishment of elastic development areas is a hot topic in the compilation of territorial and spatial master plans, and its concept has become relatively mature in the development of planning. In land use master plans, it was called conditional construction areas, referring to spatial areas identified in plans where urban and rural construction can be carried out only after meeting specific conditions.

Currently, the rules for establishing elastic development areas are finally stipulated in the Guidelines for the *Compilation of Municipal Territorial and Spatial Master Plans (Trial)*, which states that elastic development areas should not exceed 15% of the area of concentrated construction areas in principle, with smaller proportion limits for large cities and super-large cities. However, the rules for land-use conversion, positive and negative lists have not been clarified and need to be detailed by local regulations.

Elastic development areas are zones outside urban concentrated construction areas within urban development zones. In specific practice, they adopt control methods such as retaining road networks and "white" land-use. Some scholars have proposed constructing a "unit-combined" overall land use layout method system, setting "basic units" for urban expansion. With rigid control as the basis, this method aims to break the ultimate blueprint-style land use plan. As long as the development conditions are met, the construction of "basic units" can be initiated, providing adjustable flexibility for urban construction space

Table 2 Construction of expression system with different level of precision and enforcement

Category		System (from strong to weak)
Control Precision	Structure:	Layout---Schema
	Land use:	Classification---Zoning
	Control lines:	Polygon outline—Control line—Structure line
Control Intensity	Certainty:	Demarcation–Delineation
	Positive terms	Must–Should–May
	Negative terms	Strictly prohibit–Should not–May not
Transmission Mode		Implementation–Deepening–Optimization—Supplementation

Source by the author

and offering a new perspective for innovating urban overall land use layout methods.[8]

3.2 Elastic Expression Techniques in the Formulation of Territorial and Spatial Master Plans

(1) Control Expressions with Various Precision and Intensity

The content of the control system exists in various forms and levels. Adopting appropriate control forms and intensities for different planning contents is the most direct and effective way to avoid excessive control and retain flexibility. The hierarchical territorial and spatial master planning system adds a vertical dimension to traditional control tools. It is necessary to establish a conceptual system with multiple levels, various precision, and hierarchical intensity to reflect the differences in control policies at each level, enhance the operability of plans, and avoid repetition and conflicts between plans at different levels (Table 2).

(2) Progressive Expression in Texts

Whether formulating technical standards for spatial planning or compiling planning text, the main basis for judging the degree of flexibility lies in legal provisions. Therefore, it is essential to use appropriate vocabulary to distinguish between rigid and flexible content, and even necessary to form a dedicated vocabulary system for spatial planning to standardize the content and application of elastic planning.

Compared with drawings, the elastic expression through language is more direct and understandable. For example, "delineation (in a general sense)" versus "demarcation (in a precise sense)", "schema" versus "layout", "zoning (in a broad sense)" versus "classification (in a specific sense)". By using accurate terms, the control precision of different planning contents is distinguished,

[8] Zhang Ruixia and Lin Zhiming (2019).

and the progressive relationship and varying degrees of flexibility between different levels are conveyed in the transmission process. "Delineating development boundaries" has appeared many times in the guidelines for territorial and spatial master plans in various regions. In terms of enforcement intensity, legal terms are more mature: "must" and "prohibit" have the strongest mandatory nature; "should" and "should not" come next; "may" and "may not" indicate that priority should be given under permissible conditions, allowing for some exceptions and adjustment space compared to the former.

(3) Structural Expression in Drawings

The aforementioned "delineation (general)—demarcation (precise)" reflects different positioning precision and control intensity through conceptual definition. However, in terms of planning outcomes, when the "demarcation" of central urban areas and the "delineation" of peripheral towns appear in the same drawing, planners should make efforts to let readers accurately understand the rigid and flexible scopes of the controlled content. In the plans of European and American countries, as well as cities like Shanghai and Beijing in China, there are relatively mature cases of graphic expression, demonstrating that various drawing techniques can effectively assist in conveying the intentions of elastic planning.[9]

Stefanie Duhr summarized the analysis of graphic expressions in European spatial planning, established an analytical framework, and based on this framework, sorted out a set of spatial planning graphic methods ranging from diagrammatic to refined, from vague to rigorous, and from region to location (Fig. 1). After effective drawing elements being modularized and useless interfering information being removed, a networked and topological schematic space will be presented. The key of spatial structure diagrams lies in how to accurately, clearly, and effectively map planning intentions onto the relative space and arrange various policies within this space (Liang Jie, 2020).

People may question that structural expression in control drawings appears less rigorous and accurate. However, the author believes that structural expression, due to its information refinement, can more precisely convey core information and directly discard information that does not belong to the current level of planning. In some traditional drawings in the past, "approximately demarcating boundaries" that were not carefully considered due to insufficient level depth often caused confusion and misinterpretation in lower-level plans. The so-called "less is more" applies here: structural expression can focus information on control contents such as network structure, nodes, relative scales, and topological relationships. Highly abstract points, lines, and hatches obviously do not impose control on site selection and alignment, thereby granting complete flexibility to the planning and design of the next level (Figs. 2, 3).

(4) Reference-Based Alignment System.

We usually consider rigid boundaries as coordinate-based alignment systems, often ignoring that alignment systems can also be based on physical features

[9] Duhr (2007).

Fig. 1 Cross-country comparison of abstract concepts of strategic planning schemes in the Netherlands, Germany and the United Kingdom. *Source* The author redrawn from DUHR S (2007). The visual language of spatial planning-exploring graphic presentations for spatial planning in Europe. London: Routledge

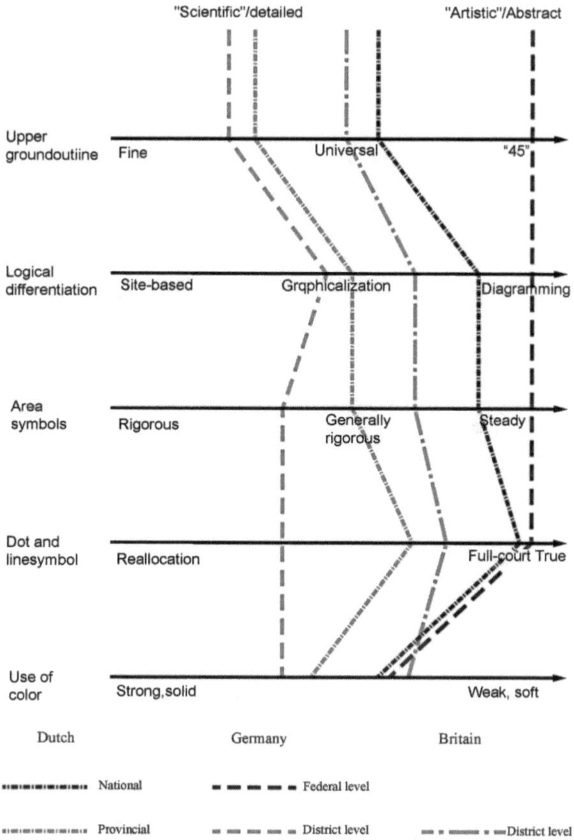

as references. In the draft discussion of the Guidelines for Urban Development Boundaries, the demarcation principles emphasize the need to make full use of natural geographical boundaries such as forests, rivers, lakes, mountains, and hills to form urban development boundaries. A reference-based alignment system, on the one hand, serves as a rigid basis for alignment; on the other hand, it avoids the difficulty of coordinate positioning (especially in upper-level plans, where precise coordinate positioning is overly challenging). It provides an option of "relative accuracy" between precision and ambiguity, while maintaining both boundary flexibility and control rigidity.

For example, the master plan text of Ottawa is accompanied by an interpretation document for the planning content. The document states that boundary alignment uses physical features as the reference system: when coinciding with physical features, it must strictly align with them; in other cases, it is an approximate boundary. This reference-based alignment system can also be extended to the relationship between planning boundaries and planning zones. Whether it is a zone or a development boundary, the upper-level delineation is not landing-specific and is an elastic boundary. However, during the implementation of

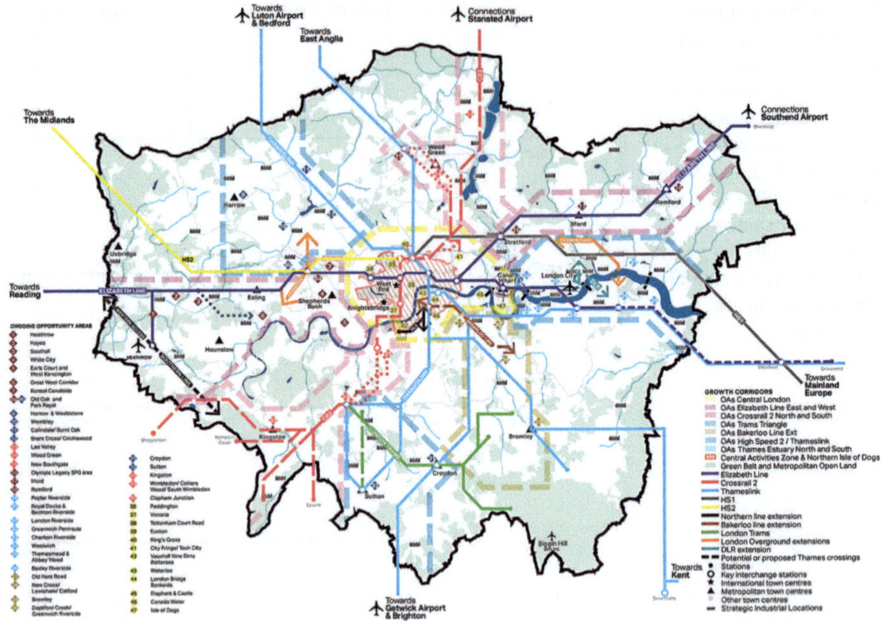

Fig. 2 London Plan. *Source* https://www.london.gov.uk/what-we-do/plan/london-plan/current-london-plan

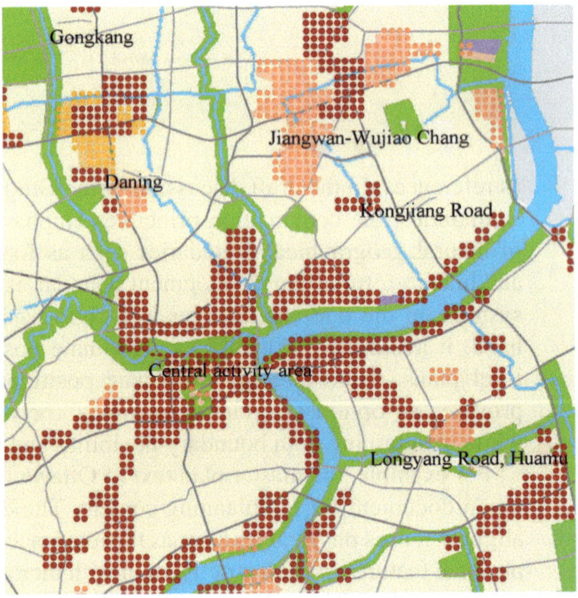

Fig. 3 Partial enlargement of land use layout plan of Shanghai main city. *Source* Shanghai Urban Master Plan (2 0 1 7–2 0 35)

lower-level plans, it is necessary to ensure that the relationship between zones and development boundaries is consistent with that in the upper-level plan. This achieves strict transmission in the relative relationship of planning content while leaving flexibility for lower-level plans.

3.3 *Elastic Mechanisms in the Implementation of Territorial and Spatial Master Plans*

(1) Achieving Institutional Flexibility through Regulations

It was necessary for the territorial and spatial planning system to emphasize rigidity and seriousness at its initial establishment. However, as the system shapes, various subdivided systems should be supplemented to the basic institutions. For example, Portland set both rigid growth boundaries and elastic growth boundaries during the compilation stage: the rigid growth boundary serves as a strict no-construction-outside-the-boundary rule to protect the ecology, while the elastic growth boundary restricts urban expansion by combining indicators and timelines. During the implementation stage, these two types of boundaries adopt graded management with different timeline and control rules, and allow revisions to the development boundaries through three modes: "minor adjustments, major adjustments, and legislative amendments".[10]

Currently, the elastic mechanisms for development boundary in the territorial and spatial planning system are mainly reflected in the compilation stage, with only relevant policies such as spot-like land supply and basic farmland occupation in the implementation stage. As territorial and spatial plans are gradually completed, the construction and implementation process will inevitably face numerous changes and new demands, such as major projects, replacement and adjustment of elastic indicators, and boundary corrections. We should not only reserve flexibility during the compilation process but also standardize elastic systems in implementation, fully consider different scenarios, and provide corresponding institutional buffers in approval and revision procedures (Figs. 4, 5).

(2) Permission for Construction Outside Development Boundaries

Unlike the stipulation in the *Urban and Rural Planning Law of the People's Republic of China* that "urban and rural planning authorities shall not grant planning permission beyond the scope of construction land defined in urban and rural plans", under the territorial and spatial master planning system, construction outside urban development boundaries is regulated through two approaches based on dominant land use zoning: "detailed planning + planning permission" and "constraint indicators + zoning access".

[10] Wang Ying et al. (2014).

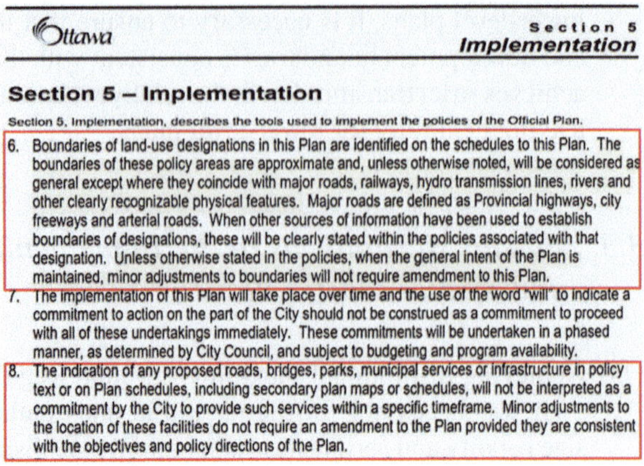
Ottawa Section 5 **Implementation**

Section 5 – Implementation

Section 5, Implementation, describes the tools used to implement the policies of the Official Plan.

6. Boundaries of land-use designations in this Plan are identified on the schedules to this Plan. The boundaries of these policy areas are approximate and, unless otherwise noted, will be considered as general except where they coincide with major roads, railways, hydro transmission lines, rivers and other clearly recognizable physical features. Major roads are defined as Provincial highways, city freeways and arterial roads. When other sources of information have been used to establish boundaries of designations, these will be clearly stated within the policies associated with that designation. Unless otherwise stated in the policies, when the general intent of the Plan is maintained, minor adjustments to boundaries will not require amendment to this Plan.
7. The implementation of this Plan will take place over time and the use of the word "will" to indicate a commitment to action on the part of the City should not be construed as a commitment to proceed with all of these undertakings immediately. These commitments will be undertaken in a phased manner, as determined by City Council, and subject to budgeting and program availability.
8. The indication of any proposed roads, bridges, parks, municipal services or infrastructure in policy text or on Plan schedules, including secondary plan maps or schedules, will not be interpreted as a commitment by the City to provide such services within a specific timeframe. Minor adjustments to the location of these facilities do not require an amendment to the Plan provided they are consistent with the objectives and policy directions of the Plan.

Fig. 4 Excerpts from explanatory documents on planning alignment in Ottawa master plan. *Source* Author redrawn from Official Plan Consolidation for the City of Ortaua

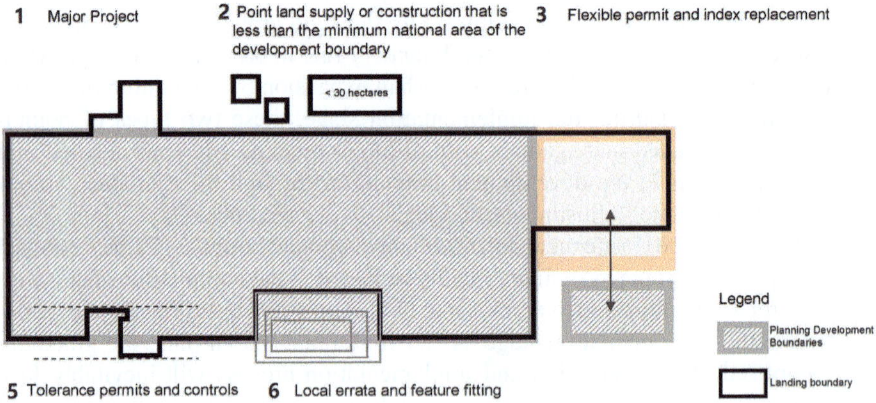

Fig. 5 Flexible revision of development boundary through regulation relaxation under different scenarios. *Source*: By the author

Multiple draft versions of the *Guidelines for Delineating Urban Development Boundaries* have listed permissible construction projects outside development boundaries, including: transportation, infrastructure, and other linear projects; special construction projects such as military and security facilities, religious sites, funeral facilities, comprehensive disaster prevention and mitigation projects, and strategic reserves; supporting facilities for country parks and scenic tourism areas; construction projects directly serving the rural revitalization strategy; and other necessary facilities and urban livelihood projects.

Although the *Guidelines for Compiling Municipal Territorial and Spatial Master Plans (Trial)* does not directly permit construction outside development boundaries, certain provisions implicitly allow dispersed urban construction of less than 30 hectares outside such boundaries, and such land use should be restricted to less than 10% of the total planned urban construction land. These provisions include: "Taking counties (districts) as statistical units, the planned urban construction land included in urban concentrated construction areas shall, in principle, account for no less than 90% of the total planned urban construction land within the county (district)"; and "The area enclosed by a single closed line of the development boundary shall, in principle, be no less than 30 hectares". Permissions for such construction obviously require stricter preconditions and construction requirements.

"Spot-like land supply" has emerged as an exploration in recent years. In its early stage, the focus was on its feature of solo, "supplying as much land as needed for construction". Investors and developers could "carve out" the exact land required for construction and obtain surrounding ecologically protected land through leasing or other means. With the establishment of the territorial and spatial planning system, it has gradually become a form of permission for construction outside urban development boundaries (Table 3). Relevant documents in Sichuan and Guangdong Provinces have clarified its connotation, but it remains largely within the framework of "rural revitalization". *The Notice of the Department of Natural Resources of Guangxi Zhuang Autonomous Region on Implementing spot-like Land Supply to Boost Rural Revitalization (draft for comments)* stipulates that "spot-like land supply policies can be applied to land for solo projects such as power transmission line towers, wind farm towers and booster substations, oil/gas pipeline valve chambers and distribution stations, and communication towers", expanding the scope of "spot-like land supply" to municipal facilities. *The Opinions of the Department of Natural Resources and Planning of Hainan Province on Implementing the spot-like Land Use System* comprehensively specifies the definition of spot-like land, positive and negative lists for the scope of implementation, and management approval procedures. Under the framework of the free trade zone, it is expected to become a experience to be replicated and recommended.

(3) Occupation of Permanent Basic Cropland and Ecological Conversation Red Lines.

Despite adhering to the "strictest land protection policies", there are unavoidable circumstances involving the occupation of permanent basic cropland or ecological red lines. *The Land Administration Law of the People's Republic of China* stipulates that: if the site selection of key construction projects such as national energy, transportation, water conservancy, and military facilities cannot avoid occupying permanent basic cropland, which involves agricultural land conversion or land expropriation, such cases must be approved by the State Council.

The Notice of the Ministry of Natural Resources on Doing a Good Job in the Pre-Examination of Land for Major Construction Projects Occupying Permanent Basic Cropland allows major construction projects under six scenarios. During the recent epidemic prevention and control period, the Ministry of Natural Resources issued a notice stating that "for projects with special site selection requirements that indeed need to occupy permanent basic cropland

Table 3 Summary of policies related to "point land supply"

Time		Document	Content
Before the establishment of the territorial and spatial planning system	March 2017	Anhui Province's 13th Five-Year Plan for Tourism Development	Adhere to economical and intensive land use, reform and improve the tourism land system, innovatively adopt various land supply modes such as spot-like, directional, and leasing. The construction land plans for key tourism projects are included in the provincial annual land use plan for overall arrangement.
	March 2017	Opinions on Supporting Land Policies for Tourism Development by Chongqing Municipal Bureau of Land and Resources and Housing, Chongqing Municipal Planning Bureau, and Chongqing Municipal Tourism Bureau	Implement the requirements of ecological civilization construction, adhere to the principle of "using land for agriculture if suitable, for forestry if suitable, and for construction if suitable". According to the regional resource and environmental carrying capacity, location conditions, and development potential, fully rely on mountain and forest natural landscape resources, combine with the topographic and geomorphic characteristics of the project block, follow the mountain terrain, and carry out spot-like layout, spot-like land expropriation, and spot-like supply for tourism project land based on the building footprint, half-building spacing range, and necessary environmental land.
	January 2018	Opinions of Hainan Provincial People's Government on Further Strengthening Land Macro-Control and Improving Land Use Efficiency	Actively explore the implementation of "spot-like land supply" in the construction of 100 towns and 1,000 villages, shared farms, and other tourism project facilities. With the land around the main projects remaining in their original state, adopt decentralized and spot-like distribution methods to further improve the refinement, precision, and intensification of land use.

(continued)

Table 3 (continued)

Time		Document	Content
	June 2018	Several Opinions of the General Office of Zhejiang Provincial People's Government on Promoting the Development and Utilization of Low Hills and Gentle Slopes and Advancing the Construction of Ecological "Slope Villages and Towns"	For blocks included in village construction through planning guidance in conjunction with relocation and new rural construction, multiple plots can be developed in combination with spot-like or belt-like layouts. For blocks developed for ecological (agricultural) tourism, leisure and vacation projects relying on mountain and forest natural landscape resources, multiple plots can be developed in combination with spot-like layouts.
	July 2018	Guangdong Province's Implementation Plan for Promoting All-for-One Tourism Development	Support rural collective economic organizations to revitalize and utilize idle rural houses and homesteads in accordance with laws and regulations to transform and build homestays, makerspaces, etc. Township land use 总体规划 can reserve part of the planned construction land indicators (no more than 5%) for scattered and individually selected rural tourism facilities. For land used for new industries and new formats in rural tourism projects, as well as land for people's livelihood that meets the requirements of targeted poverty alleviation and other policies, indicators can be resolved through overall planning from the provincial new construction land indicators.

(continued)

Table 3 (continued)

Time		Document	Content
	November 2018	Implementation Opinions on Promoting Rural Revitalization and Doing a Good Job in Planning and Land Management in this Municipality (Trial) by Shanghai Municipal Bureau of Planning and Land Resources	For new rural industries and new formats projects managed as construction land, "spot-like" and "belt-like" layouts can be implemented, with multiple plots developed in combination. Other land within the project area shall still be managed according to the original land type. All districts can carry out spot-like layouts based on the construction land specified in the suburban unit (village) plan. After going through the formalities of agricultural land conversion according to the scope of construction land plots, land can be supplied through collective construction land use or expropriation as state-owned land.
After the establishment of the territorial and spatial planning system	July 2019	Guiding Opinions on Standardizing the Implementation of "spot-like Land Use" to Boost Rural Revitalization (Trial) by Sichuan Province	In areas outside the urban development boundary (expansion boundary of urban and township construction land) that are not suitable for large-scale development and construction, according to the regional resource and environmental carrying capacity, location conditions, and development potential, combined with the topographic and geomorphic characteristics of the project block, carry out spot-like layout based on the floor area of buildings (structures), etc. Implement spot-like approval in accordance with the principle of "converting, expropriating (occupying) as much as constructed", and flexibly supply land for projects serving rural revitalization according to the planned land nature and land use.

(continued)

Table 3 (continued)

Time		Document	Content
	December 2019	Notice of Guangdong Provincial Department of Natural Resources on Implementing spot-like Land Supply to Assist Rural Industrial Revitalization	For the construction of rural industrial projects such as modern planting and breeding industry, agricultural product processing and circulation industry, rural leisure tourism industry, local characteristic industry, rural information industry, and new rural service industry, as well as their supporting infrastructure and public service facilities, if it is necessary to use scattered construction land outside the urban development boundary and the total construction land area of a single project does not exceed 30 mu, spot-like land supply can be implemented.
	April 2020	Opinions on Implementing the spot-like Land Use System by Hainan Provincial Department of Natural Resources and Planning	The scope of implementation of spot-like land use includes: 1. Land for rural infrastructure (including transportation, water conservancy, etc.) and public service facilities; 2. Land for rural leisure agriculture, tourism projects, and their supporting facilities; 3. Land for agricultural product processing, exhibition, storage, etc.; 4. Land for new industries, new formats in the integrated development of rural primary, secondary, and tertiary industries, and new tourism formats; 5. Land for production facilities serving seed breeding in the south reproduction scientific research and breeding base; 6. Land for tourist highway stations and transportation hubs; 7. Land for other spot-like projects as stipulated by laws and regulations. Individually selected projects that use land outside the development boundaries of cities, towns, and industrial parks without adopting the spot-like land use method are not applicable to these opinions. Land for facility agriculture shall be managed in accordance with the relevant provisions on facility agriculture and shall not be applicable to the relevant provisions on spot-like land use.

Source According to relevant documents

and ecological conversation red lines, they shall be regarded as major projects that allowed to occupy such areas". *The Guiding Opinions on Coordinating the Delineation and Implementation of the Three Control Lines in Territorial and Spatial Planning* stipulates that within the ecological conversation red lines, in the core protection zones of nature reserves and on the premise of complying with current laws and regulations, except for national major strategic projects, only limited human activities that do not damage ecological functions are permitted. These mainly include: moderate tourism visits that do not damage ecological functions and the construction of relevant necessary public facilities; linear infrastructure construction, flood control and water supply facilities construction, operation and maintenance that are necessary and unavoidable and comply with territorial and spatial plans at or above the county level; and important ecological restoration projects.

4 Conclusion

4.1 The Judgement of "Consistency" is the Technical Guarantee of Planning Flexibility

The judgment of consistency for flexible content is the core issue of flexible rules. In the past, master plans usually adopted subjective judgment. For example, whether a regulatory detailed plan meets the requirements of the master plan, and by what standards to judge specific content, were not specified in the plans, which often led to disputes. This is particularly true for sensitive issues such as green belt width and commercial mixing ratio. Without defining flexible boundaries, there have been many cases of taking advantage of ambiguities during construction or absurd one-size-fits-all practices.

In the "*Guidelines for Compilation of Municipal Land and Space Master Plan (Trial)*", it is pointed out that "if local corrections are needed during plan implementation due to topographic differences, land surveying and demarcation, property rights scope definition, scale coordination, etc., it will not be regarded as boundary adjustment after being determined by the municipal natural resources department". This provision puts forward the allowable elastic error in the process of delineating the development boundary, which provides a certain tolerance for the planning content. However, the judgment rules of tolerance category are still insufficiently defined, such as the lack of definition of concepts such as topographic differences and scale coordination, which leads to difficulty in judgment. In the implementation process, because the rules are too vague, managers either reject ambiguous areas or subjectively judge consistency. The former undermines the role of flexibility, while the latter increases the risks associated with flexibility—neither is desirable. In the context of building a society ruled by law, it is necessary in the future to provide clearer definitions and explanations of tolerable scenarios in technical specifications or planning outcomes, making explicit the previously unspoken hidden rules.

4.2 Spatial Planning Towards Procedural Justice

The ideal realm of management science is "self-limitation within the boundaries of capability, so as to provide abundant realistic selectivity to the most people".[11] The planning is the smae. Absolute flexibility and absolute rigidity are not difficult to realize, but it is difficult to measure and master the actual situation. The seriousness of planning is by no means based on precise control lines on drawings, but should be reflected in the stability and application of the implementation process, and the rules should be used to ensure the long-term procedural justice of spatial planning, instead of pursuing short-term coercive force with the bottom line.

The thinking of spatial planning should be changed from "short-term" to "long-term", from "engineering blueprint" to "pattern framework", from "precise layout" to "guidance, regulation and inclusiveness", and the rational core of planning should be revised from "instrumental rationality" to "limited rationality".

References

Taniguchi Anping (1996). *Procedural justice and litigation.* Beijing: China University of Political Science and Law Press.

Yang Baojun(2003). *Facing the reality, the way of change: Discussing the theoretical and practical significance of short-term construction planning.* Urban Planning, (03): 5–9.

Zhang Bing, Lin Yongxin, Liu Wan, et al. (2014). *"Urban Development Boundary" Policy and National Spatial Governance.* Journal of Urban Planning, (03): 20–27

Duhr S (2007). *The visual language of spatial planning-exploring cartoon presentations for spatial planning in Europe.* London; Routledge.

Cao Kang, Henry Hui Wang. (2009). *From instrumental rationality to communicative rationality: the change of modern urban planning thought core and theory.* Urban Planning, (09): 44-51.

Liu Kun, Tong De, & Jin Shan, et al. (2012). *Resilience planning, interval control and dynamic organization: Experience summary and method refinement of flexible planning in Shenzhen.* Planner, (05): 36–41.

Zhang Ruixia, Lin Zhiming(2019). *Method and practice of "unit combination" overall land use layout.* Planner, (03): 26–31

Zhou Xiao (2004). *The boundary of management science: can be and can not be.* Journal of Hengyang Normal University, (04): 123–126.

Wang Ying, Gu Chaolin, Li Xiaojiang. (2014). *Research progress of urban growth boundarybetween China and foreign countries.* International Urban Planning, 29 (04): 1–11.

Zheng Zhenyuan. (2004). *Reform of overall land use planning.* Chinese Land Science, (04): 14–19.

[11] Zhou Xiao (2004).

Implementation of the Territorial and Spatial Comprehensive Plan at City and County Levels

Pei Xinsheng

1 Understanding of Implementation Mechanism of Territorial and Spatial Planning (TSP)

1.1 Reform Background

The "*Several Opinions of the Central Committee of the Communist Party of China and the State Council on Establishing a Territorial and Spatial Planning (TSP) System and Steering its Implementation*" clearly proposes the establishment of a TSP system. Territorial and spatial planning serves as the spatial guide for development, the blueprint for sustainable development, the fundamental basis for land development and conservation activities, and the governing agenda for governments at all levels.

One of the fundamental purposes of this round of reform in TSP is to achieve the modernization of national governance capacity. In the practice of planning formulation, the reform process should not be simply understood as a technological change. While technology is important, it is even more important to fully reflect the policy attributes of planning in governance. Therefore, the general plan for territory space, as a crucial component of the TSP system, should embody strategic thinking, bottom-line thinking, people-oriented thinking, and governance thinking. Strategic thinking means that the comprehensive plan for territory space should not be simply and directly crafted as indicator allocation and bottom-line control. Every planning process must include strategic research and strategic guidance, in other words, "small cities also have grand strategies." Bottom-line thinking refers to the need to clearly define the boundaries of planning in order to promote the intensive and efficient use of resources in the context of ecological civilization reform. People-oriented

P. Xinsheng (✉)
Shanghai Tongji Urban Planning and Design Institute Co., Ltd., Shanghai, China
e-mail: peixs@163.com

Z. Peng (ed.), *Theories and Frontiers in Territorial Spatial Planning*, Urban Sustainability, https://doi.org/10.1007/978-981-96-9825-7_8

thinking means focusing on people and their livelihoods in order to build cities for the people. Governance thinking means that the understanding and application of planning implementation should be based on the perspective of governance systems capacity. In the practice of formulating TSP, clarifying the role of planning is one of the important changes that reflect the requirements of reform. In the previous planning system, regardless of urban–rural planning, land use planning, or development planning, there was a lack of a clear chain of formulation, implementation, supervision, and feedback. Local governments did not clearly define the roles of various plans, the specific requirements for supervision and feedback in planning, and the criteria for evaluating the effectiveness of planning implementation during the implementation of existing plans. Similarly, these elements often lack clear definitions in the planning formulation process, and planners often only fill in the planning contents according to normative documents and technical standards. Therefore, planners cannot define the implementation, control, and supervision attributes of various planning elements in the planning text. This is also a point that needs to be reflected upon in the process of TSP reform.

From the perspective of reform background, it is essential for planners to have a clear understanding of "the establishment of a TSP system as a systematic restructuring process." The practice of planning preparation needs to be recognized within the context of the planning operational system. The TSP system includes the system for preparation approval, implementation supervision, laws and regulations, and technical standards. By placing planning preparation within the broader context of reform, we can better understand the significance of planning implementation within the entire planning system.

1.2 Requirements for the Construction of Transition System

(1) Three relations of transimission

The national TSP system of "five levels and three categories" has been clearly defined. The emphasis of planning differs at the levels of national, provincial, city, county, and township. The national level emphasizes strategy. The provincial level focuses on coordination, and the city and county levels emphasize implementation. The five-level planning itself requires a vertical transimission mechanism, which includes both the direct implementation of rigid planning elements from higher levels to lower levels (the direct implementation of the intentions of higher-level planning by lower-level planning) and flexible transimission. In addition to implementing the superior strategic and bottom-line control requirements, city and county-level planning should also reflect the development demands that correspond to the respective government responsibilities at their levels. Therefore, the transimission system emphasizes both top-down vertical transimission and bottom-up feedback.

The TSP should not only be of high quality and comprehensive, but also systematic and focused. More importantly, it must be transferable and implementable. The implementation system mainly focuses on implementation, and the overall TSP of cities and counties in the intermediate stage should not only implement the higher-level planning, but also guide the lower-level planning and coordinate the specialized planning. Therefore, the implementation system of planning also includes three types of implementation relationships: vertical, horizontal, and implementation.

Vertical implementation includes the coordination and implementation of overall planning at different levels, and the vertical implementation system includes the implementation from province to city, county, and township. Horizontal implementation mainly refers to the constraints and verification of the comprehensive plan on specific plans. In principle, the current round of TSP is linked with departmental detailed planning, but significant differences in the progress among different regions in practice are witnessed. Therefore, it is necessary to reserve interfaces in TSP to reflect and accommodate major goals, strategies, and especially important projects and implementation requirements that need to be implemented in detailed planning. The implementation at the implementation level of planning involves the decomposition, implementation, and constraint of planning elements between the comprehensive plan, detailed plan, and short-term construction plan. After the completion of TSP, detailed plans are prepared within the boundaries of urban development, village plans are prepared outside the boundaries of urban development, and local governments also need to prepare short-term construction plans. All these plans need to be aligned with the requirements of the overall plan (Fig. 8.1).

(2) Principle of transition

It is generally believed that the basic principles of implementation include the correspondence of administrative responsibilities, the combination of rigidity and flexibility, the implementation-oriented and the emphasis on feedback.

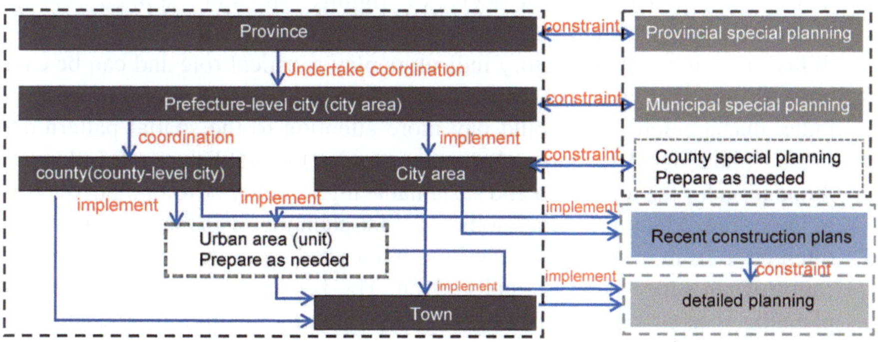

Fig. 8.1 Countries Schema of transimission system in soil spatial planning. *Source* Self-painted by the author

(1) Correspondence of administrative responsibilities

The implementation of planning should correspond to the planning administrative responsibilities of governments at all levels. It should be based on the administrative responsibilities of the Ministry of Natural Resources and other relevant departments, and clarify the administrative responsibilities of governments at all levels and departments. The implementation should be classified and the contents of implementation control should be precisely defined. In previous overall urban planning, the issue of excessive complexity and wide coverage of the content was addressed by clearly defining mandatory content. This includes emphasizing and enforcing specific elements such as control lines (green, blue and purple) in the city, as well as public service facilities and major infrastructure. The purpose of defining mandatory content is to align with the management administrative responsibilities of local governments. According to the principle of corresponding administrative responsibilities, planning texts (especially those submitted for approval) should not aim for comprehensive coverage, but rather correspond to the responsibilities of examination and approval management. The planning text submitted for approval to higher-level governments should clearly define core elements, such as development strategies, positioning and objectives, key indicators, planning patterns, bottom-line controls, and the spatial implementation of various major facilities required by national strategies. Conversely, detailed planning arrangements within the city may not necessarily require direct management and approval from higher-level policy makers, and can be excluded from the mandatory content. In summary, aligning powers with responsibilities reflects a key shift in thinking underpinning the TSP reform.

(2) Banlance between rigidity and elasticity

Planning transmission should focus on binding requirements and strategic guidance, balancing the transmission of rigid control indicators with the strategic guidance of flexible element. The planning should address factors of protection, development, and the mutual contradiction throughout the entire process of territorial and spatial development. While certain elements need to be controlled, it is essential to determine the degree of "rigidity" should be applied. As local governments also have strong development demands, to what extent can "flexibility" be allowed? Mandatory indicators play a critical role and can be categorized into two types: binding indicators and guiding indicators. In terms of index management, we should pay more attention to the spatial pattern and overall planning of resource elements, in order to establish a solid planning foundation for the long-term and sustainable high-quality development of cities in the future. Many problems in the original planning or management process often arise from a lack of supervision, while leading to developmental suffocation when supervision is finally given. The key challenge in this round of planning is to strike a balance between maintaining the bottom line and flexibly responding to dynamic development, ultimately improving the efficiency of land space utilization and achieving high-quality development.

(3) Down-to-earth Implementation

Transmission should aim to achieve effective supervision, quantification, and coordination. In the process of constructing the TSP system, the development of "one map" is one of the core components. All planning elements should be incorporated into the "one map," attaching great responsibility to planning practioners. In previous urban and rural planning system, there was a lack of top-down database management and linkage, resulting in many issues not being fully exposed. Under the context of TSP reform, it is necessary to integrate planning elements into the database during the planning process, ensuring precise placement of the planning, which will become one of the challenges faced in TSP formulation.

(4) Focuses on feedbacks.

To achieve effective planning implementation, it is essential to establish a feedback mechanism that ensures coordination and communication between upstream and downstream of planning. Planning should not be seen as a one-off task; instead, it requires continuous dynamic management. In the process of planning formulation, it is crucial to emphasize a planning path that focuses on a life cycle management. Planning is a dynamic and ongoing optimization that requires annual review, dynamic assessments, and regular monitoring of the quality and effectiveness of plan implementation. This enables further optimization and adjustment of the control measures within the planning framework.

2 Transimission Mechanism of Comprehensive Plan of Land and Space in Cities and Counties

2.1 Requirements for Compiling the Comprehensive Plan

(1) Compilation content requirements

The Municipal Guide to the Compilation of Comprehensive plan(Trial) of 2020 is a guiding framework, which serves as a guiding framework, indicating that the current compilation of comprehensive plans is a gradual exploration and a gradually unified process. It allows local governments and compilation units to conduct beneficial exploration based on local conditions.

According to the requirements specified in the "*Municipal Guidelines*" regarding target positioning, zoning, important control lines, factors allocation, and index decomposition, it is necessary to first clarify a "to-do-list" of the city and county levels in order to to determine which contents need to be transmitted and in what manner. For instance, the decomposition of binding indicators is an essential aspect of municipal planning, encompassing important indicators such as the scale of construction land, the preservation of arable land, and the identification of permanent basic farmland. Additionally, it involves zoning, where each level of planning at the city and county levels must be delineated to a certain depth. The city and county levels, as well as the municipal districts, differ in this regard. Municipal planning should coordinate the various counties, but it should not overstep its authority by determining the specific content of county-level planning. Municipal planning should adhere to the overall development pattern determined by higher authorities for the entire province, and based on the municipal development strategy and goals, establish a framework for the city's development. County-level planning then further refines this framework by delineating detailed zoning. As for the implementation of coordinates and positioning, this task is primarily carried out at the township-level planning stage. This process represents a hierarchical optimization and implementation, gradually and precisely delineating boundaries, and thus embodies the implementation of planning throughout the process.

The differences in the hierarchical levels of jurisdiction for the "three lines" are also significant. The eco-protection red line represents national or provincial authority, and therefore, it needs to be directly implemented. After multiple rounds of communication, the eco-protection red line in some areas has been largely settled, which is an important part of the ecological space corresponding to national or provincial authorities, and strict implementation is required in city and county planning. However, it does not mean that the ecological space in city and county planning is limited to the eco-protection red line. The planning of ecological space not only needs to implement the designated ecological protection red line in spatial terms, but also requires further systematic planning based on the "dual evaluation" (Fig. 8.2). Of course, there are also differences between the municipal and county levels in terms of the "dual evaluation" itself. Currently, the requirement is to carry out the "dual evaluation" at the municipal level, while the "dual evaluation" for county-level planning is chosen by the local government. However, it should not be a repetitive evaluation that is comprehensive in nature, but rather an evaluation based on the local resource endowment and specific binding requirements.

(2) Transition requirements.

Municipal planning needs to undertake provincial planning upwards, transmit it downwards to county planning, and put forward binding requirements to detailed planning.

Upward undertaking is reflected in the deepening of the clear objectives of provincial planning, implementing structural control requirements, and the arrangements for cross-city coordination in provincial planning. Downward implementation is manifested in the requirements imposed on the planning and compilation of counties, cities, and districts under overall coordination, in terms of objectives, structure, major projects, and key areas.The implementation to implementation is manifested by the binding requirements imposed on specific plans and implementation plans (Fig. 8.3).

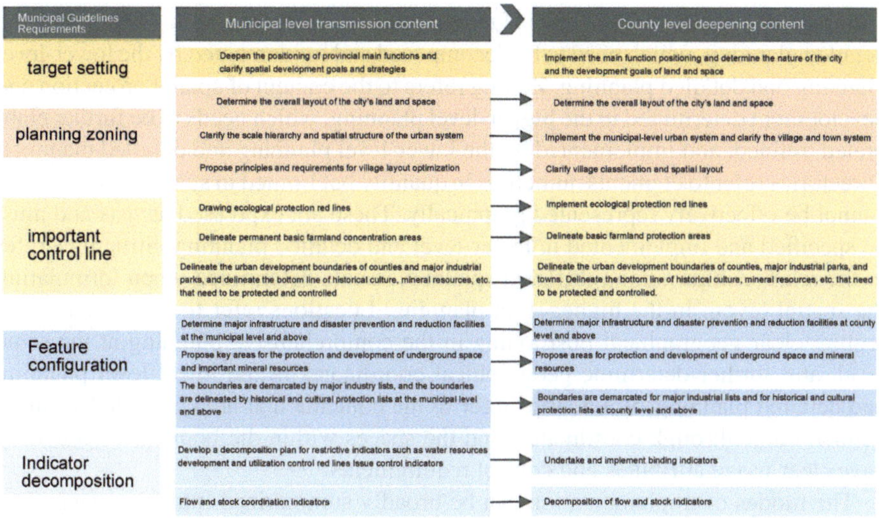

Fig. 8.2 Relationship and Content Depth Difference of Planning Implementation between Cities and Counties. *Source* Self-painted by the author

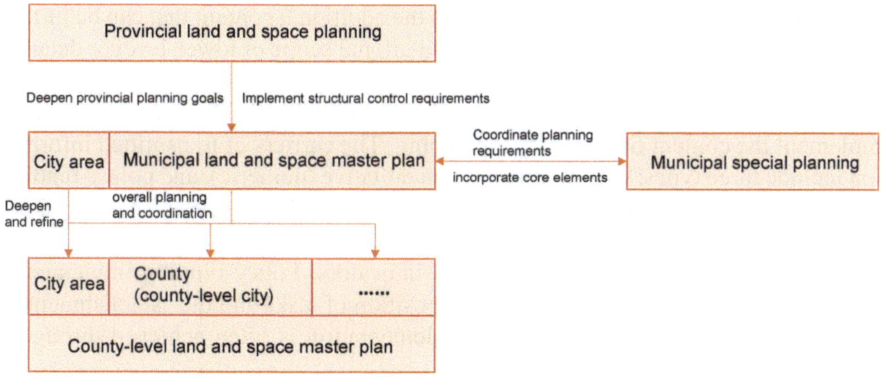

Fig. 8.3 Sketch of implementation direction of municipal planning. *Source* Self-painted by the author

2.2 *Transition Framework of City and County Comprehensive Plan*

By combining the implementation requirements of the current comprehensive planning in city and county, we will get a relatively clear implementation framework, classified into: transition elements, modes and information carriers.

The elements of implementation include indicators, zoning, directories, structures, locations, boundaries, and so on. Indicators refer to the quantifiable control requirements that the higher-level planning imposes on the lower-level planning and detailed planning, which need to be decomposed and implemented by the lower-level planning and detailed planning. Zoning refers to the content of spatial protection and development determined in the higher-level planning, which needs to be further elaborated, refined, and implemented by the lower-level planning and detailed planning. Directories refer to elements numerous in quantity but limited in spatial extent, which cannot be effectively represented graphically. These are expressed as lists and must be specified and implemented in lower-level and detailed planning. Structures refer to the relationships among different spatial elements determined when formulating the spatial layout in the higher-level planning. Locations refer to the contents that require clear geographical positioning in the comprehensive planning at the same level, and further determine geographical coordinates in the lower-level planning and detailed planning. Boundaries refer to the contents that are controlled within a defined range through coordinates, and the spaces within the boundaries generally have clear usage attributes and control requirements.

The modes of implementation can be broadly summarized into four categories: implementation, deepening, supplement, and translation. Implementation refers to the strict adherence of lower-level planning to the transmitted content of upper-level planning, without any changes during the continuous implementation. Deepening involves further decomposition and refinement of the content by lower-level planning and detailed planning. Supplement refers to the additional content that can be further included and supplemented within the jurisdictional scope of lower-level or detailed planning. Translation means that as the planning spatial hierarchy descends, the depth of planning also increases, and lower-level planning will transform and transmit or implement the content of upper-level planning. The carriers of transmitted information include three types: spatial (graphic), quantitative (numeric), and policy-binding (text). Spatial implementation (graphic) primarily entails conveying spatial characteristics and information. Quantitative implementation (numeric) mainly involves the decomposition and implementation of indicators. Policy-binding implementation (text) mainly encompasses the decomposition of tasks and the establishment of rules. In planning practices, effective implementation is often achieved through a combination of multiple modes.

3 Implementation Practice of Comprehensive Planning in Cities and Counties:Case Study on Jiaxing City and Jingzhou City

3.1 Basic Thoughts of Implementation of Municipal Comprehensive Plan

Jiaxing is a typical representative of the Zhejiang model, located adjacent to Shanghai and Suzhou yet underwent a completely different development model. Jiaxing exhibits prominent features such as a patchwork-like economy, diversified development, and strong counties and towns. It has the smallest urban–rural income gap in the country, yet its central city ranks second to last in the province. The development model of "isolated governing" has led to a mismatch among population, economy, and spatial resources, thereby impeding the overall development efficiency of the city. Against the backdrop of the rise of high-quality integrated development in the Yangtze River Delta as a national strategy, Jiaxing's comprehensive planning bears the important mission of showcasing the region's "integration" and "high quality". In 2019, Jiaxing was designated as a pilot city for comprehensive reform in Zhejiang Province, with the municipal-level coordination and management of the comprehensive planning serving as a key focus of the reform. In light of this, it is of paramount importance to explore a novel planning and coordination model that aligns with Jiaxing's development characteristics. Integrating technical logic with policy logic enhances the modernization of urban governance capabilities. Municipal planning should prioritize major issues while avoiding unnecessary involvement in minor ones. Therefore, the planning framework establishes the "five holistic plans" for the city, namely, focusing on the five key elements of "pattern, scale, bottom line, elements, and facilities" in municipal planning. On one hand, it establishes a planning technology implementation system that integrates structure, indicators, boundaries, zoning, directories, and locations. On the other hand, it establishes a planning implementation mechanism to ensure the comprehensive planning of the city, thereby realizing the planning's role in driving high-quality and coordinated development between the city and counties (Fig. 8.4) (Kai and Ying 2021).

3.2 Content and Form of Implementation

(1) Planned pattern-structural implementation

Jiaxing is not short of strategies. It is the superposition of various strategies from different levels, including the integration strategy of Yangtze River Delta region, the strategy of Zhejiang Province, and the strategic blessing of G60 and Hangzhou Bay. In Jiaxing's planning, structural research is very important to coordinate the spatial arrangement of strategies at all levels. For example, to

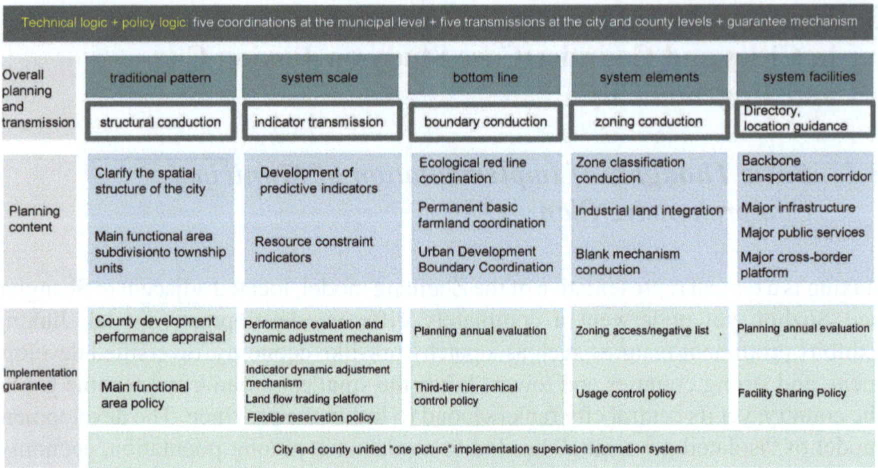

Fig. 8.4 Basic thoughts of comprehensive planningand implementation of municipal land and space master plan. Kai and Ying 2021 *Source* Revised according to reference ()

match the regional development pattern, Jiaxing needs to undertake the Yangtze River Delta's plan by shifting from Z-shaped corridor to radial network. In addition, it has to absorb G60 Science and Technology Innovation Corridor, Greater Bay Area Transformation and Development Corridor and Water Town Ecological Civilization Corridor.

Structural implementation also involves main functional zones, while the basic units of provincial main functional zone planning are generally county-level units, and hence it is difficult to accurately guarantee the implementation of policies. Zhejiang's reform and innovation in the main functional zones is worth popularizing. According to resource endowment, spatial strategy, population, economic and social development, the main functional areas are further refined to township units, so that a set of technical logic can be formed for the planning and zoning of future leading functions and differentiated policy formulation, including index allocation. For example, in Jiaxing, the provincial main functional zone planning only distinguishes urbanization areas, ecological protection areas and main agricultural products producing areas with county level as the unit. Combined with the study of urban spatial structure, it can be further refined. Strategic areas such as G60 Corridor and Greater Bay Area are defined as urbanization advantage areas and potential areas, Shenjiahu Water Township Cultural Belt is implemented as ecological functional areas, and Tangpu Agricultural Area between Bay Area and G60 is implemented as main agricultural products producing areas. This main functional zoning refined to township units can become an effective policy tool (Fig. 8.5) The main functional zoning provides important guidance, under which local governments can formulate and implement specific policies based on policy packages including fiscal policy, industrial policy, population and employment policy and ecological environment policy.

Among the prefecture-level cities in agricultural areas, such as Jingzhou in the Yangtze River Basin, it is located in the main grain producing areas of China. However, when the planning requirements are refined in cities or counties, the policy guidelines are not completely consistent. In addition to ensuring efficient grain production, some regions should also undertake the function of urbanization. Therefore, one of the upmost meanings of deepening and refining the main functional zones is to help local governments formulate targeted guiding policies according to regional characteristics, which is of fundamental significance to index implementation and zoning control (Fig. 8.6).

(2) Planned size-index implementation

Fig. 8.5 Deepening and refinement of main functional zones. *Source* According to the relevant contents of reference (Shanghai Tongji Urban Planning and Design Institute Co., Ltd. (2021) 2021)

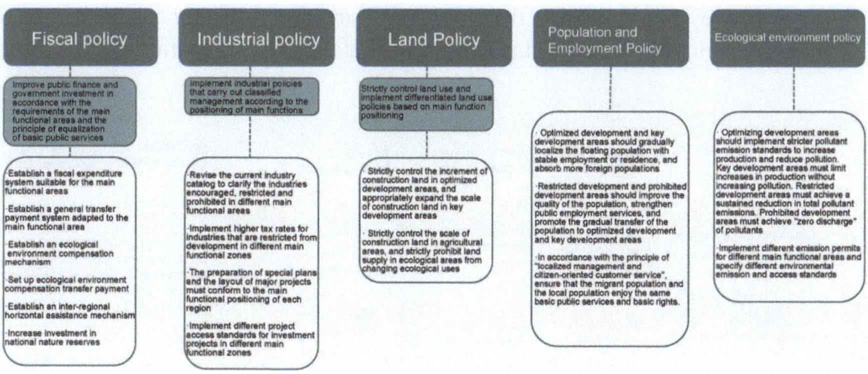

Fig. 8.6 Policy requirements of main functional areas. *Source* According to the relevant requirements of the state and some provinces and cities

Zhejiang Province has made detailed arrangements in the work of index decomposition, and determined the six-quantity overall planning ideas of "limiting capacity, framing total amount, optimizing increment, tapping stock, revitalizing flow and improving quality". Different provinces have different situations, so the policy system cannot be simply translated, but its logic is reasonable. Indicator measurement needs to be based on clear logic. First of all, it should match with the carrying capacity of resources and environment, and the index must be measured under the scientific and reasonable capacity limit. In addition, it is necessary to optimize the increment, tap the stock and revitalize the flow on the premise of locking the total amount. In this round of comprehensive planning, incremental indicators are distributed from top to bottom. Planners need to think about how to optimize the increment, tap the stock and revitalize the flow on the premise of locking the total amount. Indicators are not a simple number, and the overall planning of land space is not entirely a problem of indicators. There are also strategies and structures. Local governments need to design flexible space, blank space and other planning tools to cope with the future uncertainty.

On the index decomposition of municipal planning, the research method of forming the basic logic of spatial resource allocation through the main functional areas in Zhejiang is worth learning from. For urbanization advantage areas, potential areas and ecological function areas, the arrangement of index delivery and incremental space is different. The index of a city must not be an average distribution, but must reflect the allocation and guarantee of resources logic of important development space (Fig. 8.7).

After the completion of index allocation, we should pay attention to performance appraisal from the perspective of implementation. In the planning process, we should realize that indicators are not naturally available, result in establishment of performance appraisal standards, formation of indicators

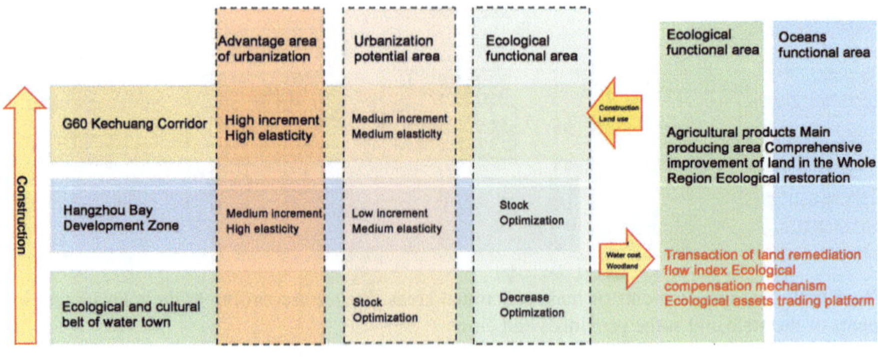

Fig. 8.7 Logical schematic diagram of index allocation based on main functional zone. *Source* According to the relevant contents of reference (Shanghai Tongji Urban Planning and Design Institute Co., Ltd. 2021)

delivery and dynamic regulation mechanism, management ofusage of indicators in the whole life cycle, and supervision the use of indicators with annual plans. In addition, it is necessary to emphasize quality and efficiency, combine the annual "physical examination", assess the efficiency of various types of land use, and supervise the use of indicators in the process of construction and management. This will ultimately determine whether the indicators can be successfully landed and used efficiently.

(3) Bottom line-boundary transimission.

Corresponding to the powers and responsibilities, the three control lines have deep differences in "delineation" and "delineation" at different planning levels, and have differences in implementation and deepening.

The implementation of ecological protection red line is a relatively rigid boundary implementation, with an upper-to-lower level's plan to delimit the boundary, locking the scale and implement in the lower levels' plan. Of course, it is not to say that the ecological space planning of county-level cities and counties has nothing to do except strictly implement the red line of ecological protection. Within the powers of the government at the same level, in addition to strictly implementing the red line of ecological protection, we should further identify the ecological space that needs to be controlled at the same level according to the "double evaluation". The formation of state network creates high-quality ecological space.

In transmitting permanent basic farmland requirements within the municipal territorial spatial master plan, it is necessary to specify protection targets (indicators) for cultivated land and permanent basic farmland, but in terms of spatial location, the municipal planning only needs to delimit centralized protected areas. County-level planning delimits the boundary of permanent basic farmland protection areas. The boundary of permanent basic farmland centralized protection area demarcated at the municipal level and the boundary of permanent basic farmland protection area demarcated at the county level can be understood as the difference between "demarcation" and "demarcation", and there will be fine-tuning on the boundary and differences in area. However, the landing of the map spot at the township planning level requires that "the scale is indispensable and the boundary can be fine-tuned" (Fig. 8.8).

The implementation of urban development boundary has been clearly defined in the "Municipal Guide". The guide clarifies that the municipal comprehensive planshould delimit the urban development boundary of municipal districts in accordance with the control requirements such as urban positioning and scale indicators determined by the superior comprehensive plan, combined with local development reality; Make overall plans to put forward the urban development boundary guidance scheme for towns (streets) and various development zones where the county people's government is located. The county-level comprehensive planshould delimit the urban development boundaries within the county according to the guidance scheme of the municipal master plan, including the towns (streets) where the county people's government is located, other towns, various development zones, etc. (Fig. 8.9).

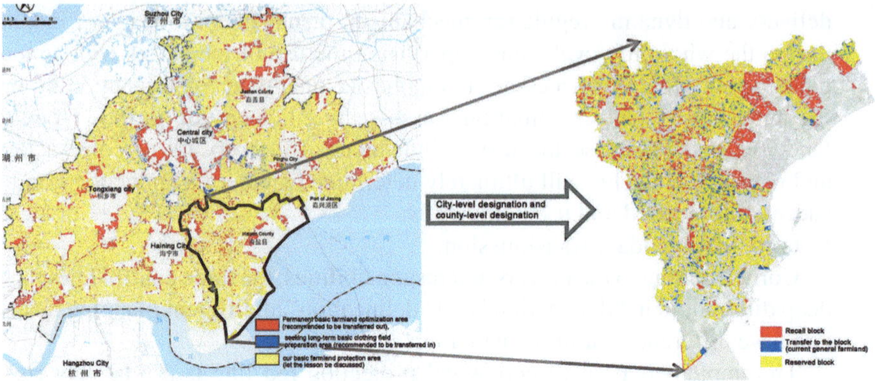

Fig. 8.8 Implementation of permanent basic farmland in cities and counties. *Source* According to the relevant contents of reference Shanghai Tongji Urban Planning and Design Institute Co., Ltd. (2021)

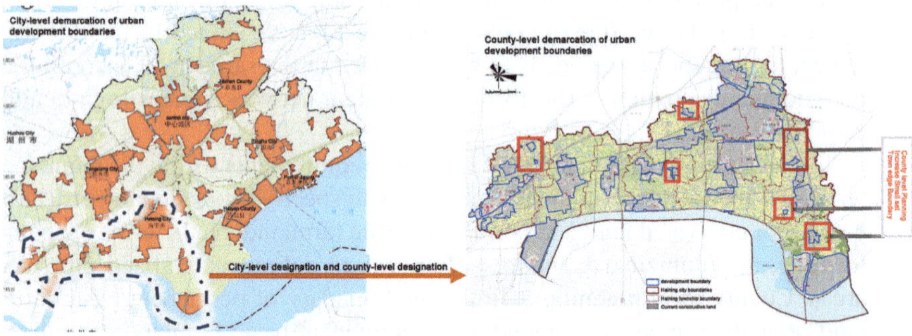

Fig. 8.9 City and county transimission of urban development boundary. *Source* According to the relevant contents of reference Shanghai Tongji Urban Planning and Design Institute Co., Ltd. (2021)

(4) Series elements-zoning implementation

(1) Hierarchical transimission of partitions.

Similar to the mapping process of different scales, zoning transition is also a process of gradual optimization and refinement. It is essental to determine the orientation of large-scale partitions based on the logic of main functional areas. Especially in some prefecture-level cities with relatively large areas, their urbanization areas, main agricultural products producing areas and ecological functional areas must be differentiated according to their respective development and protection requirements, which cannot be generalized on a single scale of the city. Therefore, in order to realize the implementation to lower-level planning, it is necessary to divide the leading functional zoning, implement the functional access-type land space control, and then determine the usage classification as the basis of usage control.

For example, in Jingzhou City's comprehensive plan comprehensive plan (under compilation), the city-level zoning (first-level zoning) serves as a guiding framework rather than a rigid control boundary, although the associated binding indicators remain mandatory; The zoning categories of municipal districts and counties are the same as those of the central city (secondary zoning/functional zoning), but the central city deepens the municipal districts, including shape refinement, function refinement and level deepening. Of course, the central city can be further refined into use classification according to the needs of various places (Fig. 8.10).

(2) Transimission from partition to classification

The focus of spatial expression in central city planning is the implementation from zoning to classification. The land-use planning map of the original urban comprehensive plan distinguishes the land use attributes

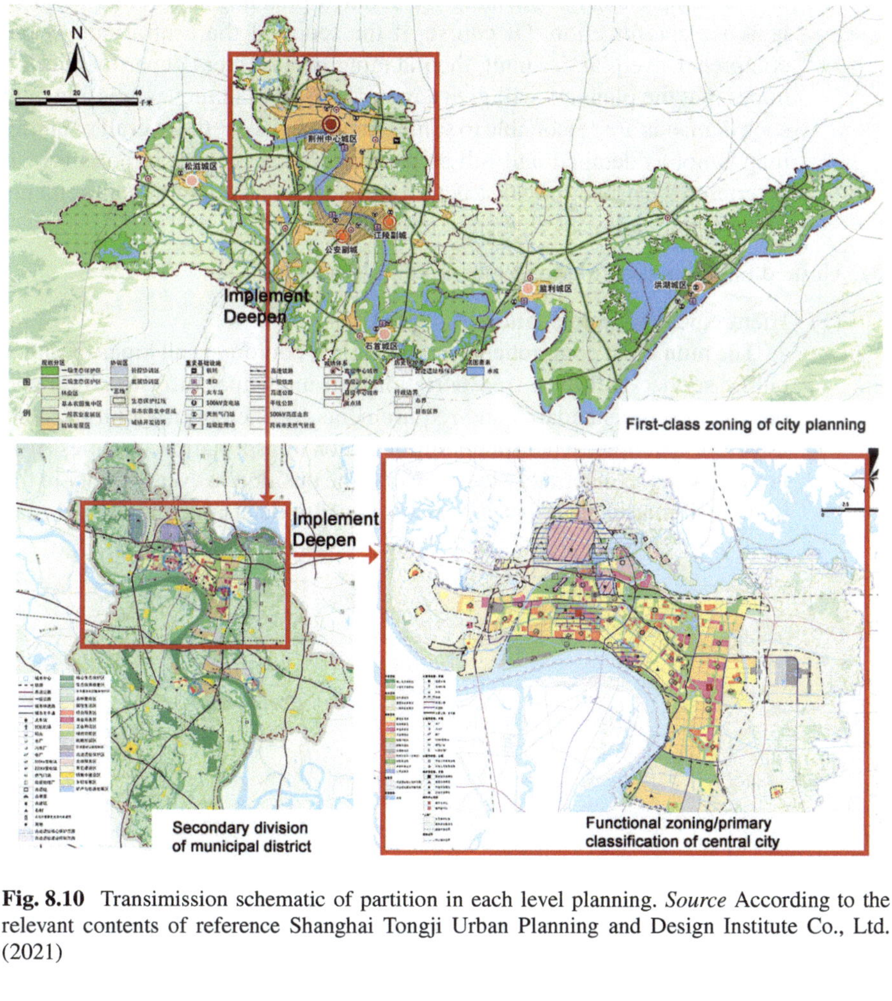

Fig. 8.10 Transimission schematic of partition in each level planning. *Source* According to the relevant contents of reference Shanghai Tongji Urban Planning and Design Institute Co., Ltd. (2021)

into middle and small categories, and the plot boundary is as detailed as possible, which is expressed according to the standard of land usage classification. In the comprehensive planning, the original operation mode is still open to question. Hence, zoning control reflecting the leading function can be adopted. For example, a certain area is identified as a commercial business area, an industrial development area or a green leisure area. This is an expression that reflects the dominant function, and is consistent with the logic of planning elements such as index control and use access, instead of directly deepening into the specific use classification that should be determined by detailed planning in the overall planning.

Therefore, it is suggested to draw the functional zoning control chart at the comprehensive planning level, put forward the control requirements in the form of zoning chart + control table, leave the flexibility of land use arrangement to detailed planning, and implement the zoning requirements of the comprehensive planning by detailed planning and refine them to land use classification. Of course, if the scope of the central city in the comprehensive plan is small, the planning management department often hopes that the planners will give a more detailed planning expression, and such demands are reasonable to some extent. However, the overall planning map is not as detailed and full as possible. What is more important is to determine the urban spatial structure and the urban structural requirements in the planning layout element control (Fig. 8.11).

(5) Unified facilities-directory and location implementation

(1) Hierarchical implementation of master plan.

The municipal comprehensive plan should coordinate all kinds of major public service facilities, transportation and municipal infrastructure, and put forward the standards and requirements for the construction of facilities. For example, when planning of major transportation facilities, it is not necessary to accurately locate every line position in every county in the municipal master plan, but it is also critical to determine the overall road

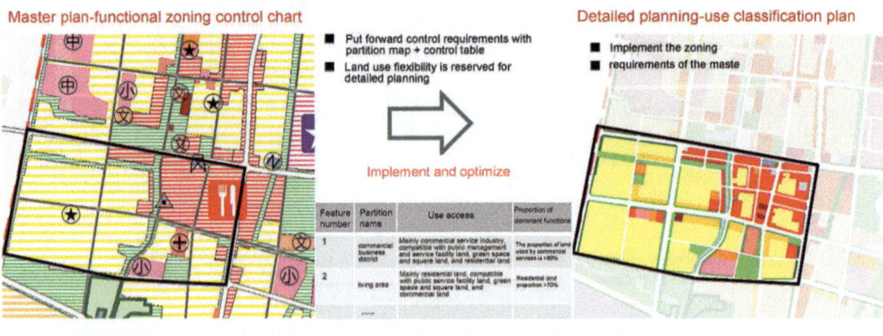

Fig. 8.11 Transimission schematic of partition-classification *Source* Revised according to reference (Yuhan 2020)

network structure of the whole city. In this process, the municipal comprehensive plan only needs to determine the macro indicators, trend and layout, and the specific line position cannot be directly defined. However, in the county-level and township-level planning, it is necessary to further connect with the detailed planning, and some major transportation facilities with clear plans can be implemented (Fig. 8.12).

In the implementation of major public service facilities, the basic idea is to put forward the allocation standards of public service facilities at different levels.In the municipal master plan, only the demarcation standards of urban circle and community living circle and the supporting requirements of facilities can be determined, and it is difficult to implement the specific spatial arrangement of various public service facilities. However, in the county-level planning, it can be further refined and implemented according to the requirements put forward by the municipal planning.

(2) Implementation from comprehensive plan to special plan

The implementation from the municipal overall plan to the special plan mainly includes that the municipal plan defines the list of special plans, and puts forward binding indicators and control requirements. Compile and review the special plan, incorporate it into the database after completion, and superimpose it on the "one map" of national spatial planning. Taking the special transportation planning as an example, the comprehensive plan determines the direction of important transportation routes and the location of important transportation facilities. In the preparation of the special traffic

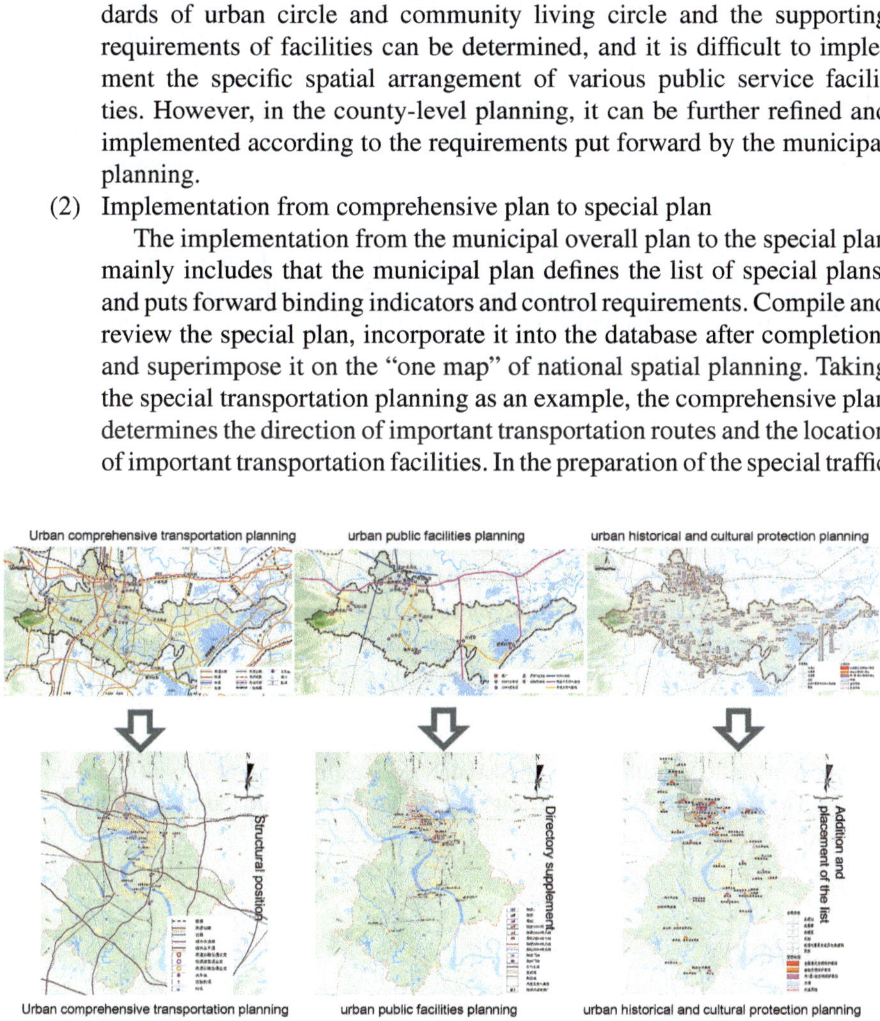

Fig. 8.12 City and county implementation schematic of various facilities (elements). *Source* According to the relevant contents of reference (Shanghai Tongji Urban Planning and Design Institute Co., Ltd. 2021)

plan, the boundary of route selection and facility land shall be determined, and it shall be reviewed with the master plan, and incorporated into the comprehensive plan after approval.

3.3 The Result Form of Implementation

Based on the working basis of city planning and related mechanism construction, the elements used for implementation can be expressed separately in the planning outcomes. In the case of Jiaxing comprehensive planning, referring to the experience of *Shanghai Urban Comprehensive plan(2017–2035)*, a county guidance consisting of "one article, one table and four maps" was formed, and a implementation mechanism to lower-level planning was formed through the mode of county guidance (Fig. 8.13). "One article" is a specific description of implementation in six major areas, "one table" is a data list of city indicators transmitted to each district, and "four maps" include four maps of bottom line control, central system and public services, infrastructure and urban design guidelines, which are the drawing expressions of specific graphic implementation. Through the combination of "text + table + map", the municipal planning needs the county level.The bottom line, central system and major infrastructure of planning implementation are transmitted to lower-level planning. At the same time, through the guidance of urban design, we can guide the construction of important districts, corridors and important scenic areas, and give full play to the structural guidance of superior planning to subordinate planning.

Of course, in the process of municipal comprehensive planning, attention should also be paid to the establishment of "one map" information platform. Through provincial summary, municipal construction and county-level access, a three-level information linkage among province, city and county can be formed to ensure that all kinds of elements can be transmitted hierarchically on the information platform, and planning assessment and supervision can be carried out in an orderly manner through the planning information platform, and an early warning mechanism can be formed. There are three key points in the construction of "one map" information platform: First, it is necessary to ensure the unified integration of the planning base map base; The second is to establish a unified database specification at the city and county levels; Third, it is necessary to clarify the database standards and data interfaces of sectoral planning.

4 Summary

First, we should fully understand the importance of planning implementation in this planning work, and establishing a implementation mechanism is an important means to ensure the accomplishment of planning objectives. Second, the comprehensive plan emphasizes the implementation of top-down planning, and at the same time, it

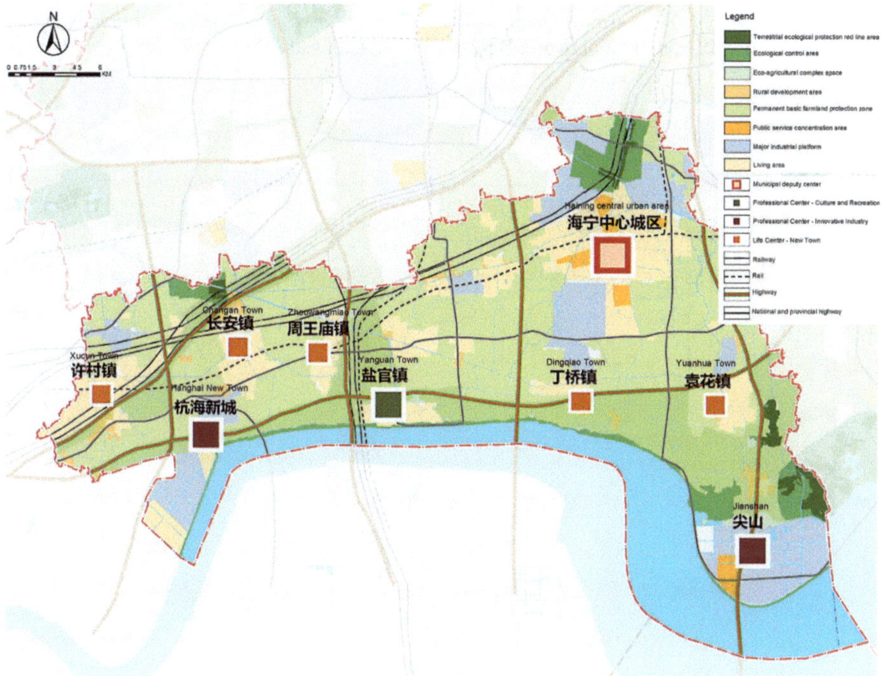

Fig. 8.13 Example diagram of "County Guidelines". *Source* According to the relevant contents of reference (Shanghai Tongji Urban Planning and Design Institute Co., Ltd. 2021)

should also reflect the overall planning of linkage from top to bottom. The effective combination of government efforts and market power is the key to the success of planning. The comprehensive plan of land and space should not only consider the technical logic transmitted by cities and counties, but also consider the policy logic of planning implementation. Third, planning is a dynamic process, and a blueprint has many levels and can never be achieved overnight. The combination of rigidity and flexibility and flexible reservation need to be considered emphatically. It is necessary to explore the planning path of assessment-strategic reservation-dynamic adjustment within life cycle management.

References

Yao Kai, Yang Ying. (2021). *Exploration on the overall planning and transmission practice of municipal land spatial planning* Southern Architecture, (02): 34–38.

Shanghai Tongji Urban Planning and Design Institute Co., Ltd. (2021). *Comprehensive Plan of Land and Space in Jiaxing City (Stage Achievements).*

Shanghai Tongji Urban Planning and Design Institute Co., Ltd. (2021). *Jingzhou City Land and Space Comprehensive Plan (Stage Results).*

Yang Yuhan. (2020). *How to draw a "map" of county-level planning for implementation* Shanghai: Shanghai Tongji Urban Planning and Design Institute Co., Ltd.

Constructing Rural Planning System

Zhang Li, Li Wenqi, and Zhang Shangwu

The legality of Village Planning can be traced back to the Regulations on the Administration of Village and Town Planning and Construction promulgated by the State Council in 1993. The initial understanding is that "village and town planning" is generally divided into two parts: "village and town master planning" and "village and town construction planning", which clearly distinguishes master planning from construction planning. The former focuses on overall deployment, while the latter is the guiding basis for construction actions. The implementation of the Urban and Rural Planning Law in 2008 marks that the traditional urban planning has officially expanded to rural areas. It explicitly stipulated that township plans and village plans should be formulated for rural regions, and village planning has risen from administrative regulations to legal level. The "Several Opinions of the Central Committee of the Communist Party of China and the State Council on Establishing a Territorial and Spatial Planning System and Supervising Implementation" (hereinafter referred to as "Several Opinions") issued in 2019 clarified that the scope of village planning is the village area and the positioning requirements of village planning in the new era: It is first a statutory planning; Second, as an integral part of the national spatial planning system, it belongs to detailed planning; Third, we should work out the practical village planning of "multi-regulation integration".

In fact, during the 11 years from the implementation of the Urban and Rural Planning Law to the promulgation of Several Opinions, it has been continuously

This paper was published in Urban Planning Forum Journal, No. 6, 2021.

Z. Li · L. Wenqi · Z. Shangwu
College of Architecture and Urban Planning, Tongji University, Shanghai, China
e-mail: leonzsh@tongji.edu.cn

Z. Shangwu (✉)
College of Architecture and Urban Planning, Shanghai Tongji Urban Planning and Design Institute Co., Ltd., Academic Committee of Rural Planning and Construction, Urban Planning Society of China, Tongji University, Shanghai, China
e-mail: zhshangwu@tongji.edu.cn

Z. Peng (ed.), *Theories and Frontiers in Territorial Spatial Planning*, Urban Sustainability, https://doi.org/10.1007/978-981-96-9825-7_9

discussed whether the working object of village planning is village or residential area, master planning or detailed planning. The "Several Opinions" generally clarify the above two core issues, but the relevant policy documents are still vague in guiding the specific contents of village planning, and there are the following key issues that need further discussion. First, the objectives and tasks of village planning are too comprehensive, including management and control, development, construction and design. For example, the Notice of the General Office of the Ministry of Natural Resources on Strengthening Village Planning to Promote Rural Revitalization not only emphasizes that village planning should bear the legal basis of land space use control, but also emphasizes the role of village planning in promoting rural revitalization and development. The content of village planning is tedious, which easily leads to heavy planning tasks, repeated results s and difficult implementation.[1] Second, the relationship between the level and system of village planning is still unclear. In the orientation of the current village planning content, it is required to have both the overall spatial layout at the village scale and the detailed construction guidance at the residential scale; There are both macro guidance at the overall planning level and use control at the detailed planning level. Third, the transmission relationship between village planning and superior planning is unclear. As a detailed planning, village planning must be guided by superior planning. However, at present, the territorial space panning at the county and city level system has not been fully complied and the compilation content is still unclear. In addition, there are various forms of territorial space panning at the township level,[2] and different types of upper-level planning are different in content, which leads to fuzzy upper-level planning of village planning and complex transmission relationship.

In the critical period of establishing the territorial space panning system, village planning shoulders the mission of territorial control and construction permit in rural areas. At the same time, the development opportunities and construction demands brought about by the implementation of rural revitalization strategy urgently need rural planning to play an effective role in guiding the investment of policy resources. However, the difference between the "full coverage" of management and control and the "targeted implementation" of revitalization should be fully understood and reflected in the planning logic. Facing the diverse and complex demands of rural development, the practice in many provinces and cities has proved that it is difficult to manage, develop and build rural areas with only one plan at the same time, and it is necessary[3,4] to establish systematic thinking to clarify the connotation of rural planning again. Therefore, while actively promoting the preparation of rural planning, it is necessary to carefully sort out and screen in detail, explore the rural planning

---

[1] Even a few years ago, it was often mentioned that "village rules and regulations can be village planning", but up to now, there is no officially approved village planning, only "village rules and regulations".

[2] "Several Opinions" clarify that township planning can be prepared separately, combined with city and county planning, or jointly prepared with several townships as a unit.

[3] Geng Huizhi and Li Kaiming (2020).

[4] Zhao Peipei et al. (2019a).

system and method suitable for today's rural areas, and enhance the scientific nature of rural planning and effectiveness.

1 Discussion on the Evolution of Rural Planning

Looking back on the course of village construction in China since 1949, it has gone through three stages: The first stage focused on the primary construction from 1950 to 1960s, which ensured that villagers' housing needs meeted and farmers had their own land through the reform of land system; The second stage focused on the improvement of village differentiation from 1980s to the end of 1990s, which improved housing conditions and agricultural production efficiency; The third stage concentrated on the implementation of new rural construction and rural revitalization strategy since the twenty-first century, so as to comprehensively improve the rural living environment. In each stage of rural construction, there are macro-policy guidance and micro-planning guidance. Although the latter may be informal practical planning,[5] objectively speaking, these informal planning have played a very good guiding role in rural construction.

1.1 Compilation and Implementation of Rural Planning

Since the twenty-first century, the development and evolution of rural planning has resulted in a variety of "disorderly" forms in the process of rapid advancement due to different compilation purposes and development stage requirements.[6] All local governments have initially formed a planning awareness from macro angle to micro angle about rural planning promoting work. There are four common problems in general. First, village planning is based on solving specific problems Problem-orientation, such as Shanghai focusing on rural construction land reduction,[7,8] Zhejiang focusing on beautiful countryside design,[9,10] Jiangsu focusing on human settlements improvement,[11] Guangzhou focusing on villagers' actual homestead construction demands,[12] etc. Therefore, rural planning often revolves around needs and policies, without a clear main line of planning system; Second, the planning focuses on the construction space, and the content is not comprehensive. Although

[5] Zhang Shangwu (2014).

[6] Wei Shuwei et al. (2019).

[7] Zhou Xiaojuan (2018).

[8] Yang Qiuhui (2019).

[9] Wu Qianbo et al. (2017).

[10] Zhang Jianbo et al. (2020).

[11] Chen Xiaohui, *Lvhai* (2021).

[12] Ye Yumin et al. (2016).

Comprehensive Regional Planning has been putting forward, it actually lacks effective management of non-construction space such as ecology and agriculture[13]; Third, many rounds of planning work have been promoted in various places, and there are obvious problems such as the absence of implementation mechanism and difficulties in connecting with villagers' governance; Fourth, in the actual implementation process, the rural planning system has not actually operated effectively.

In the process of widely promoting rural planning in various places, some provinces have reached a high planning coverage rate.[14] For example, Xinjiang, Yunnan and other provinces achieved full coverage of village planning in 2015. However, with the wide coverage of planning, there is also a negative voice like "village planning is useless", which not only contains the inherent contradiction[15] of governance logic, but also has some problems such as unclear planning level[16], improper methods[17,18] and the lack of planning implementation mechanism.[19] With the complexity of planning types, the positioning of village planning tends to be more vague, until "Several Opinions" made it clear that "village planning is detailed planning".

However, while reflecting on the diversified development of various rural planning in the past, we also proved the diversified practical demands in the evolution process of rural development (planning), which not only needs to complete the top-down bottom line control, but also needs to solve the problems of economic and social development, and also needs planning to guide the implementation of specific construction actions, which is both dynamic and complex. However, many planning practices have proved many times that the complete presentation of the planning itself cannot ensure the effective implementation of the village planning, and ultimately depends on the overall system design and operation.[20]

1.2 Rural Planning and Village Planning

Due to the long-term lack of authoritative definition of "rural planning", the academic understanding of "rural planning" is simply equivalent to "village planning",[21] or it

---

[13] Zhao Guangying and Jerry Lee (2019).

[14] According to the statistical yearbook of urban and rural construction of the Ministry of Housing and Urban-Rural Development in 2017, by the end of 2016, 61.5% of administrative villages in China had prepared village plans.

[15] Tang Yan et al. (2015).

[16] Fan Lingyun and Lei Cheng (2010).

[17] Wen Jiangang and Wen Hanzi (2015).

[18] Wei Guangjun (2021).

[19] Young Fan (2019).

[20] Ying Hongbo et al. (2019).

[21] In order not to misunderstand the two words "rural" and "village", the "rural planning" described in this paper tends to refer to the planning of rural areas, focusing on the physical system; "Village Planning", on the other hand, focuses on individual planning types.

is considered that the planning to solve the social problems and renewal problems caused by rural shrinkage should play the role of "social planning" such as comprehensive development planning, village rules and regulations, and rural governance.[22] In view of the above differences in cognition, the author believes that it is necessary to distinguish the scale and type of planning.

According to the scale of rural planning, it is generally believed that it mainly includes three levels: county, township and village, among which county planning is the key level to carry out village distribution, optimize rural spatial pattern as a whole and promote multi-regulation integration,[23,24] while township planning should cover the bottom line control of most villages.[25] Other scholars believe that there is no planning implementation ability below the county level, so the broad village planning content should be covered at the county level, and 'village planning and design conditions' should be put forward in the county planning, while effective and constructive planning can be prepared at the village level.[26] The "County Rural Construction Plan" launched by the Ministry of Housing and Urban-Rural Development in 2016 is intended to be based on the county as the basic unit coordinates the development and construction of rural areas.

In view of the horizontal planning classification, the existing research puts forward that the planning at the village level needs to be combined with the actual classification, rather than being combined into a "unified planning", which generally includes the main types such as village master planning, village construction and renovation planning, and village action planning.[27,28,29] There are also recent studies that explore the planning technology of "multi-regulation integration"[30,31] according to the requirements of territorial and spatial planning, and the research perspective mostly focuses on the technical integration of spatial planning. At the same time, some scholars believe that rural revitalization planning should deal with villages[32,33] (Land space) planning plays a guiding role and connecting relationship, making spatial planning become the guide implementation of rural revitalization resource allocation and classification.

---

[22] Sun Ying and Zhang Shangwu (2017).

[23] Gu Chaolin and Zhang Xiaoming (2016).

[24] Cao Lu (2017a).

[25] Peng Zhenwei et al. (2020).

[26] Zhang Jingxiang et al. (2019).

[27] Zhang Shangwu. *Urbanization and the transformation of planning system: An understanding based on the rural perspective.* Urban Planning Forum, (06): 19–25.

[28] Kuai Li and Liu Haiyan (2013).

[29] Ge Dandong and Hua Chen (2013).

[30] Le (2021).

[31] Li Na (2021).

[32] Hu Shou-Geng et al. (2019).

[33] Zhao Peipei et al. (2019b).

2 Construction of Rural Planning System

From the local practical experience, rural planning includes at least two levels and multiple dimensions of work. Two levels are the county Township (town) level and village level correspond to macro-scale planning control and layout planning and micro-level village level respectively. While the latter includes multiple dimensions of work, namely, industrial revitalization in village development dimension, resource control and layout in village space dimension, detailed design and land remediation in development and construction dimension, etc. Facing different dimensions such as territorial space panning control and rural revitalization strategy development, it is necessary to superimpose planning at the village scale to realize the organic integration of content, it is necessary to distinguish the planning content requirements of different levels and dimensions, instead of relying on a plan to nest complex content, which increases the complexity of planning preparation and the difficulty of control and implementation.

Therefore, it is necessary to realize that "rural planning is a system while village planning is a type of rural planning"[34]. Combined with the demands of top-down spatial management and control, the needs of rural revitalization and development and the actual needs of rural construction, this paper puts forward the framework of rural planning system, including vertical "hierarchical transmission" and horizontal "distinguishing three basic dimensions of management and control, development and construction, corresponding to three types of planning", and constructs a complete rural planning system.

2.1 *Vertical Transmission, Distinguishing Planning Levels*

As for rural spatial planning, every spatial level of territorial and spatial planning includes rural work, with varying emphases on rural planning content across different spatial levels.[35] Macro includes national, provincial and municipal levels: In the national planning, the overall strategy of rural revitalization, agricultural protection requirements and major principles and policies are deployed; In the provincial planning, policy formulation and strategy guidance are emphasized, programmatic requirements for optimizing the layout of rural residential areas are put forward, the spatial pattern of agriculture and rural areas adapted to the main functional areas is constructed, and the classification standards and planning orientation of village classification are put forward; In municipal planning, spatial guidance and zoning control are emphasized, the general principle of village layout optimization is put forward, the situation of rural construction land resources in the city is explored, and the reduction guidance strategy of zoning differentiation is formulated.

[34] Zhang Shangwu (2020). Zhang Li. *Ideal space: territorial space panning of towns and villages.* Shanghai: Tongji University Press, 6–7.

[35] He Xinghua (2020).

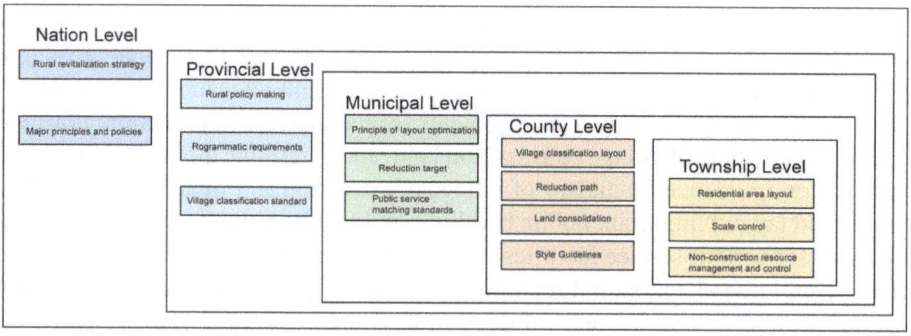

Fig. 1 Rural planning content system with vertical and vertical transmission. *Source* By the author

Spatial form control requirements such as characteristic protection, style shaping and height control are put forward by classification and zoning, infrastructure is arranged, public services are extended, and policy guidelines for activating rural vitality are put forward (Fig. 1).

From the practical demand, the transmission of county, township (town) and village is the core of the transmission of rural planning system.[36] In the past, county-level rural planning, such as county-level urban-rural overall planning or rural construction planning under the urban-rural planning system, played an important role in guiding the spatial agglomeration of towns and villages at the important level.[37] At present, the reform of spatial planning continues to advance, and further emphasis is placed on the overall role of protecting natural resources in rural areas and optimizing the overall spatial layout of villages at the county level. The "Several Opinions" pointed out that the village layout work should be basically completed at the county level in combination with the preparation of land spatial planning. The specific contents can be subdivided into: First, it is necessary to build the overall framework of rural regional spatial development, determine the key areas of rural revitalization, and put forward rural development guidance in different areas; Second, it is necessary to make overall plans to divide the types of villages in the county, and put forward the target requirements and implementation paths of rural construction land reduction; The third is to make overall arrangements for rural public service facilities and infrastructure; Fourthly, special tasks such as rural land remediation and style protection are put forward, among which the general rules of rural style protection are established at the county level based on the regional perspective, including the characteristic guidance, protection points, positive and negative lists of village style protection, etc., and a unified basic norm is established for the protection actions of regional village style through top-level design. The content of rural planning at the county level can be combined with the actual situation, or compiled as a special planning at the county level, or incorporated into the compilation results of the overall land and territorial space panning at the county level as a special content.

[36] Yuan Yuan et al. (2020).

[37] Cao Lu (2017b).

In the township-level planning, we should base on the perspective of town-level overall planning, implement and deepen the layout of village residential areas, put forward the scheme of village classification retention and relocation and the implementation timing, make overall arrangements for the increment, reduction, flow and flexibility indicators of village construction land, and determine the scale control and construction (guidance) boundary of village construction (for qualified towns); We also should lay out important public buildings and facilities, put forward supporting standards for various service facilities in villages, determine the key implementation areas of rural revitalization, put forward differentiated positioning guidance and put forward the control requirements of non-construction resources in township area.

2.2 *Horizontal Combination, Distinguishing Planning Functions*

Starting from the goal of village planning, this paper puts forward to distinguish the village (management and control) planning with the goal of resource management and control, the practical village (revitalization) planning with the goal of rural development and the village (special project) design with the goal of engineering construction Complete village planning system at village level (Fig. 2).

(1) Village (control) planning: Bottom line constraints, focusing on resource management and control in the global space

Village (management and control) planning corresponds to the management and control logic of territorial and spatial planning, focusing on the management of spatial elements, which is to carry out national spatial development and protection activities in rural areas, implement national spatial use control, issue planning permits

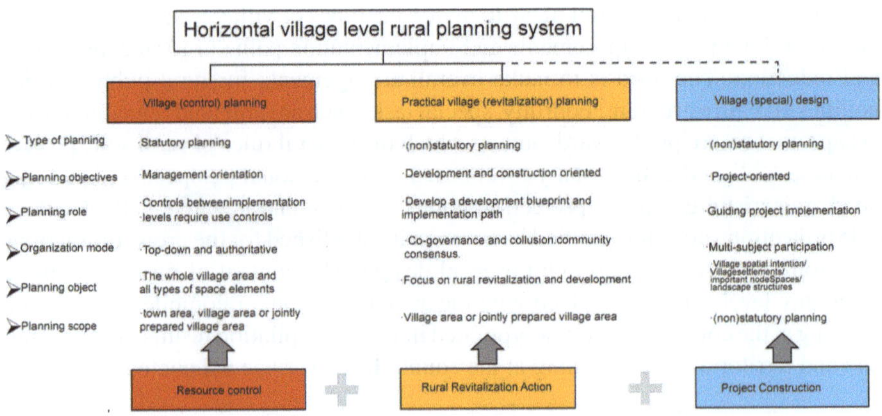

Fig. 2 Rural planning system at the horizontal village level. *Source* By the author

for rural construction projects, and serve as the legal basis for carrying out various constructions, the control of village spatial elements is solved at the micro level.

As the basic level of the territorial spatial planning system, in order to implement the control requirements of the upper planning, the technical logic of village (control) planning should be unified with that of the upper planning, namely, ① reflect the all-factor control, shift from focusing on village settlements to substantive coverage of village areas, and have more in-depth and adequate planning control over the non-construction space and land types of villages; ② reflect the bottom line constraints, implement the core control requirements and control forms issued by the superior planning, and keep the resource protection boundary and the construction and development boundary of the village; ③ Reflect planning rigidity and reservation flexibility and rigidity should reflect the bottom line constraints and the implementation of specific land use and construction intensity, flexibility means that should reserve space response for future rural development and construction uncertainty and formulate management rules to improve the operability of planning.

Based on the positioning understanding of village (management and control) planning, the content of village (management and control) planning should be inclined to the spatial control of the overall level of the village, including the overall spatial layout, the total amount of construction land constraints, and the delineation of control lines, which frame the basic rules for the rural revitalization work and project construction in the next stage.

(2) Practical village (revitalization) planning: Pursue for Efficiency and Focus on Development Orientation of Rural Revitalization

Since the implementation of the national rural revitalization strategy, the national, provincial, city and county levels have successively issued rural revitalization strategic planning or implementation planning to achieve rural revitalization "development concept layer by layer transmission" and "development goals step by step decomposition".[38] The rural revitalization strategic plan formulated at the macro level is a comprehensive and programmatic plan for local rural development. At the local level, the rural revitalization plan should be an operational planning blueprint, answer the overall idea of local village development, and plan the integration and coordination of special work from different lines to guide the rural revitalization funds, projects so as to guide the various funds, projects and plans for rural revitalization to implement more accurately and efficiently.

Different from the bottom line constraint of village (management and control) planning, emphasizes rigidity and flexibility, and strives for full coverage. Revitalization planning is a development-oriented planning, which should adapt to dynamic requirements and embody practicality, and it should be compiled as needed and highlight the unique features.[39] Therefore, the content of practical village (revitalization) planning should not be completely confused with the statutory planning. Instead, for

[38] Hu Shougeng et al. (2019).

[39] Gao Jing et al. (2020).

villages with revitalization and development needs, it is encouraged to prepare practical village (revitalization) planning at the village level as an action plan to solve local development issues, face operational implementation and implement capital projects, and at the same time, it can also become an integrated platform for rural revitalization work.

Development and control are mutually causal. Practical village (revitalization) planning and village (management and control) planning need to realize dynamic connection. Namely, take the idea of revitalization and development as the guide, and take the spatial management and control of village (management and control) planning as the spatial constraint of rural revitalization implementation. On the premise of conforming to the village (management and control) planning, the village revitalization ideas, industrial system, action plan, implementation mechanism, etc. are systematically laid out. And the spatial demands of rural revitalization are dynamically incorporated into the management and control content of the village (management and control) planning. Therefore, the practical village (revitalization) planning should coordinate the contents outside the village (management and control) planning, including village development strategy, industrial development strategy, spatial assumption, action plan and organization construction, mechanism and system innovation, etc. Also, its contents should be flexible, demand-oriented and keep pace with the time.

(3) Village (special) design: Project construction orientation, focusing on implementation and construction.

Village (special) design is a general term for detailed design based on engineering projects and aim at specific rural construction behaviors, including design types that take improving space quality as the starting point or serve the needs of specific projects. The former includes the overall village design and landscape design at the village scale, as well as the public space design, farmhouse design, street design and environmental sketch design at the micro scale; The latter includes land comprehensive improvement design, environmental improvement design, ecological restoration design, traditional village protection design, tourism planning design and so on.

Under the long-term path dependence of "project system" to develop villages (revitalize villages), there is a great demand for project-oriented village (special) design in practice, which leads to the fact that most of the achievements of legal village planning in the past are mainly at the design level, and confuse the nature of planning, construction and design.[40] Therefore, it should be pointed out that village (special) design is the end link of rural planning for implementation and construction, and it is also an important measure for fine management of villages. However, design itself cannot replace management and control, nor is it a necessary condition for legal management and control. Instead, it is compiled according to needs on the basis of village (management and control) planning and under the guidance of practical village (revitalization) planning to serve the practical needs of village construction at different stages of development.

[40] Zhang Li and He Lian (2017).

3 Differentiated Compilation Method and Achievement System

According to the positioning and actual needs of different villages, appropriate rural planning formulation methods should be matched (Table 1). In principle, legal control should achieve full coverage, and all kinds of villages should be compiled, but the forms can be different. For villages with small scale, stable development of villages and no need for development and construction, the use control rules and construction control requirements of villages can be clearly defined in the county and township land space (overall) planning, and the bottom line control of village construction can be carried out in the way of "elements + indicators + diagrams + directories", which can be used as the basis for village "construction and approval according to law".[41] Agglomeration promotion villages, characteristic protection villages and suburban integration villages belong to development-oriented villages, which are the main spatial carriers of economic and social development and population agglomeration in rural areas. For these three types of villages with clear development needs and development goals, village (control) planning and practical village (revitalization) planning are encouraged as the guide, and village (special) design is used as the implementation tool to form a complete and systematic rural planning achievement.[42]

Village (control) planning needs the approval of the people's government at or above the county level, and the results should be concise, clear and accurate, and the results can be generally formed into "one figure, one table, one book, one regulation, one database" (Fig. 3). "One map" "corresponds to the land space use control map (can be divided into a series of maps), and "one table" corresponds to the space use structure adjustment table. The "book" is the planning text, the "One principle" corresponds to the village rules and conventions for the protection and utilization of village features and resources, and the "one database" corresponds to the village planning database, which is integrated into the superior territorial spatial planning platform for unified management. Different from village (control) planning, practical village (revitalization) planning results should be presented in a form suitable for various groups of people to read, reflecting strategic transmission, policy guidelines and villagers' consensus. The village (special) design results are mainly for technical implementation.

[41] Peng Zhenwei et al. (2022).

[42] According to the Notice of the General Office of the Ministry of Natural Resources on Strengthening Village Planning to Promote Rural Revitalization in 2019, "Comprehensive Planning for Villages that Focus on Development or Need More Development, Construction, Restoration and Renovation" is proposed.

Table 1 Guide to the Compilation Mode of Rural Planning system

Type of village	Village (control) planning	Practical village (revitalization) planning	Village (special) design
Stability Improvement Category (Other category)	In principle, it should be covered in the township territorial and spatial planning, and it can also specify the village territorial space use control rules and construction control requirements in the county territorial and spatial planning	In principle, it should not be prepared	Compilation on demand
Cluster lifting class	In principle, it should be covered in the township territorial spatial planning and encouraged to prepare separately	Encourage compilation	Compilation on demand
Suburban fusion class	In principle, it should be covered in the territorial spatial planning of towns and villages, and separate preparation should be encouraged; If it is within the boundaries of urban development, a detailed control plan shall be prepared	Compilation on demand	Compile as needed
Featured Protection Categories	Support separate compilation, may also be covered in the township territorial spatial planning	Support separate compilation	Encourage establishment
Relocation and withdrawal of classes	In principle, it should be covered in the national land spatial planning of towns and townships, and it can also be clarified in the county land spatial planning of village land space use control rules and construction control requirements	Not to be compiled	Not compiled

Source Collated by the author

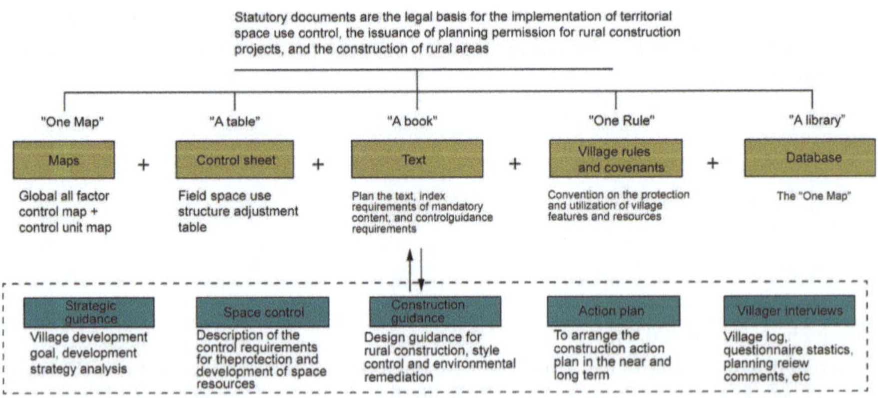

Fig. 3 Schematic composition of legal results of village planning. *Source* By the author

4 Key Contents of Village (Management and Control) Planning

Combined with the actual situation of the current reform progress of territorial and spatial planning, the key contents of village (control) planning should include: three line placement, factor control and space optimization.

4.1 Three-Line Position

In practice, the spatial data sources involved in villages are complex, including territorial survey, cadastral data of rural villages, spatial data of forest and grass departments, etc. Although the current land spatial planning at all levels is based on the data of the third Land Resource survey (referred to as the "three adjustments"), there are still many conflicts in the data information of the "three adjustments" at the village level that need to be further corrected. For example, in the process of top-down scribing transmission, there are deviations of control objects caused by differences in precision of different scales. Taking a village in a mountainous area of a province as an example, the ecological red line recognized by the provincial level is segmented based on latitude and longitude and grid. But in fact, the village's forest resources have relatively clear ownership jurisdiction boundary and protection grade boundary. In the actual implementation of control, it is obvious that the latter is more practical (Fig. 4). In addition, in terms of construction land, the rural cadastral survey data are the property rights space of the homestead and collective land that have been confirmed and registered, often including residential houses and yards (partial or partial) in front of and behind the house, while the "three tones" data are "what you see is what you get", and the scale accuracy of 1:10,000 is coarse, often expressed in the form of continuous map spots. Therefore, the obtained construction plots and property rights plots (often with a scale accuracy of less than 1:1000) have obvious spatial dislocation in different situations (Fig. 5), which needs to be corrected and confirmed in the village (management and control) planning.

To sum up, village (control) planning should pay full attention to the integration of current data, and take the most realistic spatial data as the basic base map of village land space use control. On this basis, it is necessary to check and feedback the spatial position of the control line delineated by the upper planning on a more accurate scale, and truly reflect the spatial contradiction in combination with the actual situation for the follow-up implement accurate global space use control to provide support.

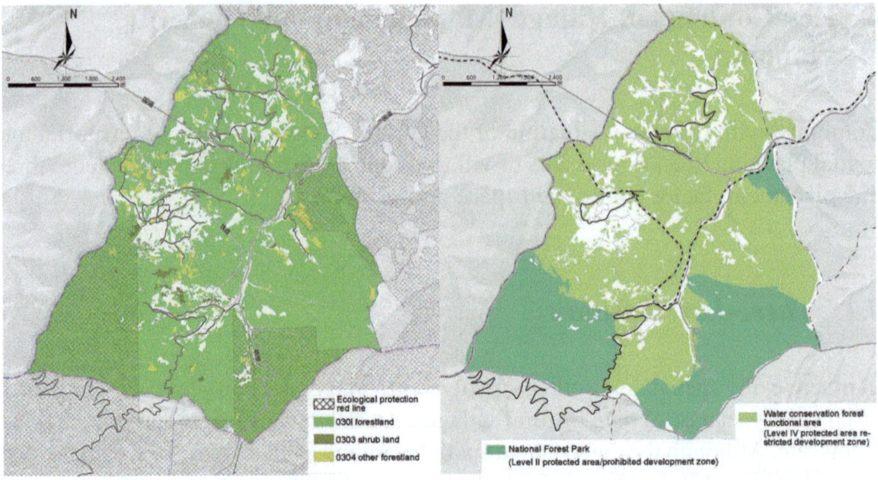

Fig. 4 Comparison of ecological protection red line (left) and forest land resource protection grade distribution (right) of a village in a province. *Source* A village Planning in a Province, 2021[43]

4.2 Factor Control

Village planning should implement the ecological protection red line, permanent basic farmland protection line and village construction (index) boundary determined by the superior land spatial planning, and delineate other control lines such as water control line and rural historical and cultural protection line according to the actual situation. In addition to the bottom line control, it is necessary to establish a control system for the spatial resources of the village, and encourage the adoption of the "Global Total Factor Control Master Map +Control Unit Plan" to control the detailed planning and form full coverage of factor control (Figs. 6 and 7).

In the general plan of total factor management and control in the whole region, it is mainly expressed as follows: ① "line elements", that is, core management and control boundaries, reflecting the rigid bottom line of planning; ② "Area elements", that is, planning zoning and use classification, reflect the land use control and construction control of the whole space elements; ③ "Point elements", that is, the distribution of public services and basic public facilities, reflect the layout requirements of facilities; ④ Relevant core indicators, reflecting the total amount and scale control of ecological and agricultural resources. In the control unit plan, the village is divided into several control units, and the land use types are further controlled according to the construction space and non-construction space. According to the non-construction space, distinguish the elements of ecology, agriculture and mineral resources, and put forward the protection and utilization requirements of different land types; According to the construction space, distinguish the homestead from

[43] Shanghai Tongji Urban Planning and Design Institute Co., LTD., Planning of a village in a province. 2021.

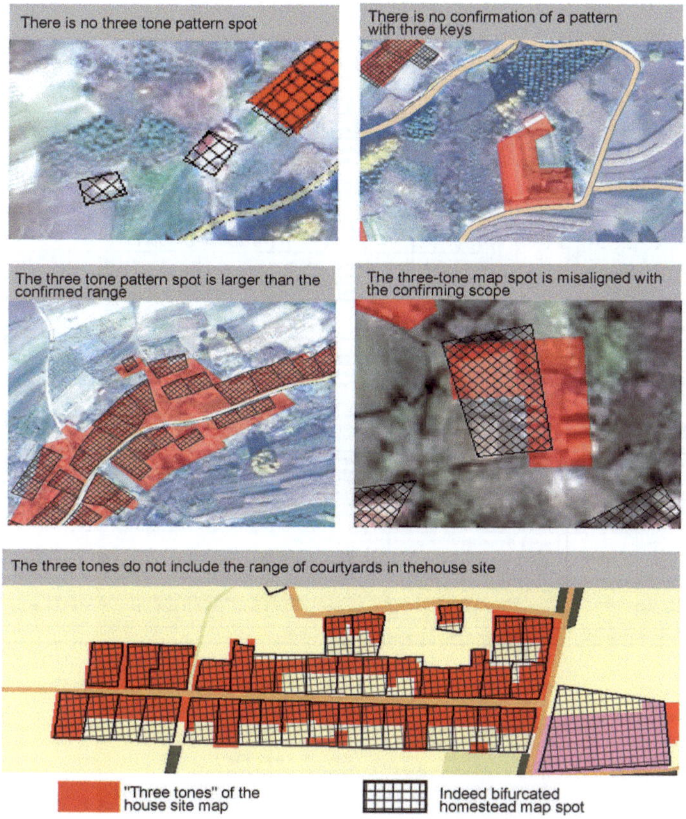

Fig. 5 Types of spatial conflicts between the scope of homestead ownership in a village of a province and the land use data of "three adjustments". *Source* Planning of a village in a province, 2021

others For construction land, the requirements of function, building density, height limit, boundary withdrawal and building quantity are put forward to form effective control for construction.

4.3 Space Optimization

From the macro trend of urbanization in our country, villages on the whole are trending toward shrinkage,[44] and most of the village planning is for "construction land reduction" planning. Therefore, the village (management and control) planning should put forward the scale, scope, timing and operational scheme of the reduction of village construction land. As for the reduction of village residential areas, it is

[44] Zhao Min et al. (2015).

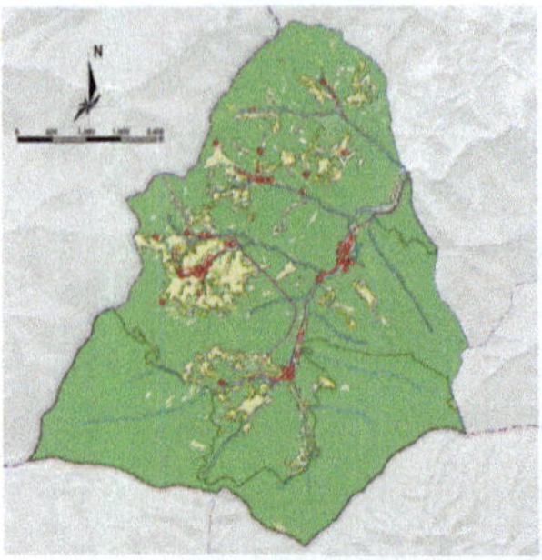

Key points of control

Control boundary (bottom line control: conduction and implementation + delineation at this level)

Legend	Name of control boundary	Basis for delineation
	Ecological protection red line	Implement the ecological protection red lines determined by the upper-level plan
	Permanent basic farmland protection red line	Implement the permanent basic farmland protection red line determined by the upper-level plan
	Village building boundaries	The map spots for village construction in Santiao are superimposed on the map spots for confirming the rights of rural homesteads and collective construction land.
	Drinking water source protection scope	The structure where 12 water intake wells are located is planned to extend 30m from the center point of the structure to the drinking water source protection area
	Flood risk control line	The center line of the main stream channel is controlled outward by 15m, the center line of the Class I tributary channel is controlled outward by 12m, and the center line of the Class I tributary channel is controlled outward by 10m
	Rural cultural protection line	Cultural protection unit: 10m extension on both sides of the railway, 5m extension on both sides of the protection range.
	Hohu reservoir management scope lines	It is recommended that the flood risk control line be used to delineate river, lake and reservoir management scope lines

Planning zoning (global zoning control)

Legend	Planning zoning name	Area(hm²)	Key points of control
	Ecological reserve	1313.89	The village has been designated as an ecological protection zone 1313.89hm² All types of development activities that are not in line with the main functional positioning are strictly prohibited. It is strictly prohibited to change the use without authorization. It is prohibited to occupy and change the nature of the land. It is encouraged to carry out activities to maintain, repair and improve ecological functions.
	Ecological control area	97.31	The village has been classified into an ecological control area of 97.31hm², which is the 10-15m range on both sides of Toudaogou and its tributaries. Strictly control the occupation and destruction of various development activities. Development and construction activities that damage the ecological landscape and pollute the environment are not allowed to be carried out without approval. Trees must be carefully planted, digging and filling are prohibited.
	Farmland protection area	239.53	239.53hm² of farmland protection zone has been designated in this village, which is the red line range of permanent basic farmland protection. It is managed in accordance with relevant regulations such as the Land Management Law of the People's Republic of China and the Regulations on the Protection of Basic Farmland. The district strictly controls the occupation of permanent basic farmland by non-agricultural construction, encourages high-standard farmland construction and land remediation, and improves the quality of permanent basic farmland.
	Village construction area	30.09	This village has been allocated 30.09hm² for village construction, which is the main area for village construction. Various spatial development activities such as rural construction must implement rural construction planning permit management within the boundaries of village construction.
	General farmland area	256.26	This village has been classified as general farmland area 256.26hm² Arable land is not allowed to be occupied at will. Those who do occupy it should submit an application. After the village committee reviews and agrees and issues a written opinion, it will be reported to the department for approval of the relevant land according to the procedures. Resolutely stop unauthorized "de-agriculturalization" of cultivated land
	Forestry development zone	2154.39	This village has been included in the forestry development zone 2154.39hm² Without approval, non-agricultural construction activities are not allowed in public welfare forests, commercial forests and other forest lands. Deforestation for reclamation and deforestation for quarrying, mining, sand extraction, earth extraction, tomb repair, house building and other illegal destruction of forest land are prohibited. Buy, sell, or otherwise illegally transfer forest land

Note: The basic farmland protection scope needs to be adjusted, and it conflicts with the village construction area by 1.08hm²

Fig. 6 All-factor control diagram of village planning. *Source* Planning of a village in a province, 2021

encouraged to combine the improvement of people's livelihood issues and environmental remediation,[45] and determine the key reduction objects. For those residential areas that are uncertain, the control method of "policy reduction" can be adopted. Namely, policy flexibility to the residential areas needs to be reduced, and start the reduction work after meeting certain policy conditions. Before that, the construction control is carried out according to the retention of village residents. In the whole space, the relationship between the reduction of construction land and the increase of non-construction territory should be handled well, and the layout of production facilities and ecological factors should be uniformly optimized to maximize the economic

[45] It can be specific to the near facilities such as highways, high-speed railways, high-voltage power corridors, as well as regional facilities, disaster protection, agricultural protection, ecological protection, historical protection, environmental governance and social governance of the villages, as the priority implementation of rural construction land reduction.

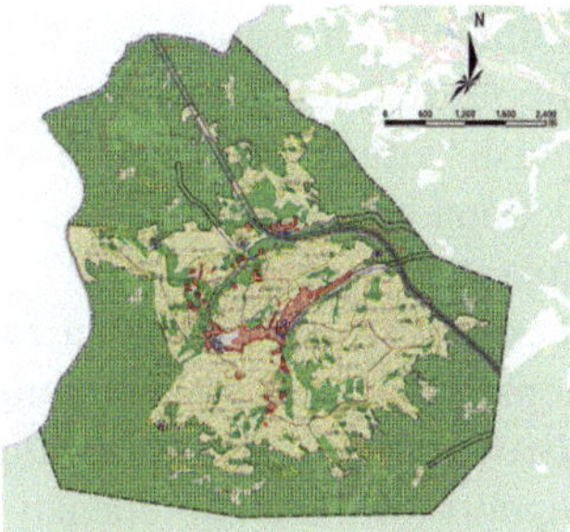

Ecological space management and control

Ecological space management and control	Lot number	Purpose Classification (Level 3)	Area (hm^2)	Control requirements
Waters	B3-01-01	Ditch	0.52	Within the scope of the ecological red line, all types of development activities that do not meet the relevant requirements for ecological protection are strictly prohibited. The forest area implements partial closure and protection management, encourages guidance and nurturing management, and commercial felling of trees is prohibited. Except for necessary occupation of highway and municipal engineering construction, it is prohibited to change the use of forest land in any form, and other land types in the ecological reserve are strictly controlled. In other areas, the occupation and destruction of various development activities are strictly controlled in the ecological control zone. Development and construction activities that damage the ecological landscape and pollute the environment are not allowed without approval. On the premise that no damage is caused to the ecological environment, activities such as sightseeing, tourism research, and education can be carried out appropriately.
Woodland	B3-0201 to B3-02-99	Arbor woodland	316.30	
	B3-02-100 to B3-02-106	Shrubland	4.61	
	B3-02-107 to B3-02-150	Other woodland	41.37	
Grassland	B3-03-01	Natural pasture	0.05	
	B3-03-02 to B3-03-29	Other grassland	8.36	

Binding indicators

Ecological protection red line (hm^2)	0	Wetland holdings (hm^2)	0
Permanent basic farmland protection area (hm^2)	112.32	Forest land holdings (hm^2)	365.88
Cultivated land holdings (hm^2)	161.27	Scale of historical and cultural protection lines (hm^2)	9.18
Construction land size (hm^2)	18.95	Scale of village construction land (hm^2)	11.53

Agricultural space management and control

Lot type	Lot number	Purpose Classification (Level 3)	Area (hm^2)	Control requirements
Agricultural facilities	B2-01-01 to B2-01-18	Village road land	4.04	Within the red line of permanent basic farmland protection, it is prohibited to build kilns, build houses, build tombs, dig sand, dig ponds to raise fish, quarry, mine, borrow soil, develop forestry and fruit industries, pile up solid waste or other activities that damage basic farmland, activities to stop idle and barren basic farmland, promote and encourage agricultural producers to apply organic fertilizers, and rationally apply chemical fertilizers and pesticides to the basic farmland they operate. In other areas, cultivated land cannot in principle be occupied at will. Resolutely stop the "non-agriculturalization" of cultivated land, prohibit the occupation of cultivated land to build kilns and tombs, or build houses on cultivated land without authorization, dig sand, quarry, mine, borrow soil, dig fields and landscape lakes, build green passages beyond standards, etc.
Construction land	B2-01-11 to B2-01-24	Internal village road land	1.06	
Arable land	B2-02-01 to B2-02-211	Dry land	161.27	
Garden	B2-03-01 to B2-03-17	Orchard	2.97	
	B2-03-18	Other gardens	0.17	

Construction space management and control

	Lot number	Usage classification	Land use function	Area (hm^2)	Volume rate	Building density	Height limit control	Green space rate	Construction guidance
Collective construction land management and control (public)	B1-02-01	Rural community service facility land	Business	0.25	≤1.5	≤50%	No more than 3 floors	30%	1. To build a complex public building, it is advisable to combine the building with outdoor activity venues to create a core space for villagers' exchange activities. 2. The design of public buildings should respect the current status of the site, encourage the restoration, transformation and improvement of original buildings and landscape spaces, and replace them with public service buildings. 3. New buildings should fully absorb simple local elements, encourage the use of pitched roofs, use red bricks or similar materials, and actively integrate into the rural
	B1-02-03	Rural community service facility land	Square	0.07					
	B1-02-04	Rural community service facility land	Culture room	0.10					
	B1-02-08	Rural community service facility land	Parking Lot	0.08					
	B1-02-11	Blank space		0.12					
Homestead control (private)	B1-01-01 to B1-01-187	Class I rural homestead	As for villagers' residences, farmhouses can be developed in compliance with laws and regulations.	Homestead control requirements 1. In a single-family rural homestead, the total hardened area of each household is 220-330m^2, which is determined according to the province's land management regulations. 2. The building height generally does not exceed 2 floors. 3. The per capita building area (based on registered population) does not exceed 50m^2					1. Residential buildings should adopt the form of pitched roofs, use tiles as much as possible, and focus on local traditional colors to avoid the use of uncoordinated color steel plate roofs. 2. The walls of residential buildings should be made of red brick or brick-red stone to continue the original village style, and avoid using bright colors or ceramic tile veneers. 3. Doors, windows, components, etc. of residential buildings should be made of wood and coordinated in color.

Fig. 7 Village planning control unit drawing. *Source* Planning of a village in a province, 2021

and ecological benefits of the village. At the same time, according to the needs of people's livelihood, perform well in the support system planning, and improve the layout of various infrastructure and public service facilities in the village.

5 Conclusion and Discussion

As China enters a new period of ecological civilization construction, more refined space use control means and tools are needed for national territory construction and development and natural resources management. The importance of village planning in the construction of land spatial planning system is obvious.

Although village planning can be setted in various forms, as a detailed planning, the basic content and flexible addition content of village planning, as well as the different functions of village planning about different construction and development conditions and different requirements are not clear at present. This has also led to widespread recognition throughout the country that "village planning is important and village planning is necessary", but it is difficult to implement it in a comprehensive and effective manner. It is very important to distinguish "village planning" from "rural planning" in the general sense to clarify the work boundary and work content of village planning under the current territorial spatial planning system.

Therefore, based on the combing of relevant domestic research and practical experience, this paper puts forward the general idea of constructing the rural planning system, and initially clarifies the logical relationship between the village planning itself and the rural planning system, so as to provide support for the comprehensive promotion and implementation of the subsequent village (control) planning. Based on this, it is advisable to carry out the full coverage of village (control) planning at the national level, and complete the control of all elements of territorial space resources; All local governments should promote the formulation of practical village (revitalization) planning for specific villages according to the actual situation; In economically developed areas and areas allowed by local financial strength, it is appropriate to combine project construction with village (special) design on demand.

References

Cao Lu (2017a). *Key points of planning County rural construction: A case study of Shexian County rural construction planning.* Urban Planning Forum, (05): 81–88.

Cao Lu. (2017b). *Thoughts on the main points of county rural construction planning: Taking rural construction planning in Shexian County as an example.* Journal of Urban Planning, (05): 81–88.

Chen Xiaohui, *Lvhai.* (2021). *Exploration of rural spatial planning under the construction of territorial Spatial Planning System: Take Jiangsu as an example.* Journal of Urban Planning, (01): 74–81.

Fan Lingyun, Lei Cheng. (2010). *On the legal implementation strategy of rural planning in China: Discussion based on urban and rural planning law.* Planners, 26 (01): 5–9.

Gao Jing, Peng Zhenwei, Gao Nan (2020). Research on the development of rural revitalization demonstration village *under the mode of co-creation: take shuiyuan village, Jinshan district, Shanghai as an example. Zhang Li. Ideal space: Land and territorial space panning of towns and villages.* Shanghai: Tongji University Press, 116–120.

Ge Dandong, Hua Chen (2013). *Construction of a new system and model of village planning adapted to the demands of rural development.* Urban Planning Forum, (06): 60–67.

Geng Huizhi, Li Kaiming. (2020). *Territorial spatial planning management and control strategy of rural areas under land spatial planning system: an empirical analysis based on Shanghai Urban planning* Journal, (04): 58–66.

Gu Chaolin, Zhang Xiaoming (2016). *On the planning of county, town and rural area.* Urban and Regional Planning Research 8(02): 1–13.

He Xinghua (2020). *On Small Town Planning and Rural Planning.* Zhang Li. *Ideal space: Territorial space panning of towns and villages.* Shanghai: Tongji University Press, 148

Hu Shougeng, Wu Si, Liu Yansui (2019). *Preliminary study on rural revitalization planning system and key technologies.* Geographical Research, 38(03): 550–562.

Hu Shou-Geng, Wu Si, Liu Yan-Sui (2019). *Preliminary study on planning system and key techniques for rural revitalization.* Geographical Research, 2019, 38(03): 550–562.

Kuai Li, Liu Haiyan (2013). *Discussion on the status quo, existing problems and Improvement Measures of village and town planning in China.* Planners, 2010, 26(06): 69–74.

Le (2021). *Discussion on practical village planning strategies under territorial spatial planning system*. Urban and Rural Planning, (S1): 69–81.

Li Na (2021). *Exploring the link path between rural space and territorial spatial planning system*. Urban and Rural Planning, (S1): 82–89.

Peng Zhenwei, Zhang Li, Dong Shuting, et al. (2020). *The necessity, orientation and key contents of township national space overall planning*. Urban Planning Forum, (01): 31–36.

Peng Zhenwei, Zhang Li, Dong Shuting, et al. (2022). *Necessity, orientation and key contents of the overall planning of land and space at township level*. Journal of Urban Planning, (01): 31–36.

Sun Ying, Zhang Shangwu (2017). *Review and prospect of rural planning research in China*. Urban Planning Forum, (04): 74–80.

Tang Yan, Zhao Wenning, Gu Chaolin. (2015). *The formation of rural governance system in China and its enlightenment to rural planning*. Modern Urban Research, (04): 2–7.

Wei Guangjun. (2021). *The Tide of Rural Planning in China; Characteristics, Dilemma and Thinking*. International Urban Planning, 08–18: 111.

Wei Shuwei, Wang Yang, & Chen Kaiyue, et al. (2019). *Evolution characteristics and enlightenment of China's rural system planning since reform and opening up*. Planners, 35 (16): 56–61.

Wen Jiangang, & Wen Hanzi. (2015). *Research on the landing of rural governance and planning in China*. Modern Urban Research, (04): 16–26.

Wu Qianbo, Yu Xiaying, & Chen Qianhu. (2017) *Development course and policy supply of rural construction in Zhejiang Province in the new period*.

Yang Qiuhui. (2019). *Unit compilation and management of territorial spatial planning in towns and villages: Development and exploration of country unit planning in Shanghai*. Shanghai Urban Planning, (04): 24–31.

Ye Yumin, et al. (2016). *Exploration on the compilation of implementable village planning in Guangzhou*. Beijing: China Building Industry Press.

Ying Hongbo, Duan Degang, & Zhang Bing. (2019). *How to do a good job in practical village planning: Observation and enlightenment based on informal circulation phenomenon*. Urban Planning, 43 (11): 103–111.

Young Fan. (2019). *Reconstruct the cognition of rural planning from three aspects: law, system and spatial implementation*. Western Journal of Human Settlements, 34 (05): 9–55.

Yuan Yuan, Zhao Xiaofeng, Zhao Tai, et al. (2020). *Study on hierarchical planning and vertical transmission of village planning under territorial space panning System*. Journal of Urban Planning, (06): 43–48.

Zhang Jianbo, Yu Jianzhong and Kong Bin. (2020). *Experience and typical techniques of village design in Zhejiang province*. Urban Planning, 44: 47–56

Zhang Jingxiang, Zhang Shangwu, Duan Degang, et al (2019). *Practical village planning based on multi-planning*. City Planning Review, 44(03): 74–83.

Zhang Li and He Lian (2017). *Examining the layout planning and extension of towns and villages from the perspective of villagers and government; Based on the Central Jiangsu region*: A case study of town. Urban Planning, 41(01): 55–62.

Zhang Shangwu (2020). *Village planning is a type of village planning*.

Zhang Shangwu. (2014). *Rural Planning: Characteristics and Difficulties*. Urban Planning, (02): 17–21.

Zhao Guangying and Jerry Lee. (2019). *Thoughts on technical reform of detailed planning under the system of land spatial planning*. Journal of Urban Planning, (04): 37–46.

Zhao Min, You Hunting, chenchen (2015). *On the "smart contraction" orientation and planning strategy of rural human settlement space*. Urban Planning, (07): 9–18 +24.

Zhao Peipei, Hu Qinggang, & Lu Dongmin, et al. (2019a). *Exploration on the planning of rural revitalization in the eastern pioneer areas; take Hangzhou as an example*. Journal of Urban Planning, (05): 68–76

Zhao Peipei, Hu Qing-Gang, LV Dong-Min, et al. (2019b). *Research on planning of rural revitalization in eastern China: A case study of Hangzhou*. Urban Planning Forum, (05): 68–76.

Zhou Xiaojuan. (2018). *Thoughts on rural revitalization strategy and planning strategy of megacities under the background of tight resource constraints: Take Shanghai as an example*. Shanghai Urban Planning, (06): 22–29.

Territorial Ecological Spatial Planning

Shen Qingji

1 Basic Concepts

1.1 Ecological Space

Ecological space can be simply divided into "natural ecological space" and "other ecological space". Natural ecological space refers to the territorial space with natural attributes and the main function of providing ecological services or ecological products, including forests, grasslands, wetlands, rivers, lakes, tidal flats, coastlines, oceans, waste land, deserts, gobi, glaciers, alpine tundra, uninhabited islets, etc. (Table 1).

"Urban ecological space" is also one of the important types of ecological space other than natural ecological space that belongs to other ecological space. Urban ecological space refers to the territorial space with natural attributes located in the urban area, whose subjectivity function is to provide ecological products or services, including important ecological areas with the functions of supply and support such as water conservation, food production, biodiversity maintenance and coastal zones stabilization, ecological restoration areas for prevention and regulation of soil erosion, land desertification, rocky desertification and salinization, as well as ecological recreation areas for providing social and cultural services such as ecotourism, outdoor activities, cultural entertainment and aesthetic value for human beings (Table 2).

S. Qingji (✉)
College of Architecture and Urban Planning, Tongji University, Shanghai, China
e-mail: sqj5688@126.com

Z. Peng (ed.), *Theories and Frontiers in Territorial Spatial Planning*, Urban Sustainability, https://doi.org/10.1007/978-981-96-9825-7_10

Table 1 Comparative analysis of definitions and connotations of ecological space

Source of concept	Connotation and inclusion categories
Definition of ecological space in *National main functional area planning*	Natural grassland, forest land, wetland, reservoir water surface, river water surface, lake water surface, barren land, sandy land, saline-alkali land, plateau desert, etc.
Definition of natural ecological space in Measures for the Control of *Natural ecological space usage (trial)*	The territorial space with natural attributes and the leading function of providing ecological products or services includes forests, grasslands, wetlands, rivers, lakes, tidal flats, shorelines, oceans, wasteland, deserts, Gobi, glaciers, alpine tundra, uninhabited islands, etc.
Shanghai ecological space connotation	The ecological space, agricultural space and green space in urban areas defined at the national level allbelong to the ecological space, including various ecological spatial elements such as offshore waters, mountains, rivers and lakes, forest land, green space, wetlands and farm land

Source Guo Chunbin. (2018). *Exploration on "Shanghai 2035" ecological spatial planning*, (05):118–124. Shanghai Urban Planning Review

Table 2 Types of ecological spatial elements in Shanghai

Numbering	Type of ecological space	Description
1	Nature reserves	National and municipal nature reserves
2	Forest parks	National forest parks
3	Geoparks	Core area of national geoparks
4	Wetland parks	National wetland parks
5	Drinking water sources	Primary and secondary protection areas for drinking water sources
6	Aquatic germplasm reserves	Conservation area for national aquatic germplasm resources
7	Natural shoreline	Natural coastline of the mainland and the islands
8	Important wetlands	National and municipal important wetlands and other wetlands with important functions except wetland parks
9	Important forest lands	Urban parks and municipal forest for public benefit
10	Important water bodies	Lakes and rivers under city and district management
11	Species distribution habitat of very small populations	Important habitats and reintroduction habitats of very small population species

Source Su Jinghua & Dongyang. (2020). Study on ecological space identification and control unit delimitation of a metropolitan city: taking Shanghai as an Example. Environmental Impact Assessment, 42 (01): 33–37

1.2 Ecological Spatial Planning

"Ecospatial Spatial Planning" can be recognized in three ways. Firstly, from the perspective of the most basic objective, ecological spatial planning refers to the planning type for maintaining and improving the function and quality of ecological space. Secondly, from the perspective of specific planning content, ecological spatial planning refers to all kinds of planning for protecting, controlling and guiding the construction of ecological space, including ecological spatial planning, ecological control line planning, ecological conservation planning, ecological network planning, and planning for development-prohibited zones and development-restricted zones. Thirdly, from the perspective of relationship, ecological spatial planning refers to the relationship between natural ecological space, urban ecological space, agricultural ecological space and facility ecological space at all levels of territorial space, as well as the relationship between ecological space and other spaces arrangement, planning and layout of coordination and integration, promotion and optimization.

2 Types of Ecological Space

2.1 Types of Ecological Space in Relevant Policy Documents

In *National plan on urbanization* (2014–2020) mentions two types: "natural ecological space" and "urban ecological space"; Several opinions on building a national territorial spatial planning system and supervising the implementation by CPC Central Committee and State Council refers that it is necessary to "scientifically lay out production space, living space and ecological space" (hereinafter referred to as "production-living-ecological space"), and appose "ecological space" with production space and living space as one of the three main spatial types of human settlements; The division of territorial space in the *National Main Functional Area Planning* includes: urban space, agricultural space, ecological space, other space (transportation facilities, water conservancy facilities, special land) and ecological space are one of the space types juxtaposed with urban space and agricultural space.

From the above information, it is evident that ecological space is explicitly listed in different spatial classifications, indicating that it is one of the important and essential spatial types of human settlement.

2.2 Ecological Space Types at Different Scales

There are different scales of territorial space, so there are different ecological space types. For example, macro-ecological space includes ecological types of national and regional spatial scales; meso-ecological space includes city cluster, metropolitan circle, city and other ecological space types; micro-ecological space includes community, block, lot and other ecological space types.

2.3 *Types of Ecological Space Based on Research Literature*

Refers to the types of ecological space mentioned in various documents on ecological space research, including: territorial ecological space; urban ecological space; natural ecological space; ecological space of town and city agglomeration; green ecological space; other ecological spaces (referring to natural spaces such as sandy land, bare land and saline-alkali land that provide less ecological products or services for human beings); forestry ecological space; water ecological space; hydrological ecological space; original ecological space; healthy ecological space; ecological space of residential area; rural ecological space; urban and rural ecological space; cultural ecological space; social ecological space; humanistic ecological space; characteristic ecological space; network ecological space; institutional ecological space; suitable ecological space for an organism; resilient ecological space; landscape ecological space; regional ecological space; urban ecological space; county ecological space; compound ecological space; key ecological space; key ecological space; symbiotic ecological space; vegetation ecological space; core ecological space; basic ecological space; fragile ecological space; ideal ecological space; institutional ecological space; expanding ecological space; indoor ecological space; ecological space of new city; sustainable ecological space; multifunctional ecological space, and so on.

Based on the research literature, the ecological space types illustrate the rich diversity of ecological space in terms of types.

3 Characteristics of Ecological Space

Understanding the characteristics of ecological space is of great significance and value for scientific ecological spatial planning and construction.

3.1 *Fundamental/Universal*

The fundamental (universal) nature of ecological space is reflected in its indispensability and carrying capacity. Necessity means that ecological space is the foundation of social and economic development of a country or region, and it is essential. If it is missing or of poor quality, it will lead to the disorder of the economic system in human society and affect its quality and sustainability. Carrying capacity means

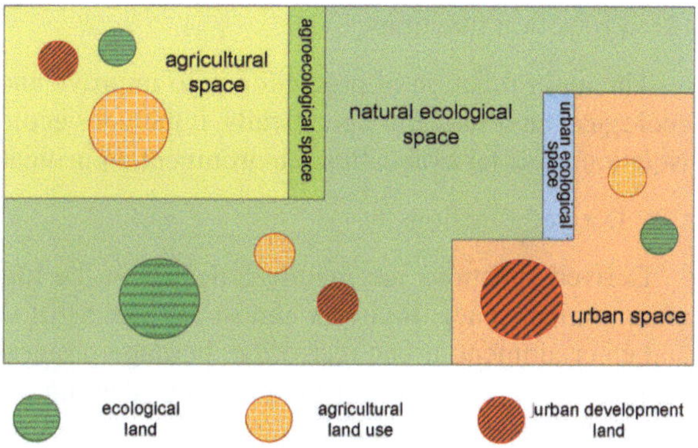

Fig. 1 Schematic diagram of national space division. *Source* Gao Jixi, Xu Delin, & Qiao Qing, et al. *Pattern construction of natural ecological space and planning theory exploration.* Acta Ecologica Sinica, 2020, 40(3): 749–755

that ecological space is the source and core carrier of ecosystem services, the material supporting foundation of ecological environment, and the bearing element of ecological civilization construction.

The most important point that can better explain the fundamental (universal) nature of ecological space is that ecological space exists in various spatial types related to human settlement. For example, as can be seen from Fig. 1, ecological land (space) exists in natural ecological space, urban space and agricultural space. To a certain extent, ecology is the root, agriculture is the trunk and town is the crown. If a tree wants to grow to a thousand feet high, it must be attributed to its roots. Without ecological space, agricultural space and urban space would be trees without roots and water without source.[1]

3.2 *Functionality*

Ecological space includes three functional types. The following takes urban areas as an example to illustrate the functionality of ecological space.

(1) Basic service functions

The basic service functions can maintain urban ecological security and keep ecological balance and stability. Maintaining biodiversity and preserving the integrity of ecological processes, which are mainly embodied in and permeate the city, specifically: improving hydrological cycle and water quality, strengthening air exchange between urban development land and peripheral ecological space, reducing urban heat island effect, ameliorating urban air quality, preventing soil erosion, lowering flood disaster risk, providing living space for various organisms and promoting energy flow circulation. The basic service is the core function of urban ecological space.

[1] Dang Shuangren (2019).

(2) Protection functions

The main function of protection is to improve and enhance the quality of urban ecological environment. Specifically, it includes ecological conservation, ecological isolation, disaster evacuation, environmental purification, etc.

(3) Derived functions

Derived functions are mainly reflected in the higher-level functional attributes of ecological space. It covers almost all aspects of urban functions such as leisure and entertainment, urban recreation, heritage protection, cultural display, industrial development, economic consumption and shaping of urban characteristics.

In urbanized areas, the dense population and compact space make the various functions of ecological space intersecting and overlapping (Fig. 2), which is the internal reason for the complexity of ecological spatial functions.

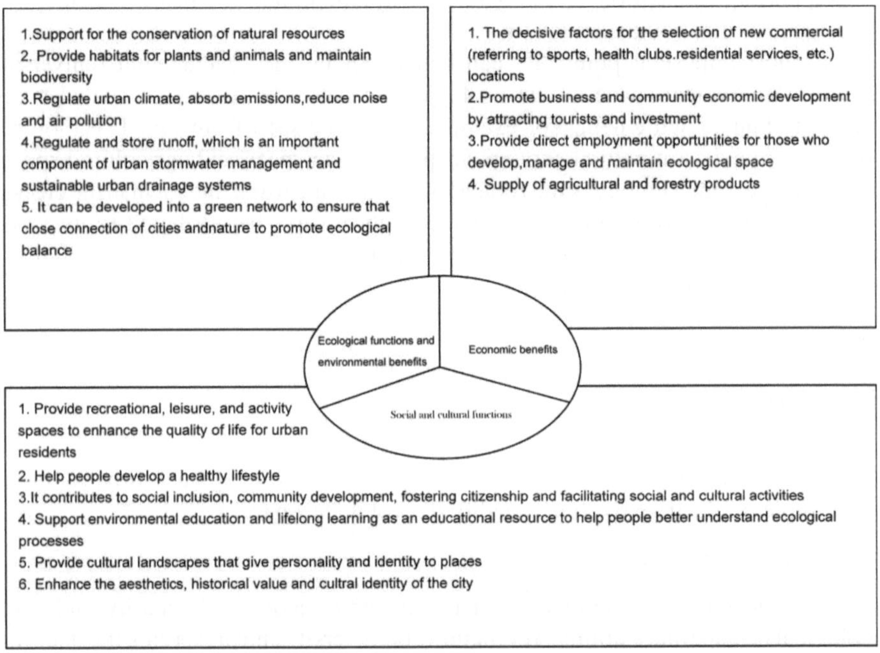

Fig. 2 Potential Functions and Benefits of Ecological Space in Urbanized Areas. *Source* Wang Fuyuan, Wang Kaiyong. (2018). *The scientific connotation of sustainable utilization of ecological space in urbanized areas.* Geographical Research, 37 (10): 1899–1914

Table 3 Basic types of ecological corridors in Suzhou

Types of ecological corridors	Manifestation forms	Resource characteristics
Vegetation	Green belts distributed along both sides of the road	Vegetation
Water body	River	Water body
Composite category	Rivers and coastal green belts	Rater body and vegetation

Source Chen Dechao, Xu Hualian, & Wang Yue. (2019). *Spatial characteristics of urban ecological resources in Suzhou.* Journal of Suzhou University of Science and Technology (Natural Science Edition), 36 (03): 54–58

3.3 Structure

The structure of ecological space is reflected by three aspects: structural elements, structural characteristics and structural effects.

(1) Structural elements

The structural elements of ecological space include point, line and area elements. Point elements refer to ecological landscape patch that are small in area. Linear elements refer to linear or strip landscape elements with channel or barrier function, and they are related point ecological elements (landscape patch), which maintains the exchange of species, nutrients and energy among landscape patch, provides special habitats for species, and increases the opportunities for species embedding, including rivers, corridors (Table 3), coasts, etc. Area elements refer to large area, best connectivity, the most complete and abundant ecological elements of the ecosystem generally include forests, wetlands, lakes, oceans, etc.

(2) Structural characteristics

It is characterized by nodes, connectivity, hierarchy and fractal. Generally, the study of ecological spatial structure requires the use of a series of indicators that reflect spatial structure. In terms of landscape patch pattern index, it includes landscape patch density, edge density, connectivity, aggregation, spread and richness. In terms of topological analysis of ecological networks, it includes network average degree, average path length, degree-degree correlation, network kernel number, etc.[2]

(3) Structural effect

Different types of ecological spatial structure will produce different ecological functions and effects, that is, structural effects. He Mei and Wang Yun summarized the urban ecological space system structure into four basic models: Eco-belt around the city model, wedge-shaped ecological belt model, ecological center core model and axial ecological belt model. The ecological, economic, social and integrated effects of these four models are different.[3] In addition, synergistic and multiplier

[2] Hou Hongbing et al. (2021).

[3] He Mei and Wang Yun (2009).

effects may be generated within ecological spaces and between ecological spaces and other spaces when they are harmonized, while conflicting and loss effect (negative external effect) in the contradictory state. In the study of Guangzhou Zhang Qifei and others found that the closer to the city center, the greater the temperature drop of the ecological space, which confirmed that even if the ecological space has the same structural characteristics, it has different cooling effects in different spatial positions. It is suggested that the cooling effect of suburban and urban central ecological space is different.[4]

3.4 Regionalism

The regionalism of ecological space includes the following aspects: Firstly, the comprehensive conditions of different regions lead to various types of ecological space (e.g. urban ecological space and natural ecological space, etc.). Secondly, the natural geographical conditions in different regions make the ecological space quality different. Thirdly, the proximity difference between ecological space and human beings in different regions leads to changes in its attributes and evolution trajectory. The ecological space near human beings produces natural-human processes due to the role of human beings in the ecological space, while the ecological space not near human beings is dominated by natural processes. Fourthly, the space utilization of natural resources by human beings in specific regions is significantly affected by regionality (Table 4); Fifthly, there are differences in the key points of ecological spatial planning in different spatial areas. For example, urban ecological space construction generally pays more attention to the improvement of environmental quality and three-dimensional and diversified green space in the central urban area, while suburban areas pay more attention to the accessibility of ecological space and the service level of recreational facilities[5]; Sixthly, different control measures of human beings on ecological space lead to different ecological space zoning schemes (Table 5).

3.5 Quality

The report of the Nineteenth National Congress of the Communist Party of China points out that "It is necessary to provide more quality ecological goods to meet people's growing needs for a beautiful ecological environment." Quality refers to the different conditions and levels of ecological environment quality and ecological service functions in the region where the ecological space is located, which are affected by various factors. For example, the quality of barren ecological space is

[4] Zhang Qifei et al. (2017).

[5] Shang Yongmin (2018a).

Table 4 91 valid expert questionnaires ranked the selection of space utilization methods of snow resources in Changchun

Space Utilization Mode of Snow Resources (N = 91)	Frequency	Percentage
1. As an ecological space utilization: Through the infiltration of snowmelt water, groundwater source can be conserved and the urban microclimate can be adjusted	81	89.0%
2. As an urban landscape space: Enhance the image of the city	69	75.8%
3.As a space utilization for water resource: Snowmelt water is used as urban domestic and production water	65	71.4%
4. As a space utilization for natural cold source: Refrigeration energy for urban living and production	57	62.6%
5. As a venue for urban ice and snow sports: Enrich winter activities	51	56.0%
6. Others	3	3.3%

Source Hou Qingyang. (2017). Study on preference and willingness to pay oriented space utilization mode of snow resources. Harbin: Harbin Institute of Technology

Table 5 Ecological spatial zoning

Ecological spatial zoning	Inclusion range	Regulatory requirements
Primary ecological space	Ecological conservation redline areas	In principle, it shall be managed in accordance with the requirements of prohibited development zones and implement mandatory ecological conservation measures
Secondary ecological space	Important ecosystem functional areas or ecological fragile areas	In principle, it is managed in accordance with the requirements of restricted development zones, and projects that meet the regional industrial and project access conditions are allowed to move in
Tertiary ecological space	Other ecological space	Implement looser management requirements, formulate access conditions, and allow more kinds of development and utilization

Source Wei Ziqian, Xu Zengrang, & Mao Shiping. (2019). Classification, scope and the impact of human activities of ecological space in Tibet Autonomous Region. Journal of Natural Resources, 34(10): 2163–2174

very different from that of non-barren ecological space. The quality of ecological space is related to the level and quality of ecological services provided by it, such as supply, regulation, support and culture, as well as the diversity of ecological space in terms of the composition of elements and the richness of the combination of elements. Of course, the quality of ecological space is also a reflection of the quality of ecological space management.

High quality ecological spatial plays an important role in the maintenance and stability of ecological space system, and the decline and weakening of ecological space quality will have a negative impact on the sustainable development of cities.

Ecological space quality model is one of the methods to express ecological space quality quantitatively and locatively (Fig. 3), and it is one of the basic tasks to improve ecological space quality (Fig. 4).

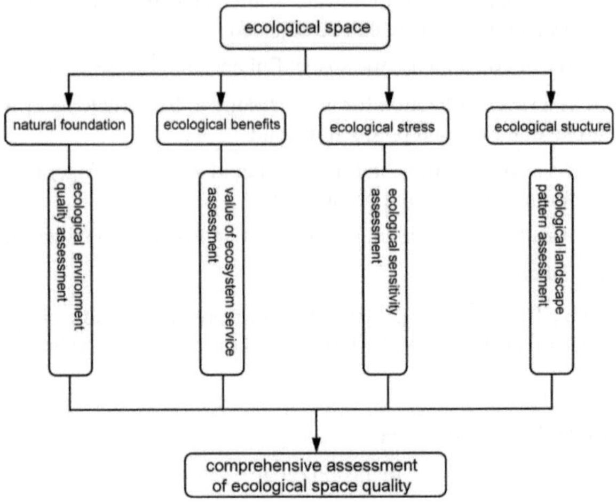

Fig. 3 Ecological space quality model. *Source* Wu Yanchang. (2019). *Study on Evaluation of ecological space quality of Chongzuo.* Nanning: Nanning Normal University

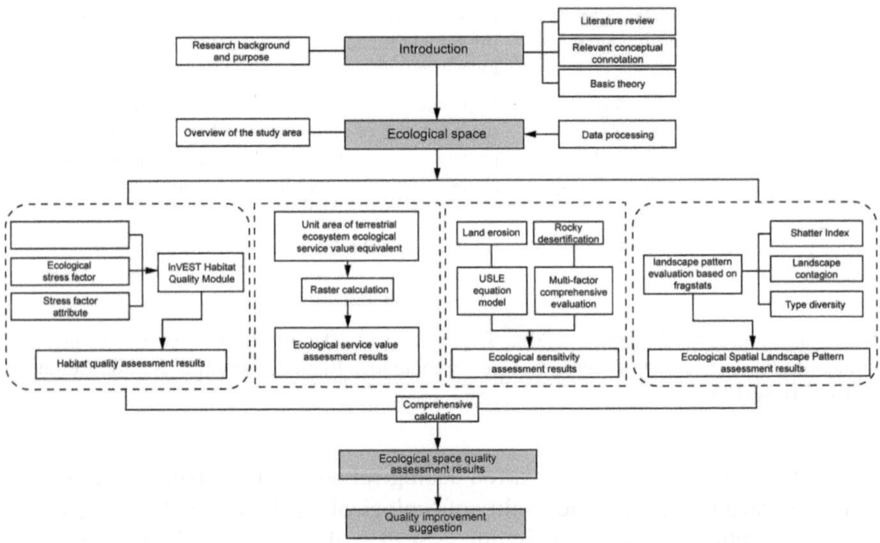

Fig. 4 Technical route of land ecological space quality evaluation in Chongzuo. *Source* Wu Yanchang. *Study on evaluation of ecological space quality of Chongzuo.* [D]. Nanning: Nanning Normal University, 2019

Fig. 5 Clean water of Lake Zurich. *Source* Author's self-photo

3.6 Dynamics

According to Conzen's periodicity theory, the transformation of space use is an unsteady process with accelerating, decelerating and stabilizing states of change.[6] The dynamic nature of ecological space can be understood through the change of ecological space quality, the type transformation and mode transformation of ecological space.

(1) The change of ecological space quality

In the 1960s, Lake Zurich became a heavily polluted area due to industrial development. The pollution was so severe that it was almost visible to the eye and detectable by the nose. Coupled with the frequent outbreaks of floods, the animal population gradually disappeared, leaving this once-popular tourist destination devoid of human visitors. [7] The quality of Lake Zurich has improved considerably as a result of the governance (Fig. 5), and 85% of Zurich's tap water is now sourced directly from Lake Zurich, which is a case of the quality change from poor to good of the water ecological space.

(2) Type transformation of ecological space

The types of ecological space transformation include: ① The transformation between ecological space and non-ecological space; ② the transformation between

[6] Conzen (2004).

[7] Shen Qi (2010).

Table 6 Landscape index changes of ecological spacein nanjing city from 1986 to 2002

Project	Main city			Urban development area			Peripheral urban area		
	1986	1996	2002	1986	1996	2002	1986	1996	2002
Number of landscape patches	61	69	67	664	759	1736	1463	1529	2336
Averagelandscape patch area (hm^2)	79.96	70.47	67.46	100.23	89.06	40.04	43.30	41.97	26.50
Boundary density	13.75	14.29	14.19	13.19	13.95	16.57	13.36	13.68	14.80

Source Chen Shuang, Liu Yunxia, & Peng Lihua. (2008). Dynamics of urban ecological space evolution and policy responses: a case study. Acta Ecologica Sinica, (05): 2270–2278

ecological space and agriculture-rural spaces; ③ Mutual transformation of various types of land in ecological space.

(3) Pattern transformation of ecological space

It includes two basic modes: contraction and expansion. The main modes of ecological space contraction are: Erosion of urban scenic forest land; Natural-near-natural land development; disappearance of water surface or agricultural development. The main modes of ecological space expansion include: grain for green project, newly building reservoirs and ponds, regreening mines, etc. Table 6 shows the dynamic changes of ecological space in Nanjing during 1986–2002 in terms of landscape patch number and area, which reflects the transformation state of ecological space in this period. Figure 6 shows the spatial changes of Nanji wetland in Poyang Lake during the wet season and the dry season ("rising water is a lake, falling water is a continent"), revealing one of the reasons for the ecological space dynamic changes is the change of natural environment.

3.7 *Publicness/Commons*

Holmes Rolston refers to the earth, air, water, environmental quality, ozone layer, wild animals and plants, the eagle as national symbols, and endangered species as public goods.[8] Many of the public goods mentioned by Rolston here are elements of ecological space, which to a certain extent indicates that ecological space has a public nature.

As for the nature of commons, it can be illustrated by "tragedy of the commons" first proposed by British scholar Garrett Hardin in 1968. Hardin believes that "commons" as a resource has many owners, each of whom has the right to use it, and none of whom has the right to prevent others from using it, which soon causes the overuse and depletion of commons resources. Just as on a public pasture, each herder wants

[8] Holmes Rolston (2000).

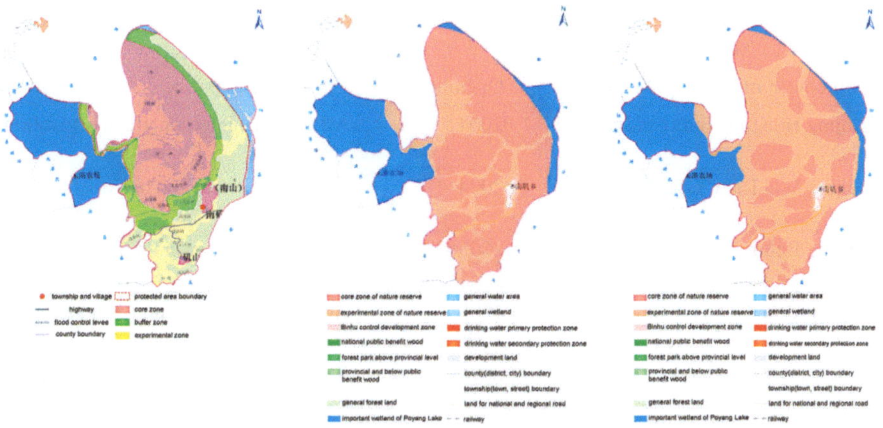

Fig. 6 Spatial changes of Nanji wetland in Poyang Lake in newly-built area of Nanchang during wet season and dry season. *Source* Deng Hongdi, Yuan Hong, & Qi Fan. (2020). *Thoughts on territorial development regulation based on natural ecological.* Urban Planning Forum, (01): 23–30

to graze the largest number of sheep to get the maximum profit. Since there is no limit to the number of sheep, the pasture is quickly overused and the quality of grassland deteriorates rapidly. Excessive deforestation, overfished fishery resources and heavily polluted rivers and air are typical examples of the "tragedy of the commons" The reason why it is called tragedy is that everyone knows that resources will be exhausted or damaged due to overuse, but everyone feels powerless to prevent the situation from deteriorating, and they all aggravate the situation with the mentality of "grab a share while it's still available".[9]

The resource attribute of ecological space determines its economic value, while the publicity of ecological space determines its openness and accessibility, when the relevant regulations are lacking or unreasonable, will eventually lead to tragedy of the commons of ecological space vulnerability to ecological space (Table 7).[10]

3.8 *Landscape/Aesthetic*

The types of ecological spaces listed in *National main functional area planning* include natural grassland, forest land, wetland, reservoirs, rivers, lakes, plateau desert and so on. The landscape of these ecological spaces (elements) are natural, which is in strong contrast with the landscape of non-ecological spaces, satisfying the pursuit of nature by human beings. In addition, it has its own landscape, thus showing its aesthetic value and characteristics (Fig. 7).

---

[9] Chen Xingang (2005).

[10] Peng Jianbo (2009).

Table 7 Privatization of public ecological space in the construction of official golf course in Beijing from 2004 to 2008

Golf course name	Opening time	Private use of public ecological space
Beijing Honghua International Golf Club	2004	All encroachment on green separation areas
Beijing Win River Golf Club	2004	Encroachment on the banks of the Wenyu River
Beijing Bohing Golf Club	2005	Some sites encroach on the riverbed of Yongding River
Beijing Grand Canal Club	2005	Adjacent to The Beijing-Hangzhou Grand Canal
Tian'an Holiday Sports Club	2005	Some sites encroach on green and isolated areas
Shunfeng Golf Country Club	2005	Occupy the shoreline of Ming Tombs Reservoir
Beijing Xinghewan Golf Course	2005	Embezzlement of public green space and private park
Beijing Jiawei Golf Club	2005	All sites encroach on green and isolated areas
Beijing Bayhood International Golf Club	2006	All sites encroach on green and isolated areas
Beijing Renji Golf Club	2006	All sites encroach on the shoreline of Huairou Reservoir
Beijing Xiangshan International Golf Club	2006	Private Huaxiangshan Ecological Sports Park
Beijing Oriental Shuangying Golf Club	2006	All sites encroach on the riverbed of Yongding River
Beyond Champion Golf Club	2007	Some sites encroach on the shoreline of Wenyu River
Beijing California County Golf Club	2007	All sites encroach on the riverbed of Yongding River
Topwin Golf And Country Club	2007	Some sites encroach on the shoreline of Yanqi Lake
Beijing Qinghebay Golf Club	2007	Some sites encroach on Qinghe River
Beijing Daning International Golf Club	2007	Some sites encroach on Daning Reservoir
Gold Century International Golf Club	2007	Adjacent to Qinglong Lake Reservoir
Beijing Junshan International Golf Club	2008	Adjacent to water source
Beijing Yanzikou International Golf Club	2008	Some sites encroach on forest land
Beijing Yintai Hongye Golf Club	2008	All sites encroach on the riverbed of Yongding River

(continued)

Table 7 (continued)

Golf course name	Opening time	Private use of public ecological space
Beijing Aoyang Property Golf Club	2008	Some sites encroach on the shoreline of Wenyu River
Beijing Huake International Golf Club	2008	Adjacent to the World Park
Beijing Ginkgo Park Golf Club	2008	Embezzlement of farmland
Pine Ridge Country Club & Resort	2008	Adjacent to Miyun Reservoir

Source Peng Jianbo. (2009). Study on the privatization of public ecological space. Beijing: Tsinghua University

Fig. 7 Landscape of waterfront populus euphratica forest. *Source* http://k.sina.com.cn/article503571802612c26f58a00100bmls.html

Different types of ecological spatial elements have unique aesthetic characteristics. For example, in Li Bai's poem "With northern grass looks like green silk thread", Li Shangyin's poem "The secluded grass may get favor from God", Han Yu's poem "Green grass can be perceived afar but not nearby", and the song of Chi-Le "The boundless grassland lies. Beneath the boundless skies. When the winds blow. And grass bends low. My sheep and cattle will emerge before your eyes." It's all about the aesthetics embodied in the grassland as an ecological space. Li Bai's poem "The Yangtze River is like a giant ax to split the male peak of Tianmen mountain, and green water runs east and back here." and "His lessening sail is lost in boundless azure sky. Where I see but the endless River rolling by." as well as Du Fu's poem "Stars hang low over the wide plain, the moon pours its light on the flowing water" and "The boundless forest sheds its leaves shower by shower. The endless river rolls its waves hour after hour." and Zhang Ji's poem "Spring water of the Yangtze River is dyed green, lotus leaves emerge like coins from water" and Su Dongpo's poem "Rocks tower in the air and waves teat on the shore. Rolling up a thousand heaps of snow." all express the aesthetic qualities of river ecological spaces. Similarly, Wang Changling's poem "The desert wind and dust blot out the sun" and Wang Wei's poem "In boundless desert lonely smokes rise straight. Over endless river the sun sinks round." and "The sun fell on the edge of the desert, and the desert was filled with the sound of fighting soldiers." and Gao Shi's poem "In the desert, wind and sand swirl, along the Great Wall rain and snow fall" as well as Li Bai's poem "Let's go to the desert to explore the tiger's lair, whip the horse across the Yellow River." and Li He's poem "The sand in the Gobi desert swirls like snow. A hook-shaped moon hangs over Yanshan Hill." These vividly express the unique artistic value of the desert landscape.

As for the charm of a city and the charm of its ecological space are causal and complementary to each other, often mutually influenced by each other and bring out the best in each other. For example, Han Yu's poem "The river is like a green gauze belt, and the mountain is like a green hairpin" records the scenery of Guilin's excellent landscape space. *Guilin* by Li Shangyin: "The mountains around Guilin seem to be falling down, wide river as if floating on the ground. To the southeast, it reaches the remote areas, to the northwest, there are tall buildings. The trees on both sides of the green Maple have the protection of the gods here: and at the bottom of the deep white stone pool, the sound of the flute and drum on the giant dragon boat never stops. In this unique land, What else can people pray for?" vividly portrays the aesthetic charm of the ecological spaces (elements) found in Guilin, such as mountains, rivers, land, shores, stones, trees, and shoals. It establishes the high aesthetic value and artistic status of Guilin's landscape.

3.9 Emotional

Emotional nature of ecological space refers to the function and role of ecological space and its constituent elements in carrying and hosting certain emotional factors of people. The emotional nature of ecological space is closely related to the "imaginability"[11] of the space, which is the basis of the emotionality of ecological space, which can explain the phenomenon that homesickness is often associated with some landscape elements in hometown (trees, mountains, rivers, streams, old houses, etc.) in hometown. Therefore, it can be considered that ecological space (elements) is an important carrier of "homesickness".

Residents' attachment to urban ecological space can reflect the emotional characteristics of ecological space. Arnberger and others found that although urban residents have less green space, they have higher utilization rate of green space and stronger ecological local attachment by studying the perceived difference in the impact of green space on local attachment between urban and suburban residents[12]; Peters and others found that residents" leisure and recreation in urban parks can

[11] Wang Mengyao (2014).

[12] Arnberger and Eder (2012).

enhance their sense of local attachment to the whole city.[13] Yin Duo and others took Ordos Kangbashi as an example to study the local attachment characteristics of residents in urban ecological recreation space in ethnic minority areas, and found that Ordos urban residents' sense of identity with Kangbashi is greater than their sense of local attachment, and residents' attachment to Kangbashi mainly comes from their emotional attachment. In addition, Mongolian residents' attachment to Kangbashi's ecological recreation place and local identity are different from those of Han residents (Table 8).[14]

Human factors in the composition of ecological spatial play a key role in spatial affection. He Mei believes: "Urban ecological space is a comprehensive multi-ecosystem composed of natural factors such as soil, water body, animals and plants and people."[15] It proves that natural interaction between man and ecological space contributes to the emergence and obvious of emotional nature of ecological space.

There are examples around the world of human factors contributing to the emotional nature of ecological spaces. Including the belief in sacred trees in Japan, the blessing culture of feng shui trees in Korea, and the worship of plants in China. For example, Cao Songye investigated and counted more than 20 tree temples in Jinhua urban area, Zhejiang Province in the late 1920s. and in his article *A simple statistics of some tree temples in Jinhua* in Folk Weekly, he recorded the following: "There are 2 pagoda tree temples, 2 soap tree temples, 1 cypress tree temple, 1 cypress tree with white pagodas temple, 1 Xiantao temple, 1 jujube tree temple, 1 tongmu temple, 1 palm temple, 1 willow (yangliu seven ladies) temple, 1 sophora pagoda temple, 1 ancient wood temple and 1 luohan cypress temple." At that time, the urban area of Jinhua city was less than 2 km^2, but there were as many as 15 tree temples, which shows that the custom of respecting tree gods in Jinhua flourished.[16] The reason for this, if explained from the perspective of ecological space emotionality, can be concluded that trees as an ecological element, people place their emotions through the spatial form of temples.

For another example, the big pagoda tree in Hongdong County, Shanxi Province is a holy place for people's sacrifices with the theme of "root exploration" and "ancestor worship" in China, and key cultural relic protection units in Shanxi Province. In 2008, the custom of worshipping ancestors with the great pagoda tree was listed as a national intangible cultural heritage, and the great pagoda tree is considered as the birthplace of "root and ancestor culture". In the history of Chinese immigration, the great pagoda tree may belong to the most extensive radiation range and the greatest influence of an immigrant birthplace. "Ask me where my ancestors came from, the great pagoda tree in Hongdong, Shanxi Province". This folk song, which has been circulating for hundreds of years, is almost well known among the older generation in North China. The great pagoda tree has become the spiritual root of all the descendants of Shanxi immigrants, no matter where their ancestors came

[13] Peters et al. (2020).

[14] Yin Duo et al. (2014).

[15] He Mei (2010).

[16] Tai Jie (2010).

Table 8 Comparison of perceived differences of local attachment of ecological recreation between mongolian residents and han residents

	Residents' local attachment factors and their items	Mean		Standard deviation		Approval rate		Objection rate	
		Han	Meng	Han	Meng	Han	Meng	Han	Meng
Factor 1	Local dependence of ecological recreation	3.65	3.97	–	–	–	–	–	–
	Other places in Ordos can't be compared with Kangbashi	3.67	4.03	1.244	1.109	56.4%	73.0%	16.2%	8.3%
	To appreciate the ecological landscape, the first thing that comes to mind is Kangbashi	3.73	3.88	1.157	1.039	59.7%	68.8%	14.9%	11.5%
	This urban area provides leisure and recreation conditions that cannot be provided elsewhere in Ordos	3.72	3.96	1.100	1.004	59.7%	70.8%	13.3%	8.4%
	You like Kangbashi more than any other place in Ordos	3.54	3.95	1.253	1.040	51.6%	69.8%	20.7%	8.3%
	Compared with other places in Ordos, Kangbashi is more worth visiting	3.61	4.03	1.209	1.051	56.2%	74.0%	17.7%	10.4%
	Local identity of ecological recreation	3.73	4.00						
Factor 2	When talking to friends from other places about Ordos, you always mention Kangbashi	3.77	4.07	1.137	1.018	62.3%	77.1%	15.1%	9.4%
	When traveling, you often think of Kangbashi in Ordos	3.85	3.92	1.148	1.111	53.4%	69.8%	16.6%	14.6%
	Kangbashi in Ordos is very special to you	3.64	3.97	1.147	1.100	54.9%	68.8%	15.4%	10.4%
	You have a deep attachment to Ordos Kangbashi	3.52	3.83	1.214	1.185	51.0%	69.8%	20.2%	16.7%
	Kangbashi makes you feel proud of living in Ordos	3.86	4.20	1.180	1.022	65.4%	77.1%	14.1%	7.3%

Source Yin Duo, Ni Meng, & Wu Tiehong, et al. (2014). Study on the characteristics of residents' local attachment in urban ecological recreation space in ethnic areas: a case study of Ordos Kangbashi. Journal of Northwest Forestry University, 29 (05): 249–255

from in Shanxi.[17] This is closely related to the fact that the great pagoda tree is the distribution center for Shanxi immigrants when they leave their hometown, and it is the most impressive symbol that can be seen when they leave their hometown. From an academic point of view, the huge size of the great pagoda tree (the ancient big locust tree is located two miles north of Hongdong County, on the bank of Fenhe River. The strong "image" caused by dozens of trees and several acres of shade.[18]) and the psychological shock caused by the group grief atmosphere when immigrants were forced to migrate are the important reasons for the great pagoda tree to become the ecological space element that sustains the feelings of immigrants' hometown.

3.10 Human Factors

The various functions and dynamics of ecological space are all related to human factors, and are all influenced and constrained by human behavior. It is particularly evident that the change and transfer (dynamic) of ecological space are related to human construction behavior, residential space expansion, social and economic system and policy orientation, etc. The aesthetics and emotional qualities of ecological space are related to people's subjective feelings towards ecological space, which can not be separated from the interaction between people and ecological space.

The characters of human factors in ecological space are closely related to the proximity between ecological space and human beings. In areas with a high degree of proximity between the two (such as cities), ecological space tends to be more susceptible to the influence of human factors. All kinds of technical indicators, planning and construction norms and standards related to ecological space are formulated and implemented by human beings. The comprehensive quality of human beings determines the level of ecological spatial planning and the results of ecological space protection and construction.

In terms of human attributes, there are space owners and administrators of spatial governance. In terms of human behavior towards ecological space, there are space utilization and space control behaviors. In addition, people's cognition of ecological space, the attributes and characteristics of people's behavior towards ecological space, and the intensity of use control have short-term and long-term impacts on the evolutionary state and quality of ecological space.

The human factors of ecological space is also reflected in the different demands of different people for ecological space. For example, highly educated people may pay more attention to the quality of ecological environment and hope that the ecological space will be richer in temperature; People with low educational background may pay more attention to the entertainment and leisure function of ecological space.

---

[17] An Jiesheng and Ge Jianxiong (1997).

[18] Yang Anxiang (1984).

Teenagers have strong demand for sports leisure space and public participation atmosphere, middle-aged people pay attention to the opportunity to get close to nature, and elderly people pay more attention to the accessibility of ecological space.[19]

4 The Significance, Role, and Planning Response to the Characteristics of Ecological Space

Cognition and revelation of ecological spatial characteristics are of great significance and function to ecological spatial planning and territorial ecological spatial planning. At the same time, the characteristics of ecological space need to be responded and reflected in the process of ecological spatial planning at all levels and types, which is the necessary way for ecological spatial planning to meet the requirements of the reform of territorial spatial planning, and for ecological space to realize its own value (Table 9).

5 Some Important Theoretical Issues of Ecological Spatial Planning

5.1 Optimality: Theory of Optimal Construction of Natural Ecological Spatial Pattern

One of the main tasks of ecological spatial planning is to build a natural ecological spatial pattern with excellent state. Gao Jixi et al., believe that ecological space in a broad sense refers to the space with natural attributes and the leading function of providing ecosystem services, including ecological space, urban ecological space and agricultural ecological space. Natural ecological space refers to the land space which is juxtaposed with urban space and agricultural space and distributed outside them, and has integrity and continuity in distribution. Based on regional ecology and landsenses ecology, the theories related to the construction of natural ecological spatial pattern (optimization) mainly include: ecological donor-receptor dual coupling theory, ecological pattern-process-function trinity theory, ecological carrying capacity and ecological suitability theory, threshold theory, ecosystem service trade-off theory (Table 10).[20]

---

[19] Shang Yongmin (2018b).

[20] Gao Jixi et al. (2020).

Table 9 Significance of ecological spatial characteristics to land (ecological) spatial planning and planning response

Ecological spatial characteristics	Significance and function of ecological spatial planning and its response		Significance and function of territorial spatial planning
	Significance and function of ecological spatial planning	Response of ecological spatial planning	
Fundamental/ universal	The main objects of ecological spatial planning; One of the embodiment of ecological space value	Attach great importance to total factor planning	One of the backgrounds and bases of territorial space development; It embodies the position of ecological space in theterritorial space development system
Functionality	Provide positive service for human beings; An important part of human well-being	Theories and methods of improving undesirable functions, maintaining positive functions and optimizing overall functions	Improve the functional system of territorial space; Improve the charm level of territorial space
Structural	One of the perspectives of representing ecological space; The basis of maintaining and optimizing ecological functions	Maintain excellent ecological network and structure, adjust defective structure and optimize structural effect	The carrier of coordinating the relationship between ecological space and other spatial types; One of the basic conditions for optimizing the overall quality of territorial space
Regionalism	Adjust measures to local conditions, reflect regional characteristics and enrich ecological space types	Theory, Method and control system of local lanning	Enhance the regional characteristics of territorial space and increase its diversity

(continued)

Table 9 (continued)

Ecological spatial characteristics	Significance and function of ecological spatial planning and its response		Significance and function of territorial spatial planning
	Significance and function of ecological spatial planning	Response of ecological spatial planning	
Quality	Improve the quality of ecological environment and enhance the overall quality of ecological space	Theory and method of ecological space quality identification and theory and standard of optimization	Promote the overall level of production-living-ecological space
Dynamics	The quality, type, combination and transformation mode of ecological space are in an unstable state, so it is particularly important to regulate them	Theory, method and management system of controlling adverse changes of ecological space and maintaining and optimizing benign dynamics of ecological space	Control the dynamic cycle of production-living-ecological space; Scientifically regulate and control the evolution of production-living-ecological space in dynamics
Publicness/ Commons	Scientifically formulate the property right system and utilization regulation of ecological space resources to avoid The Tragedy of the Commons	Through a scientific management and control system, we can avoid the phenomenon of commonality of ecological space and maintain its quality while improving its publicity	Improve public welfare and limit resource plunder
Landscape/ Aesthetic	Enhancing the value of ecological space is one of the carriers of national self-confidence	Protect the landscape carrier and elements of ecological space; Excavate and enhance the historical and humanistic value of ecological space landscape	One of the bases and foundations for enhancing the charm of territorial space
Emotional	It is suggested that the social and humanistic dimension of ecological spatial planning should be paid attention to	Theories and methods of maintaining and cultivating positive emotional factors in ecological space	Attach importance to the social psychological attribute of land ecological space
Human factors	The comprehensive quality of ecological spatial planners, managers and users plays a key role in the results of ecological spatial planning and construction	Theory and method of ecological space quality promotion and simultaneous human capital optimization	Take the improvement of human capital and three basic space (living, producation and ecology) quality as a whole

Source Self-made

Table 10 Theory of optimal construction of natural ecological spatial pattern

Names of theory	Explanation of related concepts	The content of theory
Ecological donor-recipient double coupling theory	Ecological donor refers to the ecological function bodies that provides ecosystem services in the region. Ecological receptor refers to the ecological functional bodies that receive ecosystem services in a region. Donors and receptors are linked through water, wind, resources, etc.	(1) The double coupling theory of ecological donor-receptor refers to the ground-ground coupling and human-ground coupling between ecological donor and ecological receptor, in which the ground-ground coupling refers to the spatial configuration and mutual influence of ecological donor and receptor. Man-earth coupling refers to the coupling between natural ecosystem and socio-economic system in a region; (2) Natural ecological space is the ecological donor within the territorial space, providing ecological services and ecological products for larger-scale space and human survival; (3) The construction of natural ecological spatial pattern should not only consider the ecological importance of the region itself, but also consider the mutual influence with agricultural space and urban space in spatial layout, as well as the supply capacity and coupling relationship of materials, energy and information provided by it to regional social economy
Trinity theory of ecological pattern-process-function	Ecological pattern refers to the spatial pattern and interrelationships of different ecological functions within the ecological domain. Ecological process refers to the path and process of material circulation and energy flow of different ecological functions within the ecological domain. Ecological function refers to the ability of ecological domain to provide services and products in various ecological processes based on its ecological structure	(1) The construction of natural ecological spatial pattern requires the identification of ecological functions that play an important role in maintaining the ecological process and ecological function in the ecological domain, and analyze their spatial positions and correlation, so as to build the natural ecological spatial pattern with complete functions and continuous structure; (2) The identification of natural ecological space should not only consider areas rich in biodiversity, but also consider areas that can provide important ecological services, such as the headwaters of greatr rivers and areas with fragile ecology and prone to disasters; (3) Stable pattern, smooth process and function matching between ecological donors and recipients are the basis of constructing natural ecological spatial pattern

(continued)

Table 10 (continued)

Names of theory	Explanation of related concepts	The content of theory
Ecosystem carrying capacitysuitability theory	Ecological carrying capacity refers to to the following three aspects: First, the self-regulation and self-maintenance ability of various ecosystems (ecological donors) in the region, as well as the ability to provide resources and environmental capacity for the socio-economic subsystems in the region, are the supporting parts of ecological carrying capacity; Second, the demand for ecosystem services and ecological products by economic society subsystems in the region is the pressure part of ecological carrying capacity. Third, ecological suitability refers to the suitability status and degree between social and economic development activities and ecological environment in the region	(1) The unique ecological endowmenst and resources and environmental conditions in the natural ecological space, which are binding on the living environment, industrial layout and industrial structure in the whole region, but they are also the basis for supporting social and economic development; (2) The industrial layout and structure in urban space and agricultural space, as well as the degree of population aggregation will cause the spatial circulation of ecosystem services in natural ecological space; (3) The construction of natural ecological spatial pattern should be based on the evaluation of regional ecological carrying capacity and the evaluation of the suitability of territorial spatial development. Based on these two types of evaluation, the areas facing or already overloaded ecological carrying capacity, as well as unreasonable areas such as industrial layout and population concentration, should be identified, and more ecological space should be planned and rationally constructed in these areas to match the needs of regional social economic development for ecosystem services and ecological products
Threshold theory	Threshold vaule refers to the existence of one or more interactive negative feedback mechanisms existing among the various components of the system, which will inhibit its continuous development when the system deviates to a certain extent, and make reverse adjustment to make the system return to the equilibrium state. The negative feedback mechanism will lose its function when the deviation range of the system is too large, and finally the system will be move towards imbalance and collapse. The amplitude that causes the negative feedback mechanism to lose its function is the threshold of the system	(1) The construction of regional ecological conservation spatial pattern should mainly consider the minimum protection area, the minimum ecological security distance, the radiation effect of ecosystem services, and the potential degradation possibility, etc. (2) It is necessary to establish the technical method of ecological threshold judgment and identification from three important links: pattern-process-degradation; (3) The threshold value of natural ecological space area should be determined by the matching degree between the ability of ecological donors to provide ecosystem services and the carrying capacity of receptors

(continued)

Table 10 (continued)

Names of theory	Explanation of related concepts	The content of theory
Trade-off theory of ecosystem services	Ecosystem services refers to all benefits that humans obtain from ecosystems, including provisioning services (e.g. provision of food and water), regulating services (e.g. control of floods and diseases), cultural services (e.g. spiritual, recreational and cultural benefits) and supporting services (e.g. nutrient cycling that sustains the environment in which life lives on Earth)	(1) Because of the diversity of ecosystem services, the heterogeneity of spatial distribution and the preference of human selection, ecosystem services often have a complex nonlinear relationship, including trade-off (negative relationship) and synergy (positive relationship) and compatibility (no significant relationship), etc.; (2) According to the scale of analysis and whether it is reversible or not, the trade-offs of ecosystem services include spatial trade-offs, temporal trade-offs and reversible trade-offs; (3) The trade-off of ecosystem services exists both within and between regions; It covers not only the trade-off between current ecosystem services, but also the conflict between current and future ecosystem services; (4) According to the trade-off theory of ecosystem services, we can effectively identify the natural ecological space which is very important to maintain regional ecosystem services and need to focus on protection, and build a natural ecological spatial pattern which can avoid ecological risks to the maximum extent and ensure regional ecological security

Source The author collates from Gao Jixi, Xu Delin, & Qiao Qing, et al., (2020). *Pattern construction of natural ecological space and planning theory exploration.* Acta Ecologica Sinica, 40 (03): 749–755

5.2 Co-Integration: Multi-type and Multi-level Ecological Spatial Consistency Theory

Different spatial classifications and hierarchies of human settlements are related to cognitive objects and specific objectives of planning, construction and management. Each spatial type has hierarchy. Taking ecological space as an example, it can be divided according to habitat type, landscape structure, block (subdivision) composition and hierarchy composition, etc. In the *National main functional area planning*, the ecological space is divided into natural grassland, forest land, wetland, reservoir water surface, river water surface, lake water surface, barren land, sandy land, saline-alkali land, plateau desert, etc. In *Shanghai ecological space connotation* the definition and connotations of ecological space is that the ecological space, agricultural space and green space in urban areas defined at the national level allbelong to the ecological space, including various ecological spatial elements such as offshore waters, mountains, rivers and lakes, forest land, green space, wetlands and farm land.[21] Although no matter which division method, all kinds and levels of ecological space elements show clear differences in quality function and other aspects due to their own characteristics, the consistency of classification and levels of each element is also obvious.

Consistency refers to the high degree of identity (synergy) in content, process, characteristics, connotation, essence, standards, etc., among multiple elements, objects or various parts of the system, as well as with external systems. Consistency is also one of the characteristics of validity, accuracy, logic and scientificity. Consistency can be divided into internal and external. The former means that there is no conflict or contradiction between objects, and the state and relationship are self-consistent. A complex system composed of multiple elements and multiple levels can generate synergy and innovation to effectively respond to competition. The latter refers to the synergy of the relationship between a certain element or system and external elements/external systems.

The principle of resilience consistency of multi-type and multi-level ecological space reflects people's hope that all levels of ecological space and various types of functional states and value levels have identity and similarity, so that it is easier to obtain efficient improvement and optimization. The specific contents includes the following three aspects:

First, the definition and connotation of different types and levels of ecological space should be consistent. This can make all kinds of ecological space types have a unified theoretical concept category, a unified evaluation system and technical standards, which has positive significance for ecological space research, planning, construction and management.

Second, The planning standards and construction policies of different types and levels of ecological space should be considered uniformly. The basic principles is that the micro has characteristics, harmony as a whole, and consistent logical relationship.

---

[21] Guo Chunbin (2018).

Third, The planning and construction of different types and levels of ecological space should be guided by a unified idea, and a complementary symbiotic sharing mechanism should be established, so as to optimize the health level of the overall ecological space.

5.3 Toughness: Robustness-Ductile-Elastic Coupling Theory of Ecological Spatial Structure

There are multiple perspectives in defining spatial structure, including systemic, economic, functional and integrated, which embodies the richness of spatial structure as an important concept of human settlement discipline and practice. Born described the concept of urban spatial structure from the perspective of system theory, pointing out that urban spatial structure is a set of organizational laws that connect the interaction between urban form and urban elements, and integrate them into an urban system. This definition not only points out the constituent elements of the urban spatial structure, but also emphasize the interaction among the elements.[22] For ecological spatial structure, robustness ductileness and elasticity (ductileness) are important contents to improve the health level, vitality and vitality of ecological space.

Robustness refers to the "fault tolerance" of some nodes in the system when random faults occur, and the "resistance" that can continue to maintain their functions when they are deliberately destroyed. It also refers to the ability of the system to maintain its functions when facing changes in internal structure or external environment.[23] Robustness is also called "tenacity". The robustness of ecological spatial structure refers to the "firmness" of ecological spatial structure, which can resist the negative interference of external environment; The second is that the ecological spatial structure has the ability to resist unreasonable external interference for a long time, and maintains the original scientific, reasonable and adaptable for a long time Spatial structure mode state of sex.

Resilience of ecological spatial structure, taking cities as an example, emphasizes an ideal state and ability that people hope for in urban ecological spatial structure. It means that the ecological spatial structure can not only continue the reasonable good pattern relationship between the city and the natural geographical environment endowed by the historical ecological environment, but also meet the needs of various construction and development of the city today, and continuously provide various spatial supports for the optimization of the quality of human settlements and the improvement of people's living environment. However, more importantly, the resilience of ecological spatial structure means that it can meet all kinds of extremely complex uncertainties and randomness of urban future development The disturbance and influence of events on cities.

---

[22] Zhou Chunshan (2007).

[23] Xiong Wenhai (2009).

Under certain circumstances, there is a certain contradiction between robustness and resilience of ecological spatial structure. The former expects that the ecological spatial structure will be stable once and for all, and does not want to change. The latter emphasizes the process of adapting to change and seeking stability. Obviously, the latter is more in line with the reality of the development of human settlement space system, but it will bring greater change costs.

One of the purposes of the robustness-ductile-elastic coupling theory of ecological spatial structure is to solve this contradiction, including the following content:

First, we should deeply understand the huge economic, social and economic benefits that can be obtained by the coupling of robustness-toughness and elasticity of ecological spatial structure.

Second, while insisting on the robustness of ecological spatial structure, we should consider the toughness and elasticity of spatial structure; And vice versa. The purpose is to make them reach a relative balance in different states.

Third, adopt various ways, such as planning, policy, financial policy, etc., design the flexible mechanism of ecological spatial structure, reserve flexible land, improve the inclusiveness and flexibility of ecological spatial structure, and make it adapt to the high uncertainty in the future. For example, the "*Shanghai Urban Master Plan (2017–2035)*" puts forward "flexible adaptation" and "establishment of space blank mechanism" in the "transformation of development model" (planning a strategic reserved area of 200 km^2). Including: Combine far and near, leave enough scarce resources and strategic space for future development, prepare for unexpected major events and major projects, and improve the inclusiveness of space. Ensure the implementation of important regional passages and major infrastructure by reserving mobile indicators. At the same time, the plan also proposes to innovate the flexible mode of function layout, mainly including: Strengthening the spatial development pattern of multi-center networking and adopting distributed and unified spatial layout to diversify urban functions, and at the same time, it deals with the impact of major technological changes on urban spatial structure and utilisation. Form elastic and friendly ecological separation (ecological space) in production and living space. Shanghai overall planning "elastic adaptation" and "space blank" can be regarded as one of the cases of robustness-ductileness-elasticity coupling of ecological spatial structure.

5.4 Sustainability: Sustainability Theory of Ecological Space System

In 1987, the World Commission on Environment and Development (WCED) published the report "*Our Common Future*", which defined sustainable development as: "development that meets the needs of the present without compromising the ability of future generations to meet their own needs" This definition succinctly captures the idea of sustainable development. However, in many cases, it is necessary to express the degree of sustainable development clearly and quantitatively.

At present, "sustainable power" can do this job well. Sustainability is a theoretical judgment and quantitative expression that clearly expresses the degree and level of sustainable development of system, field, region and object.

Foreign literatures of Sustainability pay more attention to sustainable resources,[24] sustainable supply,[25] sustainable design,[26] and sustainable production[27] and so on; Domestic literature mainly focuses on sustainability evaluation.[28]

Based on the discussion of sustainability theory in academic circles,[29] if ecological space is considered as an ecosystem, the sustainability of ecological space system can be defined as: Ecological space has the ability to maintain the stability of its structure, function and quality for a long time. Including the following aspects: availability, accessibility, affordability, acceptability and continuity.

Availability principle. To realize the synergy between resilience and sustainability, we should ensure the availability of ecosystem elements and their ecological services, including that ecological resources should be safe (such as rivers and soils should be unpolluted), ecosystem structure should be strong (robustness), ecosystem background resources or space should be abundant (redundant), and ecosystem genes, species and habitat types should be diverse (replaceable). In addition, the availability of the ecosystem can be monitored and improved to manage the condition of the ecosystem and ensure the provision of stable and sustainable ecological services.

Accessibility principle. It is mainly reflected in the spatial distribution of ecological resources and services. Although it may not be possible to guarantee that each community or office can have sufficient ecological resources (redundancy), we should try our best to ensure that urban residents can have relative social equity (diversity) in contacting or reaching ecological space to enjoy the welfare of ecological services. Specific evaluation indicators include the coverage rate of green space within 500 m, per capita green space per unit area, etc. Building accessibility to urban ecosystems contributes to social resilience and integrated resilience in cities.

Affordability principle. It is mainly reflected in the affordability of residents in enjoying ecological resources or services, ensuring the affordability of low-income groups or people in high-density areas in terms of economic costs or expenses (such as transportation costs) in enjoying high-quality ecological resources or services, or investing in vertical greening and other forms of ecological facilities construction in their communities, and so on. On the one hand, the affordability of ecological space system services provides an economic guarantee for the funds needed to protect the ecological system and enhances the inherent resilience of the ecological space system. On the other hand, it helps to alleviate and adjust the inequality in the use of ecological resources and services in human society which is helpful to enhance social harmony.

---

[24] Indira et al. (2019).

[25] Goel and Sharma (2019).

[26] Panyam et al. (2019).

[27] Rivera-Gonzalez et al. (2019).

[28] Xue Liming et al. (2018).

[29] Sharifi and Yamagata (2016).

Fig. 8 Sign of isolating visitors and biological district areas on Lake Zurich, Switzerland. *Source* Author's self-photo

Acceptability principle. It is mainly reflected in the social ethics level to realize the right of human beings to enjoy ecological space resources and the synergy of self-healthy development of the ecosystem. It not only guarantees the development of human beings, but also does not cause irreparable damage to the ecosystem on which other creatures depend, and realizes the social environmental ethics of harmonious coexistence between man and nature. Specifically includes: Manage ecological space resources according to the degree of being affected by human activities, and set aside certain areas to prohibit/restrict human access or entry (Fig. 8); Create ecological corridor and guarantee the area needed for ecological landscape patch function, and increase the connectivity and integrity of ecological space system Encourage social activities to participate in ecological conservation, and form a social consensus of harmonious coexistence between man and nature, and so on.

Transitivity principle mainly reflected in the continuity of resource characteristics and ecological functions in ecological space on the time scale. First of all, it is necessary to emphasize that when people make decisions on the planning and construction of human settlements, they should recognize the ecological relationship, the trend of ecological chain and the law of interaction among various elements. Avoid some measures that mean to improve the living environment, but actually it destroys the internal mechanism of ecological operation among various elements. For example, the unique aquatic communities and markets that coexist with the water on the canal in Bangkok, Thailand were originally very prosperous and were also the most popular tourist attractions in the local. However, in order to protect residents from flooding, the government built many gates at the mouth of the canal, which stopped the flooding and also sacrificed the convenience of sailing, so tourists stopped patronizing and residents relocated. In addition, the development of land transportation and automobiles has replaced water transportation in the city. The water communities and markets that were unique to Bangkok have been abandoned, and the city's traffic congestion and air pollution are increasing day by day.[30] Second, the development and construction of human settlements should leave a blank area or buffer zone (predictability, redundancy) of natural ecological functions, to ensure that the ecosystem can maintain the space for self-continuation and restoration (self-organization, adaptability, etc.) in the event of disasters or other disturbances, and ensure the regeneration of ecological resources without affecting people's rights to ecological resources in the future period.

5.5 *Symbiosis: Theory of Harmonious Coexistence Between Humans and Organisms*

From the human perspective, it is easy to regard ecological space as the sphere of human influence and the exclusive territory for human beings to enjoy ecological services, so as to place human beings above ecological space. In fact, this concept reflects the "arrogance" of human beings to a certain extent, which is bound to have a negative impact on ecological spatial planning and construction, and is also one of the manifestations and reasons for the low level of ecological spatial planning and construction. It should be recognized that taking the coordinated symbiosis between human beings and ecological space (especially natural ecological space) as the guiding theory of ecological spatial planning is the core representation of high-level and high-level ecological spatial planning and construction. For example, the early agricultural development in the Netherlands focused only on production, but with the export of Dutch agricultural products became the second exporter in the world after the United States, the biodiversity of the Netherlands also decreased to one of the low-level countries in Europe. In this case, the Netherlands has changed its agricultural development purpose and reconsidered and attached importance to the positive impact and role of agriculture in improving biodiversity. In the process of planning and construction of farmland and agricultural system, various measures are taken to improve the biodiversity level of birds, animals and insects in the Netherlands. They found that the proportion of natural landscape elements added to farmland (areas for biological foraging and habitats outside farmland) increased by 7%, and the number of butterfly and bird species increased by 64% and 74% respectively.[31] For another example, Suzhou Industrial Park has planned a coherent biological channel within the park, and designed the biological migration paths in the north and west directions within the industrial park, so as to effectively improve the biodiversity level of the park. In addition, Chongming Island in Shanghai specially identifies the living and habitat migration channels of hedgehogs on the island, and satisfies and optimizes them to the maximum extent in the planning. The forest land planning of Molonglo Valley in Australia incorporates bird protection into urban design (Table 11). An isolated area on the shore of Lake Zurich in Switzerland is specially

[30] Yu Kongjian (2020).

[31] Data source: https://www.sohu.com/a/282261019_657688.

Table 11 Recommendations for incorporating forest land birds into urban design in Molonlo valley, Australia

Recommendations	Measures		Enumeration of beneficial species
	Urban areas	Semi-urban area	
Retain eucalyptus forest land Retain high-quality riparian areas	Design and develop around the existing large-scale eucalyptus forest land; Incorporate small forest land into urban open space Plan sufficient buffer zones between rivers and urban areas; Restrict recreational activities in riverside areas	Actively manage forest land to improve its health level and function Restore the original woody tree and shrub cover; Replace weeds with indigenous plants	Brown Thornbill, Crimson Rosella Grey fantail

opened for the protection of living things; The "forbidden zone" and "access zone" designed in the green spatial planning of Vancouver, Canada. The above practice of taking the symbiotic development of human beings and organisms as planning considerations reflects the value of pursuing harmony between humans and nature, which is far higher than simply taking the satisfaction of human needs as the sole and ultimate goal of ecological spatial planning in terms of thoughts and values to the level of planning. From the perspective of interests, to some extent ecological spatial planning and construction demand to balance the relationship between man and nature; Under specific circumstances, one of the manifestations of scientific ecological spatial planning and construction is the active concession of human beings to other biological interests, which embodies the level of human thinking and values, and finally determines the quality of ecological spatial planning and construction.

6 Some Important Practical Issues of Ecological Spatial Planning

6.1 Identification of Key Ecological Space

Critical ecological space is the basic land space for guaranteeing the safety of regional water resources, biodiversity conservation, geological disaster protection, soil and water conservation, and maintaining the integrity and continuity of regional landscape patterns. In addition, critical ecological space has the important mission of maintaining the health of regional life and land. Any unit of land may undertake various ecological processes. By using the identification indicators of important ecological service functions and major ecological crises, critical ecological space

Table 12 Identification results of key ecological spaces in Poyang Lake ecological economiczone

Key types of ecological spaces	Area (km^2)	Percentage	Cumulative Percentage
Bottom line ecological space	17,125.31	32.43%	32.43%
Crisis ecological space	4431.19	8.39%	40.82%
Buffer ecological space	6194.75	11.73%	52.55%
Non-critical ecological space	25,051.63	47.45%	100.00%

Source Xie H L, Yao G, He Y F, Zhang D B. (2018). Study on spatial identification of critical ecological space based on GIS: a case study of Poyang Lake Ecological Acta Ecologica Sinica, 38 (16): 5926–5937

of the region can be comprehensively reflected. The identification of key ecological space has great reference significance for regional land use planning, delineation of permanent prime farmland and establishment of ecologically secure land use pattern. In foreign countries, the identification of regional key ecological space mainly focuses on the identification of biodiversity protection space, water security protection space, soil erosion protection space and green space. The research on the identification of regional key ecological space in China mainly focuses on landscape pattern development, index system evaluation, planning regulation and ecological security pattern.[32]

Taking the Poyang Lake eco-economic zone with fragile natural ecological environment as the study area, Xie Hualin and others combined with the relevant practical experience and research results of scholars at home and abroad, based on RS and GIS and other related spatial information technologies, selected six single-factor ecological processes, such as water conservation, soil conservation, flood regulation and storage, biodiversity protection, soil and water loss and geological disasters, from the two aspects of ecosystem service function importance and ecosystem sensitivity, respectively, and constructed a regional key ecological space identification method. Based on RS and GIS and other related spatial information technologies, the research area was carried out with grid (250 m × 250 m) the unit to identify key ecological space, the results are shown in Tables 12 and 13.[33]

It can be seen from Table 13 that in 2010, more than 90% of the construction land in the two types of land use with intensive human activities in the Poyang Lake Ecological Economic Zone was in an ecologically safe or ecologically safe space. About 6% of the development land is in a fragile or even extremely fragile ecological environment. On the one hand, the construction land in these areas poses great risks to life and property, and on the other hand, their existence poses a great threat to the regional natural environment and ecological security. In the farmland use, nearly 90% of the farmland is in the safe or relatively safe area, and about 10% of the farmland is in the fragile area. Even in the extremely fragile ecological environment, the existence of this part of the farmland poses a great threat to the regional ecological environment.

[32] Xie et al. (2018).

[33] Xie et al. (2018).

Table 13 Conflict zone of ecological security of current land use in the ecological and economic zone of Poyang Lake

Key types of ecological spaces	Development land (hm ®)	Percentage	Arable area (hm ®)	Percentage
Non-critical ecological zone	145,947.49	90.69%	1,776,972.53	85.71%
Buffer ecological zone	5465.78	3.39%	103,083.95	4.97%
Crisis ecological zone	2427.03	1.51%	42,377.08	2.04%
Bottom line ecological zone	7082.38	4.40%	150,698.68	7.27%
Total	160,922.68	100.00%	2,073,132.24	100.00%

Source Xie H L, Yao G, He Y F, Zhang D B. (2018). Study on spatial identification of critical ecological space based on GIS: a case study of Poyang Lake Ecological Acta Ecologica Sinica, 38 (16): 5926–5937

Taking Hangzhou as the study area, Zhang Liang and Yue Wenze selected 12 evaluation factors from three aspects of natural, artificial and natural-artificial interaction characteristics of urban ecological landscape patchs to establish an identification index system (Fig. 9 and Table 14), evaluated the intensity of ecological space by spatial weighted overlay method, and identified four types of ecological space: core type, auxiliary type, bottom line type and non-ecological space (Table 15).[34] The research shows that: ① The established identification system well reflects the urban ecological spatial distribution, which is suitable for urban ecological space identification. ② The core ecological space, which takes forest land as the main land use type is concentrated in the west wing of the city, and the auxiliary ecological space surrounds the core ecological space. The two types of ecological space account for 21.28% and 20.33% of the total area, respectively. ③ The distribution of ecological space in Hangzhou is characterized by local aggregation. Yuhang District in the north is the most concentrated, and the landscape patch connectivity is high. The ecological spatial distribution characteristics of different urban areas are quite different. According to their spatial distribution and ecological functions, different management and construction measures are proposed, which can provide scientific basis for the future construction and maintenance of ecological infrastructure in the study area.

[34] Zhang Liang and Yue Wenze (2019).

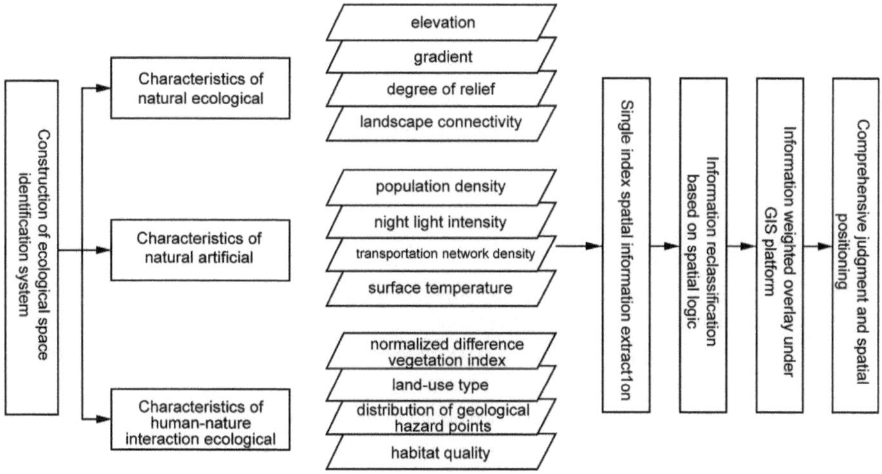

Fig. 9 Technical route of ecological space identification in Hangzhou city. *Source* Zhang Liang, Yue Wenze. (2019). *Integrated recognition of urban ecological space: a case study on Hangzhou. City*. Acta Ecologica Sinica, 39 (17): 6460–6468

6.2 *Optimal Construction of Natural Ecological Spatial Pattern and Key Points of Planning Technology*

The ultimate goal of the optimization and construction of the natural ecological spatial pattern is to scientifically control the development of national land space, rationalize the relationship between protection and development, and ultimately achieve the goal of protecting important ecological space and promoting sustainable development in the region, with the following key points for the optimization and planning techniques.[35]

(1) Construct and plan the natural ecological spatial pattern according to the needs of ecological conservation and the suitability of land spatial development, rather than the present situation

China's land space development mode has changed from focusing on social and economic development to the coordinated symbiotic development of urban space, agricultural space and ecological space. The current situation of the quantity, scale and spatial distribution of ecological land and nature reserves in China may not meet the requirements of future-oriented land spatial structure optimization and spatial planning system construction under different regional and socio-economic development conditions. Ecological spatial planning is an important opportunity to change this situation. As a forward-looking planning, ecological spatial planning is not based on the current situation, but needs to vertically optimize the current situation and the future natural ecological spatial allocation based on the ecological conservation needs

[35] Gao Jixi et al. (2020).

Table 14 Ecological spatial identification index system and weight in Hangzhou

Criteria	Indicators	Index extraction or calculation formula	Weight
Influence of natural factors	Elevation	Digital elevation model DEM	0.01
	Slope	Obtaining slope information according to DEM	0.03
	Topographic relief	Obtaining terrain relief information according to DEM	0.01
	Landscape connectivity		0.11
Human Influence	Population density	Calculate population density by street population and street area	0.03
	Night light intensity	Cutting DMSP/OIS night light data at the boundary of the study area	0.05
	Traffic network density	Linear density analysis based on traffic lines	0.03
	Surface temperature	Lands at 8 remote sensing image is calculated by radiative transfer equation method	0.03
Nature and human beings influence together	Normalized vegetation index	Calculation of landsat8 Multivariate Spectrum Remote Sensing Image Data	0.05
	Type of land utilization	It is divided into seven types: farmland, garden land, forest land, development land, water area, wetland and unused land	0.60
	Distribution of geological disaster points	According to the geological disaster survey data, the geological disaster points are mapped and then the point density is analyzed	0.03
	Habitat quality	Calculation of Habit Quality Module in INVEST Model	0.02

Note Slightly simplified
Source Zhang Liang, & Yue Wenz. (2019). Integrated recognition of urban ecological space: a case study on Hangzhou. Acta Ecologica Sinica, 39 (17): 6460–6468

and the suitability of land space development, and finally realize the unity of social, economic and ecological benefits.

(2) Planning the natural ecological spatial pattern with ecological function rather than ecological land type as the guide

The core characteristic of natural ecological space is ecological function, and the establishment of "picturesque scenery" natural ecological space is essentially in response to people's demand for ecological function. Since ecological land is relatively easy to measure in terms of quantity and space, the current natural ecological spatial planning is often guided by ecological land types, and ecological land such as forest land, grassland and wetland is classified as natural ecological space. However, planning natural ecological space from the perspective of ecological land use will leave out some potential areas that can play ecological functions and provide ecosystem services. Therefore, the natural ecological spatial planning should not

Table 15 Types and composition of urban ecological space in Hangzhou

Ecological space type	Types, states, properties and functions of various ecological space land use				
	Main land use types	State description	Nature generalization	Main functions	Proportion of total area
Core ecological space	Forest land (ecological conservation area)	The distribution of ecological landscape patch is concentrated, and the natural appearance is well protected	Key areas of ecological conservation	Water conservation, forest ecological conservation, air conditioning, water supply, biodiversity conservation	21.28%
Auxiliary ecological space	Farmland, waters, wetlands, forest lands, etc.	Mainly urban wetland conservation areas and high-quality farmland areas	Urban ecological control and coordination and an important area for developing ecological agriculture	Provide protection barrier and buffer zone for the core ecological space	20.33%
Bottom line ecological space	Mainly farmland, garden land, etc.	Affected by human activities, the ecological function is weak	The main interaction areas between ecological factors and human activities	Provide ecological agricultural products production, sightseeing, leisure and entertainment functions	23.39%
Non-ecological space	Mainly for development land	Affected by human activities influence	Parks, green spaces, green belts	Serve the construction	35.00%

Source Organized from Zhang Liang & Yue Wenze. (2019). *Integrated recognition of urban ecological space: a case study on Hangzhou.* Acta Ecologica Sinica, 39 (17): 6460–6468

stick to the current situation of land use types, but should include all areas that can play important ecological functions and provide ecosystem services, and gradually withdraw industries that do not conform to ecological functions after the conditions are allowed.

(3) To maintain the relative integrity of ecology, rather than pursuing pure ecological land to delimit the natural ecological spatial pattern

Due to the lack of proper planning of natural ecological space in China, there are non-ecological land such as urban development land and agricultural land of different scales in natural ecological space. It is difficult to maintain pure natural ecological space for areas that have been developed by human beings for a long time. However, in order to maintain its continuity and integrity, it is inevitable for some scattered non-ecological land to coexist in the natural ecological space, and there is no need to forcibly demarcate these non-ecological land in the process of demarcating the natural ecological space. At the same time, taking into consideration the coordination between ecological conservation and social and economic development, non-ecological land such as permanent prime farmland, which is easy to cause social contradictions, can also be temporarily excluded from the natural ecological space, and can be classified into the ecological space after the withdraw of industry and population when the conditions are ripe.

(4) Layout the natural ecological spatial pattern from the donor-recipient relationship

The natural ecological spatial pattern with complete structure and optimized function plays an important role in maintaining regional ecological security and establishing a reasonable and optimized land spatial development pattern. In the land spatial planning, the quantity ratio and spatial allocation of natural ecological space, urban space and agricultural space are of special importance. Guided by the donor-receptor theory of regional ecology, the coordination and suitability of natural ecological space, urban space and agricultural space can be determined through the comprehensive evaluation of ecological function donors (natural ecological space) and ecological function receptors (urban space and agricultural space) in terms of structural integrity, functional matching and process continuity, which can make the layout and scale of natural ecological space meet the needs of social and economic development and provide scientific basis for the rational construction of natural ecological spatial pattern.

(5) Comprehensive construction of natural ecological spatial pattern from the perspectives of important ecological function maintenance, human settlement barrier and biodiversity maintenance

From the perspective of maintaining important ecosystem services, mitigating natural disasters and protecting biological resources, the natural ecological space is divided into important ecological function areas, ecologically fragile zones, key species habitats and other types. The evaluation, analysis and simulation of maintaining important ecological functions, human settlement barrier and biodiversity are carried out around different natural ecological space types. Based on the ecological pattern-process-function relationship, various methods such as ecological corridor analysis and protection gap identification are integrated, and various patterns such as important ecological function maintenance pattern, human settlement ecological barrier pattern and biodiversity maintenance security pattern of natural ecological space are constructed with the help of GIS spatial analysis technology. On this basis, the natural ecological spatial pattern is formed by comprehensive overlapping.

(6) Constructing natural ecological spatial pattern based on ecological conservation redline

The ecological conservation redline is a demarcated area in the natural ecological space that has special important ecological functions and must be strictly protected. It is the most important and core part of the natural ecological space. *Several Opinions on delineating and strictly observing the ecological conservation redline* clearly requires that before the end of 2020, the delineation and demarcation of the national ecological conservation red line will be fully completed, establish the basic ecological conservation red line system, effectively protect the land ecological space, keep the ecological function stable, and perfect the national ecological security pattern.

After the ecological conservation redline is demarcated, the relevant planning should meet the spatial control requirements of the ecological conservation redline, and those that do not meet should be adjusted in time. The preparation of land spatial planning and the planning of natural ecological spatial pattern should take the ecological conservation redline as an important foundation, and give full play to the bottom line role of ecological conservation redline for land spatial development.

6.3 Ecological Spatial Planning of Coordination, Competition and Cooperation

There is a competitive (conflict) relationship between ecological space and human production and living space, ecological space and urban space, and agricultural space types (Fig. 10). It is one of the important contents of ecological spatial planning under the background of ecological civilization to effectively coordinate the competitive relationship between ecological space and other spaces and make the three ecological spaces have a good cooperative relationship.

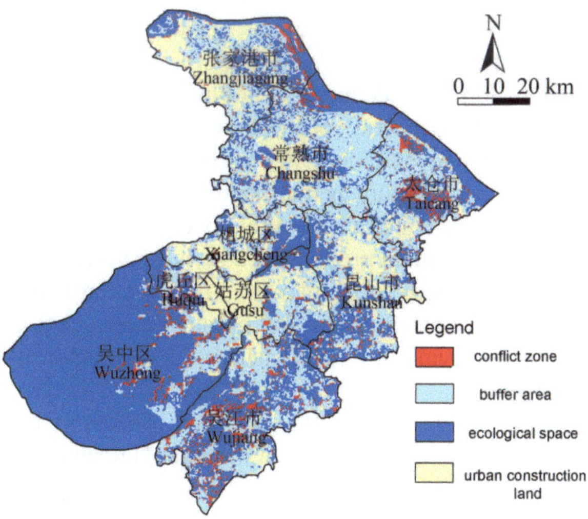

Fig. 10 Conflict between ecological space and urban expansion in Suzhou. *Source* Xu Hualian, Gao Weilong, & Chen Si, et al. (2020). *Identification of urban ecological space in South Jiangsu based on GIS: a case study of Suzhou.* Chinese Journal of Ecology, 39 (02): 614–624

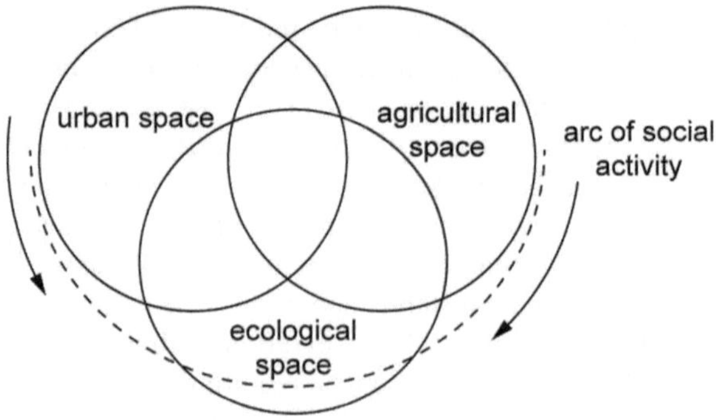

Fig. 11 The transmission mechanism of three kinds of space function. *Source* Shen Yue, Liu Tianke, & Zhou Pu. (2017). *Theoretical analysis and strategies of natural ecological space use control.* China Land Science, 31(12): 17–24

The crowding out of natural ecological space by urban space and agricultural space is an important manifestation of spatial competition. According to the meaning of various kinds of land space and the morphological characteristics of spatial evolution, Shen Yue expressed the land space as three circles with intersecting ranges, and the sum of the areas of the three circles was the total scale of land space (Fig. 11).[36] Human development or expansion will make urban space and agricultural space have an obvious situation of encroaching on natural ecological space; The increase of urban space and agricultural space (occupied side) is bound to be accompanied by the decrease of ecological space (occupied side). The degree of reduction is mainly influenced by topography, landscape type, natural resource endowment and development demand. In general, the smaller the topographic changes, the faster the downward movement speed of social activity arc, The possibility of encroachment of natural landscape space in mountainous area is obviously lower than that in plain area, and compared with grassland and tidal flat, the possibility of encroaching forest landscape is low, the ecological risk of mixed landscape is the highest, the larger the proportion of mixed space in the regional scope, the more obvious the influence of regional rapid urbanization process, and the faster the downward movement speed of social activity arc; The higher the endowment of natural resources, the more obvious the dependence of economic development on resources, and the faster the downward movement speed of social activity arc; The more vigorous the demand for regional development, the faster the downward movement of social activity arc. Without correct human intervention, ecological space will continue to be eroded by urban space and agricultural space until population growth stops or population decays, and human production and living needs weaken.

However, ecological space does not necessarily mean that it is always crowded out. Under the scientific control of human beings, ecological space can also have cooperative and symbiotic relationship with other space types, thus realizing sustainable development.

You Shang analyzed the evolution process of living, production and ecological space of representative shared farms in Hainan Province.[37] It is considered that the

[36] Shen Yue et al. (2017).

[37] You Shang et al. (2019).

living space is the place where residents in homestay settlements meet the basic necessities of life, and it is also the field where foreign tourists live and move. Life can not be separated from the material foundation. In order to keep the consistency between material and life, production space appears, including farming land, tourist accommodation, catering, shopping and entertainment places. Production space is not only a material output place, but also a spiritual output field to meet people's spiritual needs. Ecological space is the prerequisite for life and production; Ecological space defines the scope of activities of residents and tourists, which not only provides habitat for their lives and agricultural land for production, but also forms their unique experience space due to specific topography and local original ecological culture. At the same time, the living, production and ecological space of rural homestay settlements influence and restrict each other, forming an inseparable comprehensive system. The internal system of production-living-ecological space in rural homestay settlements shows the phenomenon of self-derivation, mutual adaptation and coordinated development, which embodies the characteristics of self-organization and has an obvious trend of symbiosis and cooperative development (Fig. 12).

The spatial evolution of production-living-ecological space in rural homestay settlements in Hainan Province has three modes: transition, aggregation and integration (Table 16). Its spatial evolution has external motives such as government policies, tourism market demand and production technology, and internal motives such as natural geographical conditions, traffic factors and local residents' self-awareness. It can be seen from Fig. 13 that the main elements of the optimization structure of the production-living-ecological space co-evolution optimization model of rural residential settlements are "composite space", including ecological-living composite space, production-living composite space and ecological-production composite space. The optimization measures of spatial cooperation among the three students mainly include creating ecological environment and ecological industry, expanding public

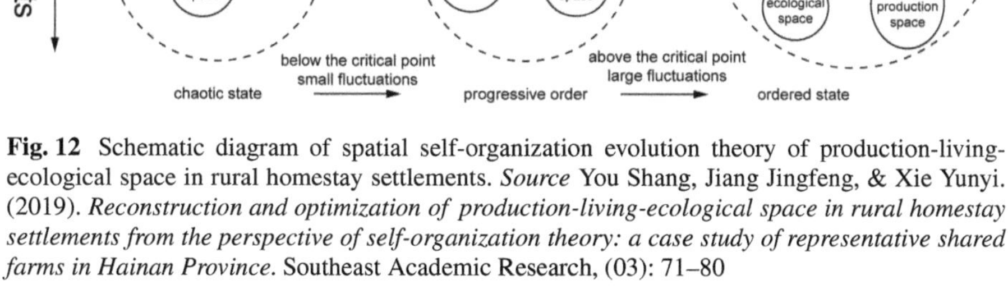

Fig. 12 Schematic diagram of spatial self-organization evolution theory of production-living-ecological space in rural homestay settlements. *Source* You Shang, Jiang Jingfeng, & Xie Yunyi. (2019). *Reconstruction and optimization of production-living-ecological space in rural homestay settlements from the perspective of self-organization theory: a case study of representative shared farms in Hainan Province.* Southeast Academic Research, (03): 71–80

space, industrial transformation, creating ecological culture, establishing ecological warning line and so on.

Ecological spatial planning based on competition-cooperation thinking suggests that we should control the conflict relationship and intensity between ecological space and other spaces, change resistance into thrust, give full play to the positive effect of cooperation and symbiosis between ecological space and other space types, and provide a solid foundation for improving and optimizing ecological space.

6.4 Ecological Spatial Planning Based on Landsenses Ecology

(1) Brief introduction of landsenses ecology

Table 16 Spatial evolution model and process of production-living-ecological space in rural homestay settlements in Hainan province

Evolutionary model	Evolution order parameter	Evolutionary space	Evolutionary process	Evolution result	Typical case
Transition type	Livable orientation	Ecology → Living	Mountain—plain/ lake	Mountain settlements gradually disappeared, Plain settlements are gradually expanding	Fengtang village/ Fengtang green park shared farm
Aggregation formula	Point centrally	Living → production	Natural Village—Central village	Settlements gather from scattered distribution to functional center	Renli village/ Shihu rueal homestay settlement
Fusion type	Function orientation	Production → Ecology	Production function area—Ecological function area	Integration of settlement production space and ecological space	Shiqiao village/ Garden dream shared farm

Source You Shang, Jiang Jingfeng, & Xie Yunyi. (2019). Reconstruction and optimization of production-living-ecological space in rural homestay settlements from the perspective of self-organization theory: a case study of representative shared farms in Hainan Province. Southeast Academic Research, (03): 71–80

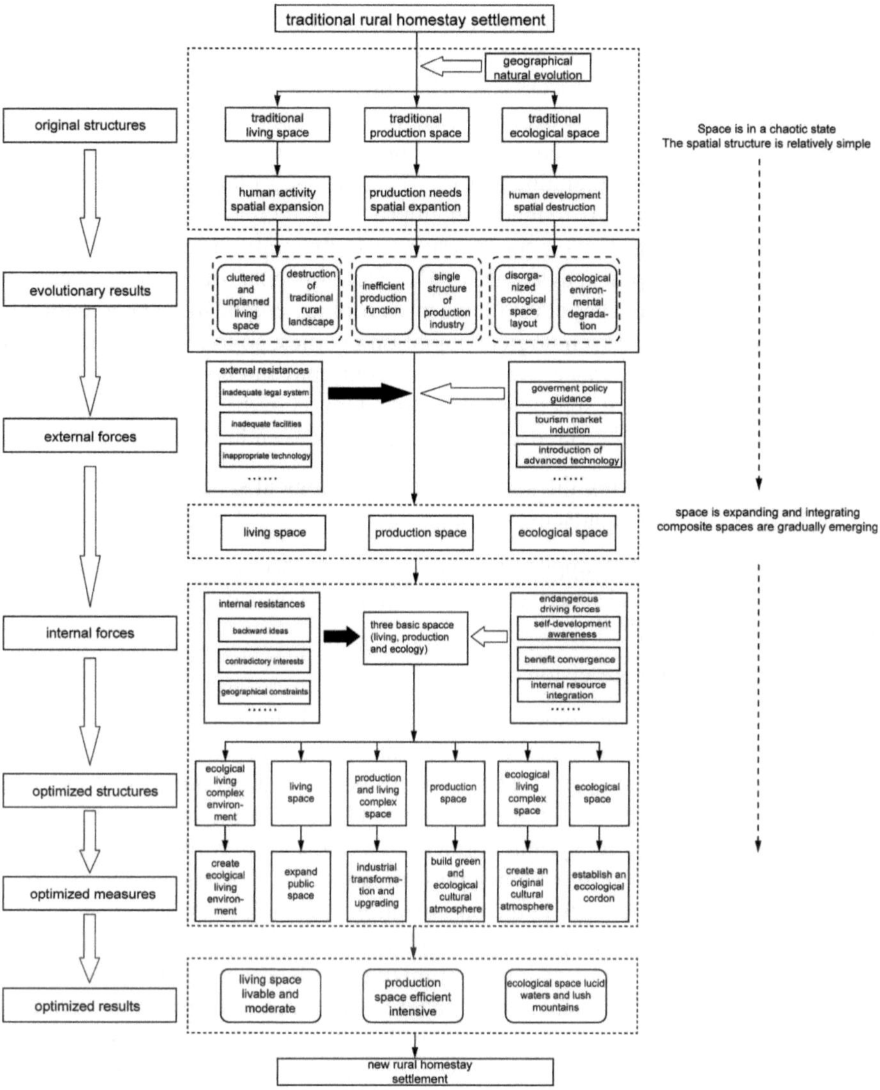

Fig. 13 Spatial co-evolution optimization model of production-living-ecological space in rural homestay settlements. *Source* You Shang, Jiang Jingfeng, & Xie Yunyi. (2019). *Reconstruction and optimization of production-living-ecological space in rural homestay settlements from the perspective of self-organization theory: a case study of representative shared farms in Hainan Province.* Southeast Academic Research, (03): 71–80

Landsenses ecology is a discipline that studies land use planning, construction and management from the aspects of natural elements, physical perception, psychological cognition, social economy, process and risk, which aims at sustainable development and is based on the basic principles of ecology.[38] Landsenses ecology advocates not only paying attention to the spatial pattern, but also paying attention to the feelings of people in the city in the planning process. The purpose of landsenses ecology and landsense creation is not only to maintain, improve and increase ecosystem services, but more importantly, to make people consciously maintain, improve and increase the "sustainable development awareness and related concepts" of ecosystem services while enjoying ecosystem services. Landsenses ecology emphasizes the attention of physical perception and psychological cognition in cities, so to improve the service level of urban ecological space, we should not only pay attention to the landscape pattern of urban ecological space, but also pay attention to the visual spatial pattern of cities.[39] Some contents of landsenses ecology are shown in Table 17.

(2) Planning for improving the service level of ecological space in Shunyi district of Beijing based on landsenses ecology

Zhao Meng et al. (2020) applied the theory of landsenses ecology to the research and planning of ecological space service level in Shunyi district of Beijing. The main contents are as follows.[40]

(1) Determination of core vision

Vision is a goal that can guide or regulate people's behavior. It can be seen from Table 17 that the designer integrates the vision into the landscape carriers by landsense creation, and realizes the sympathetic response between users and landscape, thus guiding people to actively maintain, improve and enhance the level of ecosystem services while enjoying ecosystem services. Two of the three ways of landsense creation are related to vision, which shows the importance of vision. The determination of vision is a basic work in the optimization of ecological space service level. Taking the urban ecological space in Shunyi district as the carrier, Zhao Meng and others put forward four core visions: optimizing the landscape pattern of ecological space, optimizing the visual spatial pattern, scientifically planning and controlling the ecological space, and comprehensively improving the ecosystem services. The two-way feedback relationship between these four core visions and landsense creation methods is generated through existing carriers, transformed carriers and new carriers, which lays a foundation for the application of landsenses ecology to improve the service level of ecological space.

(2) Visual spatial pattern analysis

In this case, the visual spatial pattern is expressed by the green visual ratio (Fig. 14). Green visual ratio refers to the proportion of green vegetation in the normal

[38] Zhao et al. (2016).

[39] Zhao Meng et al. (2020).

[40] Zhao Meng et al. (2020).

Table 17 Some contents of landsenses ecology

Concepts related to theory	Paraphrase
Physical perception and Psychological cognition	In landsenses ecology, physical perception and psychological cognition are interactive, and they influence and promote each other
Vision and carrier	The goal that can guide or regulate people's behavior is called vision, and the elements that can carry one or some visions are called carrier
Bidirectionality of vision	Designers integrate their vision into the landsenses carriers by landsenses ecology, so as to realize the sympathetic response between designers and people, thus guiding people to actively maintain, improve and enhance ecosystem services while enjoying ecosystem services
Landsense creation	The whole process of conceiving and building a sense of scenery is called landsense creation. The purpose of landsense creation. is to maintain, improve and enhance the ecosystem services of "sustainable development awareness and related concepts"
Double pattern	Landsense ecology emphasizes the attention to physical perception and psychological cognition in cities; To improve the service level of ecological space, we should not only pay attention to the landscape pattern of ecological space, but also pay attention to the visual spatial pattern. On the one hand, visual space has the function of representing the appearance and overall style of the city, on the other hand, it also provides an ecological space in which man and nature coexist harmoniously. Landsense pattern has objectivity, while visual spatial pattern has subjectivity
Principles of landsense creation	Bidirectional, homeopathic, combined, physical perception, psychological cognition, cultural and gradual
The way of landsense creation	① With the help of the existing carrier, integrate the vision into it and make them into landsense; ② Transform the existing carrier according to the needs of the vision and integrate the vision into it and make them into landsense; ③ Newly constructed carrier

Source Organized from Zhao Meng, Zhang Xueqi, & Zhang Yonglin, et al. (2020). Study on the enhancement of urban ecological space services based on landsenses ecology: a case study of Shunyi district, Beijing, 40(22): 8075–8084. Acta Ecologica Sinica

visual field of pedestrians, which reflects the perception degree of pedestrians to the surrounding green environment. Green visual ratio plays an important role in enhancing residents' perception of the quality of the ecological environment. Green vision Rate plays an important role in improving residents' feelings of ecological environment quality.

As can be seen from Fig. 14, the green visual ratio of some grades of roads in Shunyi district is relatively low, and there is much room for improvement. Highways and lower-grade roads have higher green viewing rate; The spatial distribution of green visual ratio of county roads is unbalanced. The green visual ratio of roads in the central area of the district is generally low, and most of them are below the middle level (green visual ratio is less than 0.35), while the roads in the eastern and western

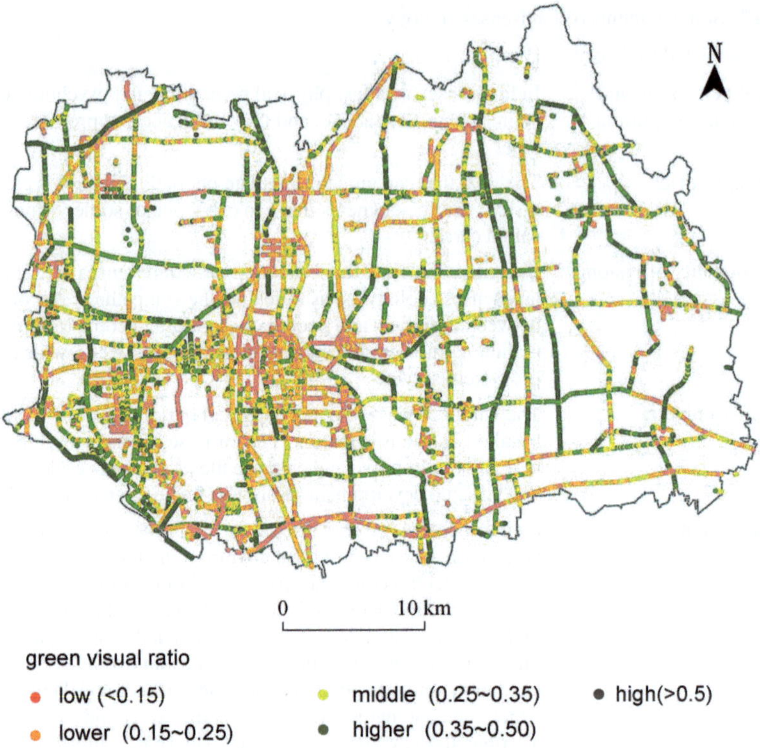

Fig. 14 Calculation results of green visual ratio in Shunyi district, Beijing. *Source* Zhao Meng, Zhang Xueqi, & Zhang Yonglin, et al. (2020). *Research on the improvement of urban ecological space service base on landsenses ecology: a case study in Shunyi District, Beijing.* Acta Ecologica Sinica, 40 (22): 8075–8084

regions have higher green visual ratio, and most of them are above the middle level. The roads with green viewing rates in (0, 0.15), (0.15, 0.25), (0.25, 0.35), (0.35, 0.5) and (0.5, 1.0) are divided into five grades: high, high, medium, low and low, and the corresponding road sections account for 20% of the total road length respectively.

(3) Optimization of visual spatial pattern

As an index to evaluate the visual perception of human settlements, the green visual ratio belongs to the physical perception in landsenses ecology. The improvement of green visual ratio can reduce negative emotions, relieve stress and improve memory, thus affecting people's psychological feelings. The analysis results of the green visual ratio for the main roads in Shunyi district urban area (Fig. 14) show that the central area especially the west of East Sixth Ring Road and Jingshen Road, the east of Youdi Road and Zuo Di Road, the north of Outer Ring Road, and the south of Baima Road and Changjin Road, appear contiguous and low green visual ratio of the road along Chaobai River Belt, which should be paid attention to. It is suggested to transform Oligomerize the Green Visibility Rate of Main Roads in

Shunyi district and improve ecological function. Replant street trees in sections with low green viewing rate to improve the quality of street greening. By employing ground greening methods such as "seizing every opportunity to plant greenery" and wall greening, to create micro green spaces with a sense of scenery, as well as encouraging the construction of tree-lined avenues, the layout of green corridors is improved. At the same time, some visual and cultural landscapes are created, such as designing Shunyi symbols, landscape sculptures and visual axes in combination with local traditional culture, so as to enhance cultural identity and scene awareness and enhance the visual space scenery quality, improve the leisure service level of ecological space in Shunyi district.

(4) Technical route to improve the service level of ecological space

It is an important feature of this case to apply the theory and method of landsenses ecology to improve the service level of ecological space in Shunyi district. The technical route of improving the service level of ecological space in Shunyi district includes four sections. One is vision analysis, the second is the principle of landsense creation, the third is the way of landsense creation, and the fourth is the measures of landsense creation (Fig. 15).

6.5 *Improve the Scientific Nature of Ecological Spatial Planning*

Scientificalness has clarity, self-consistency, mechanism, science and rationality. Scientificalness is actually the crystallization of "natural truth" and "human wisdom".[41] Scientific-oriented ecological spatial planning should closely link ecological connotation with scientific connotation, to make the ecological spatial planning results can not only accurately and completely reflect the ecological laws and principles, but also have the characteristics of high efficiency, intelligence and scientificity.

In February 2014, President Xi Jinping pointed out: "Planning science is the greatest benefit, planning mistakes are the greatest waste, and planning tossing is the biggest taboo."[42] The scientific nature of ecological spatial planning is important step to enhance the value and quality of ecological space, with rich content, is described below.

(1) Carry out ecological spatial planning according to natural geographical conditions

The regional characteristics of ecological space determine that ecological spatial planning must attach great importance to and make full use of local natural geographical conditions for ecological spatial planning, and determine planning principles

[41] Shen Qingji et al. (2016).

[42] Quote from https://www.chinanews.com.cn/gn/2015/10-29/7594594.shtml.

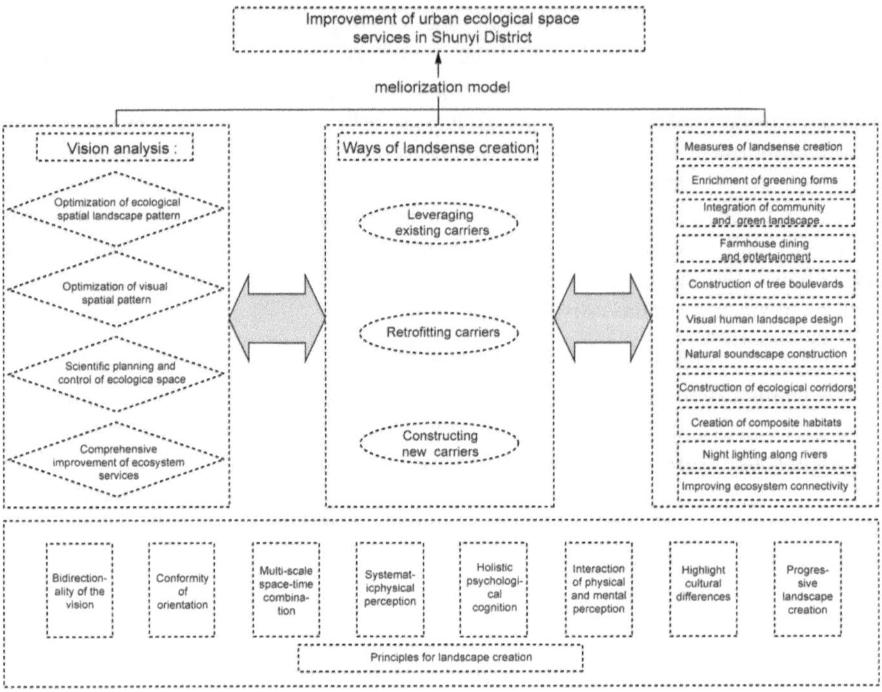

Fig. 15 Concept map of urban ecological space service promotion in Shunyi district based onlandsenses ecology. *Source* Zhao Meng, Zhang Xueqi, & Zhang Yonglin, et al. (2020). *Research on the improvement of urban ecological space service base on landsenses ecology: a case study in Shunyi District, Beijing.* EcologyActa Sinica, 40(22): 8075–8084

according to the characteristics of landscape system. This belongs to the "scientific layout… ecological space" in the "*Several opinions on building a national territorial spatial planning system and supervising the implementation by CPC Central Committee and State Council*" category. The former can be taken as an example to restore the planning and construction of high-quality original landform in Wuhan, while the latter can take the principle of spatial planning in Yichun an example.

High-quality primitive landform are characterized by landscape, regionality, scarcity and unique excellent ecological effects, and are an important part of urban's character charm. In today's highly urbanized world, it is quite difficult to restore the pure and complete urban high-quality original landform. However, there is still a great possibility of restoring the urban high-quality landform pattern. Wuhan city proposes to revive the high-quality original landform of Wuhan on top of the destroyed landform resources, to maintain the landform pattern of "mountains in the city", "lakes in the city" and "rivers in the city", and to utilize the unique natural landform of Wuhan to improve the urban construction quality of Wuhan. Among them, how to introduce water from the Yangtze River into the East Lake, so that the water of the East Lake and the Yangtze River can be connected to form a circulating water system has become an important research topic. In September 2002, the "Major Science and Technology Program for Water Pollution Control and Treatment" was launched, providing technical possibility for the lake in Wuhan to reproduce its clarity. Its main idea is to shift from simple biological pollution control to ecological environment control, the lakes are connected to each other first, and then connected to the Han River and the Yangtze River, so as to achieve the effect of "running water never turns foul". In 2005, the "Six Lakes Connection" project in Hanyang area, with a total investment of 430 million yuan, started, connecting MoShui Lake, Nantaizi Lake, Beitaizi Lake, Longyang Lake, Triangle Lake and Houguan Lake scattered among high-rise buildings in Hanyang area into a whole. It is expected to make use of the geographical gap to let the water of Yangtze River and Hanjiang River enter the lake and realize the wonderful idea of "Row a boat triping all over Hanyang".[43] In addition, Wuhan Snake Mountain is named for its shape like a winding snake, and the Yellow Crane Tower, the symbol of Wuhan, stands on the top of the Wuhan Snake Mountain. However, various constructions and buildings have make Wuhan Snake Mountain "the mountain is there, but it is not visible". To restore the charm of Wuhan Snake Mountain, Wuhan began to implement the project of "revealing the mountain and greening" in 2004. According to the relevant planning, except for a few historical buildings such as the Red Mansion, the Provincial Library and the Longhua Temple which will be retained, the rest of the buildings on the southern slope of Wuhan Snake Mountain will be demolished in phases to reproduce the high-quality natural appearance of Wuhan Snake Mountain.[44]

Yichun, on the other hand, determines relevant spatial planning principles according to the characteristics of its landscape system, including: "mountain protection and water conservation" in land planning; "embracing mountains and connecting water"in road planning; "revealing mountains and exposing water" in architectural layout; "utilizing mountains and utilizing water" in landscape design; "be near the mountains and rivers" in spatial organization, etc.[45]

(2) The analysis of ecological spatial evolution trajectory provides support for the scientific nature of ecological spatial planning

The multi-type evolution of ecological space and its elements is not only the embodiment of natural laws, but also the comprehensive result of the interaction among natural laws, economic laws, social laws and urban and rural development laws. The cognition of this kind of evolution provides a strong support for the scientific nature of ecological spatial planning.

[43] Zhao Jing (2007).

[44] BI Wei et al. (2007).

[45] Nie Huoyun and Zheng Lihua (2008).

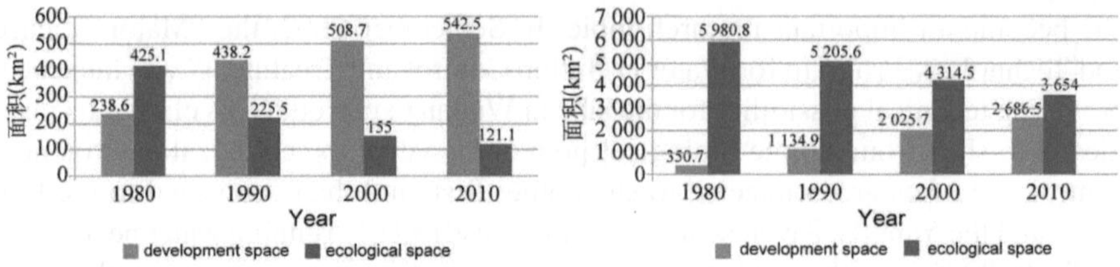

(a) Evolution of the scale of ecological space in central city area

(b) Evolution of the scale of ecological space at the external level

Fig. 16 Evolution of urban ecological space scale in Shanghai. *Source* Xu Yi, & Peng Zhenwei. (2016). *Study on spatial evolution and dynamic mechanism of Shanghai urban ecospace in 1980–2010*. Urban Development Studies, 23 (11): 1–10+59

For example, Xu Yi and Peng[46] Zhenwei analyzed the evolution of Shanghai's urban ecological space scale (Fig. 16) and the quantitative changes of several ecological environment elements (Fig. 17), and summarized the spatial evolution characteristics of Shanghai's ecological space structure system (Fig. 18) and the evolution of the relationship between Shanghai's ecological space supply and demand from 1980 to 2010 (Fig. 19).

(3) Using local ecological resources to carry out efficient construction of ecological space

Zhang Yun and Chen Mei selected 1022 protophyte in Jinan as objects, and screened out 34 protophyte adapted to Jinan ecological environment, which played a positive role in greening Jinan ecological environment efficiently.[47] Through the sulfur dioxide stress test of related tree species, Shandong Forestry Bureau screened out five tree species, such as Fraxinus tomentosa and Ailanthus altissima, as pollution-resistant tree species for urban ecological greening construction in Jinan; Based on the investigation and analysis of the present situation and patterns of greening in Jinan, three patterns of ecological greening were put forward, which improved the comprehensive benefits of urban ecological environment greening construction.[48] Table 18 shows the available pollution-resistant tree species for nitrogen oxides resistance in Zhaoqing urban area, which belongs to the utilization of local ecological resources with two cases in Jinan. By analyzing the botanical and ecological characteristics, economic value and ecological value of fast-growing Sophora japonica in Datong, Shanxi Province, Zhang Wenfang found that fast-growing Sophora japonica, as an excellent legume feed variety introduced from Korea in China, is cold-resistant, drought-resistant, barren-resistant, desertification-resistant, saline-alkali-resistant, and suitable for growing in harsh natural environment.[49] Vigorously popularizing planting can not only develop animal husbandry and increase income, but also windbreak and sand fixing and water conservation. The good ecological and economic

[46] Xu Yi and Peng Zhenwei (2016).

[47] Zhang Yun and Chen Mei (2014).

[48] Shang Jinjie (2008).

[49] Zhang Wenfang (2018).

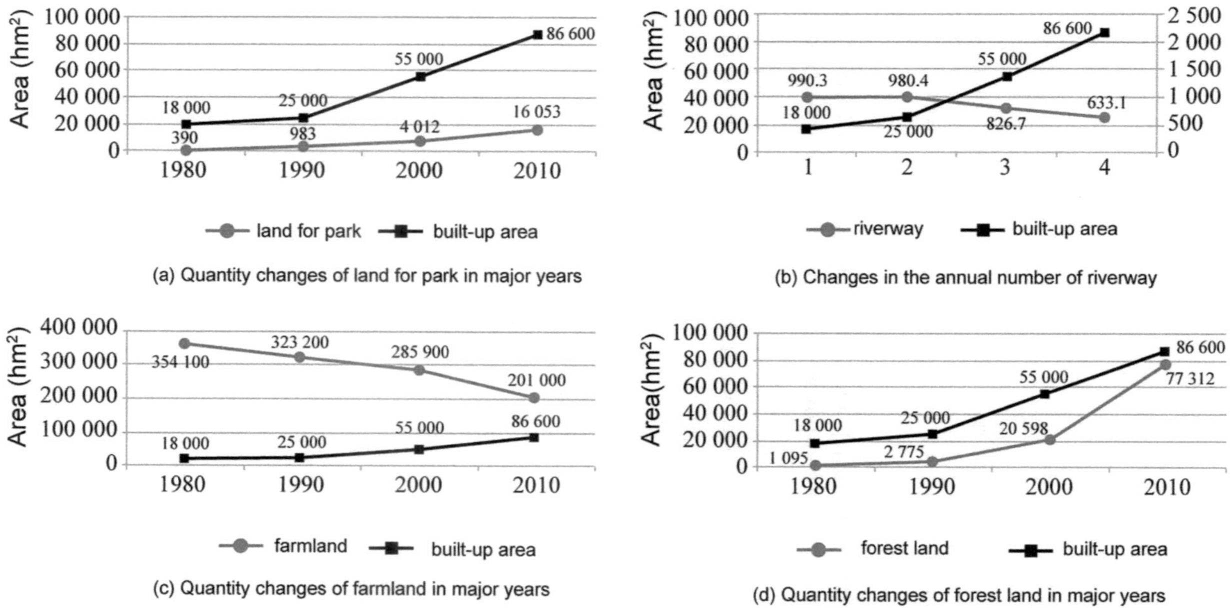

Fig. 17 Quantitative evolution of ecological spatial types in Shanghai. *Source* Xu Yi, & Peng Zhenwei. (2016). *Study on spatial evolution and dynamic mechanism of Shanghai urban ecospace in 1980–2010*. Urban Development Studies, 23 (11): 1–10+59

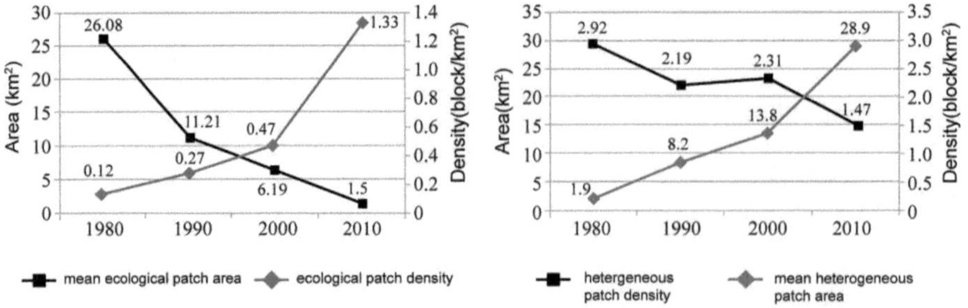

(a) Evolution of the structure characteristics of ecological space in central city area

(b) Evolution of thestructure characteristics ofecological space at the external level

Fig. 18 Evolution of urban ecological spatial structure characteristics in Shanghai. *Source* Xu Yi, & Peng Zhenwei. (2016). *Study on spatial evolution and dynamic mechanism of Shanghai urban ecospace in 1980–2010.* Urban Development Studies, 23 (11): 1–10+59

value of fast-growing Sophora japonica have dual positive effects in promoting the ecological environmental and poverty alleviation in poverty-stricken counties of Datong.

(4) Using ecological laws to improve the performance of ecological spatial planning

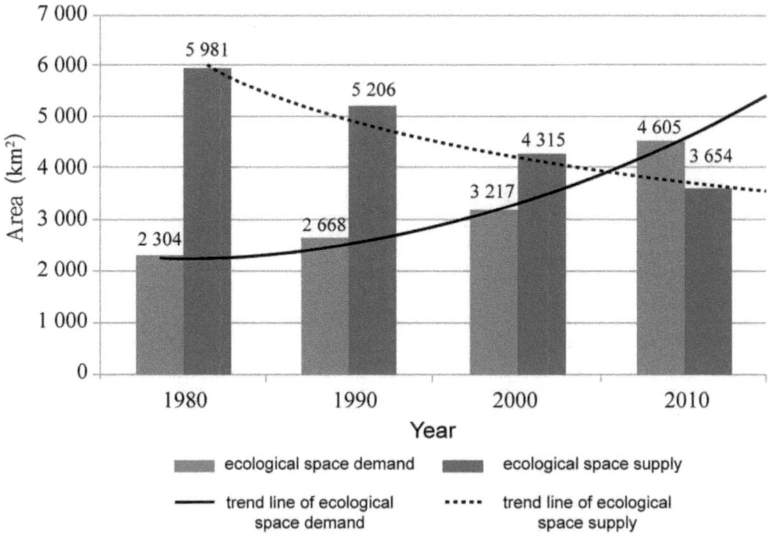

Fig. 19 Evolution of ecological space supply and demand in Shanghai. *Note* The ecological space demand is calculated according to 200 m per capita.[50]. *Source* Xu Yi, & Peng Zhenwei. (2016). *Study on spatial evolution and dynamic mechanism of Shanghai urban ecospace in 1980–2010.* Urban Development Studies, 23 (11): 1–10+59

[50] Wang Shaozeng and Li Min (2001).

Table 18 Anti-pollution tree species for nitrogen oxide resistance in Zhaoqing urban area

Tree species name	Tree species category	Resistance to air pollutants and basis for evaluation					
		SO_2	HF	CL	NO	Acid fog	Acid rain
Ficus alba	evergreen tree	Strong	Strong	Strong	Strong	Strong	Strong
Ficus chinensis	evergreen tree	Strong	Strong	Stronger-strong	Strong	Strong	Strong
Ficus microcarpa	evergreen tree	Strong	Strong	Stronger-strong	Strong	Strong	Strong
Banyan tree	evergreen tree	Strong	Strong	Stronger-strong	Strong	Strong	Strong
ficus religiosa	Tree	Stronger-strong	Stronger	Stronger	Stronger	Stronger	Stronger
Ficus altissima Blume	Tree	Stronger	Stronger	Weak	Stronger	Stronger	Stronger
dalechampia	shrub	Strong	Strong		Strong	Strong	Strong
bischofia javanica	Tree	Stronger			Stronger		
swietenia mahagoni	Tree	Strong			Strong		Strong
mimusops elengi	Tree	Strong	Stronger	Strong	Strong	Strong	Strong
tristania conferta	Tree	Strong			Strong		
syzygium hainanense	Tree	Strong	Strong	Strong	Strong		
lagerstroemia speciosa	Tree	Stronger			Stronger		
pittosporaceae	dungarunga	Strong	Strong		Strong		
bombax malabaricum	Tree	Stronger-strong	Stronger	Strong	Stronger		–
mygdalus communis	Tree	Strong	Strong		Strong		Strong
Bodhi Ficus	Tree	Stronger-strong	Stronger	Stronger	Stronger	Stronger	Stronger

Source Gao Donglian. (2019). Selection of greening tree species for anti-pollution and improving ambient air quality in Zhaoqing urban area. Journal of Green Science and Technology, (10): 135–139

Ecological law has rich in content. Ma Shijun summarized the ecological laws between organisms and the environment, including human beings, as follows: The law of synergy of mutual restriction, the law of material circulation and transformation, the law of balance between input and output, the law of net value of biological productivity and the law of development and succession.[51] Among them, "the relationship between organisms including human beings and the environment" shows that ecological laws are the mechanism reflection of the interaction between organisms (human beings) and the environment. Therefore, if some results produced by the interaction between ecological spatial elements and social economy elements are actively applied in planning (that is, the performance of related systems is adjusted by controlling variables), it will play a positive role in improving the performance of cological spatial planning.

For example, Hu Mingxiu and Zhou Huanzhen's correlation analysis results of water quality and per capita GDP in Wuhan section of Hanjiang River reveal the interaction type between them, which belongs to the category of cognition and utilization of ecological (scientific) laws, and will play an indicative and positive role in improving the quality of water ecological space in this area (Table 19).[52] Another example is Zhang Xuemei's regression analysis of the influencing factors of Shanghai's ecological space area change (Table 20), which provide a perspective to reveal the relationship between Shanghai's ecological space and economic development, and internal causes Shanghai's ecological space evolution and give basic information for future targeted regulation.[53]

For another example, the ecological green functional land area of urban non-buildable land before and after the planning of Guangzhou Haizhu eco-city is respectively 2015 hm^2 and 1499 hm^2, it decreased by 25% compared to before planning;. The urban development land area before and after planning is 2756 hm^2 and 3844 hm^2 respectively. After planning, it has increased by 40% compared to before. Although the ecological green area within the planning scope has reduced and the development land area has increased, through the adjustment of ecological green space structure and the improvement of the overall forest planting rate.The total carbon sink function of Haizhu Eco-city increased from 3834.74 tC/a to 4670.7 tC/a, with an increase of 21.8%. It can be seen that the function of green carbon sink is not only determined by the area of green land, but by the ecological function of green land unit. Increasing the rate of planting forest is the key measure to greatly improve the ecological function of unit green land, which contains the application of ecological laws.[54]

[51] Ma Shijun (1981).

[52] Source: Hu Mingxiu and Zhou Huanzhen (2007).

[53] Source: Zhang Xuemei (2017).

[54] Zhu Junhua et al. (2012).

Table 19 Correlation analysis between per capita GDP and water quality Index of Hanjiang river (Wuhan section)

Water quality index of Hanjiang River (Wuhan section)	pH value	Total hardness/ German degree	Average Water Quality Indices/(mg L-1)						
			Suspended solid	Dissolved oxygen	Permanganate index	Biochemical oxygen demand	Ammonia nitrogen	Nitrite nitrogen	Nitric acid Salt ammonia
Pearson correlation coefficient	−0.367	0.096	0.324	−0.657*	0.920**	0.454	0.682*	0.509	0.316
Two-tailed test value	0.297	0.792	0.361	0.039	0.000	0.187	0.030	0.133	0.374
N	10	10	10	10	10	10	10	10	10

Note ① "Per capita GDP" is the data from 1991 to 2000; ② ** in the table indicates that there is a very significant correlation between water quality index and per capita GDP when the two-tailed test value is 0.01, and * indicates that there is a very significant correlation between water quality index and per capita GDP when the two-tailed test value is 0.05

Source Hu Mingxiu, & Zhou Huanzhen. (2007). Study on the relation between the varlational trend of water qyality ndexes of Hanjiang (the area of Wuhan) and Donghu—based on the theory of environmental Kuznets curve(EKC). Journal of Wuhan Polytechnic University, (01): 68–72

Table 20 Regression analysis of influencing factors of ecological space area change in Shanghai

Influencing factors	R Square	F	Significance (Sig)		Coefficient of regression
			Value	Is it significant	
Gross Domestic Product (GDP)	0.905	28.647	0.013	Significant	−0.081
Primary industry	0.918	33.516	0.010	Significant	−23.437
Secondary industry	0.971	101.458	0.002	Highly significant	−0.267
Tertiary industry	0.862	18.722	0.023	Significant	−0.114
Industry	0.972	105.247	0.002	Xtremely significant	−0.299
Building industry	0.953	60.272	0.004	Highly significant	−2.432
Residential consumption level	0.889	23.916	0.016	Significant	−0.046
Gross capital formation	0.954	62.267	0.004	Highly significant	−0.218

Source Zhang Xuemei. (2017). A study on the ecological spatial evolution in Shanghai. Shanghai: Tongji University

References

An Jiesheng, & Ge Jianxiong. (1997). Hongdong Sophora japonica is the sixth birthplace of immigrants in Chinese history. Root Exploration, (06): 12–14.

Arnberger A, & Eder R. (2012). The influence of green space on community attachment of urban and suburban residents. Urban Forestry & Urban Greening, 11 (01): 41–49.

Bi Wei, Huang Jun, & Shi Jinwei, et al. (2007). The project of revealing the mountain and greening in Wuhan speeds up the Snake Mountain within months to blast another building. Chutian golden news, 03–12.

Chen Xingang. (2005). "The tragedy of the commons" and "The tragedy of the anti-commons" theory: applied research in China. Shandong Social Sciences, (03): 75–78

Conzen MRG. (2004) Thinking about urban form: papers on urban morphology. New York: Peter Lang Publishing Inc, 1932–1998.

Dang Shuangren. (2019). Promoting the construction of beautiful China by the rule of ecological space. West China Development, (12): 78–81.

Gao Jixi, Xu Delin, & Qiao Qing, et al, (2020). *Pattern construction of natural ecological space and planning theory exploration.* Acta Ecologica Sinica, 40 (03): 749–755.

Goel S, Sharma R. (2019). Optimal sizing of a biomass biogas hybrid system for sustainable power supply to a commercial agricultural farm in northern Odisha, India. Environment Development and Sustainability, 21: 2297–2319.

Guo Chunbin. (2018). Exploration of "Shanghai 2035" ecological spatial planning. Shanghai Urban Planning Review, (05): 118–124.

He Mei & Wang Yun. (2009). Construction and protection strategies of urban Eco-spatial system in Wuhan city Planners, 25 (09): 30–34

He Mei. (2010). Study on ecological spatial system planning and control of megacities.Beijing: China Architecture&Building Press.

Holmes Rolston. (2000). *Environmental ethics.* Yang Tongjin, translated. Beijing: China Social Sciences Press.

Hou Hongbing, Guo Hongqiong, & Yu Qiang, et al. (2021). *Ecospatial network optimization in Ordos Based on LMBA strategy* Transactions of the Chinese Society for Agricultural Machinery, 52(01): 219–227

Hu Mingxiu, & Zhou Huanzhen. (2007). Study on the change trend of water quality index in Wuhan section of Hanjiang River and East Lake; environment-based KuzNietz curve (EKC) method. Journal of Wuhan Polytechnic University, (01): 68–72.

Indira SS, Vaithilingam CA, & Oruganti KSP, et al. (2019). *Nanogenerators as a sustainable power source: State of art, applications, and challenges.* Nanomaterials, 9:773–808.

Ma Shijun. (1981). The role of ecological law in environmental management; on the development trend of modern environmental management. Acta Scientiae Circumstantiae, (01): 95–100.

Nie Huoyun, & Zheng Lihua. (2008). *Study on co-city construction and Sustainable development in Yichun.* Journal of Yichun University, (05): 175–179.

Panyam V, Huang H, & Davisb K, et al. (2019). *An ecosystem perspective for the design of sustainable power systems.* Procedia CIRP, 80:269–274.

Peng Jianbo. (2009). Study on the privatization of public ecological space. Beijing: Tsinghua University

Peters K, Elands B, & Bjijs A. (2020). *Social interactions in urban parks: Stimulating social cohesion?* Urban Forestry & Urban Greening, 9(02): 93–100.

Rivera-Gonzalez L, Bolonio D, & Mazadiego LF, et al. (2019). Long-term electricity supply and demand forecast (2018—2040): A LEAP Model application towards a sustainable power generation system in Ecuador. Sustainability, 11(19):5316–5335.

Shang Jinjie. (2008). Anti-pollution greening tree species were selected in Jinan, Shandong province. China Flower & Gardening News, 12–18.

Shang Yongmin. (2018) Analysis of residents' perception and influencing factors of urban ecological quality construction; Analysis of 576 questionnaires in Shanghai//Shanghai Federation of Social Science Associations. Socialism with Chinese characteristics: Practice Exploration and Theoretical Innovation—Commemorating the 40th Anniversary of Reform and Opening Up (Collected Works of the 16th Academic Annual Conference of Shanghai Federation of Social Science Associations - 2018) Shanghai Federation of Social Science Associations: Shanghai Federation of Social Science Associations, 2018: 17.

Shang Yongmin. (2018). Analysis of residents' perception and influencing factors on urban ecological quality construction: based on a survey of 576 questionnaires in Shanghai //Shanghai Federation of Social Science Associations. Socialism with Chinese characteristics: Practice Exploration and Theoretical Innovation—Commemorating the 40th Anniversary of Reform and Opening Up (Collected Works of the 16th Academic Annual Conference of Shanghai Federation of Social Science Associations - 2018) Shanghai Federation of Social Science Associations: Shanghai Federation of Social Science Associations, 2018: 17.

Sharifi A, & Yamagata Y. (2016). *Principles and criteria for assessing urban energy resilience: A literature review.* Renewable and Sustainable Energy Reviews, 60:1654–1677.

Shen Qi. (2010). *Switzerland was once polluted.* World Market, (08): 57.

Shen Qingji, Xiang Weining, & Cheng Xiangzhan, et al. (2016). Tongji declaration on ecological wisdom *and ecological practice.* Urban Planning Forum, (05): 127–129.

Shen Yue, Liu Tianke, & Zhou Pu. (2017). Theoretical analysis and strategies of natural Eecological space use control. China Land Science, 31(12): 17–24.

Tai Jie. (2010). From 'The Golden Bough' to 'Fengshui Forest': related folklore interpretation based on'Fengshui Trees' and 'Fengshui Forest'. Masterpieces Review, (23): 106–109.

Wang Mengyao. *Analysis on space construction of folk belief places in Chaozhou ancient city//* Architectural History Branch of Chinese Architectural Society, Architectural History Professional Committee of Chinese Science and Technology History Society. Proceedings of 2014 Annual Meeting and Academic Seminar of Chinese Architectural History Society.Architectural History Branch of Chinese Architectural Society and Architectural History Professional

Committee of Chinese Science and Technology History Society:Architectural History Branch of China Architectural Society, 2014:6.
Wang Shaozeng, & Li Min.(2001). Study on ecological mechanism of urban open spatial planning (part two). Chinese Landscape Architecture, (05): 33–37.
Xie HL, Yao G, He YF, Zhang DB. (2018). Study on spatial identification of critical ecological space based on GIS:a case study of Poyang Lake Ecological Acta Ecologica Sinica, 38 (16): 5926–5937.
Xiong Wenhai. (2009). Study on Structural Characteristics and Dynamic Behavior of World Shipping Network. Qingdao: Qingdao University.
Xu Yi, & Peng Zhenwei. (2016). Study on spatial evolution and dynamic mechanism of Shanghai urban ecospace in 1980–2010. Urban Development Studies, 23 (11): 1–10+59.
Xue Liming, Liu Baokang, & Zheng Zhixue, et al. (2018). *Sustainable evaluation of coal resources based on ideal point-extension cloud model*. Journal of Natural Resources, (09):1657–1665.
Yang Anxiang. (1984). Ask me where my ancestors came from the great pagoda tree in Hongdong, Shanxi Province: A preliminary study on the historical position of the great pagoda tree moving people. Journal of Shanxi Teachers College (Social Science Edition), (03): 44–48.
Yin Duo, Ni Meng, & Wu Tiehong, et al. (2014) Characteristics of residents' attachment to urban ecological recreation space in ethnic minority Areas:a ase study of Ordos Kangbashi. Journal of Northwest Forestry University, 29(05): 249–255.
You Shang, Jiang Jingfeng, & Xie Yunyi. (2019). Reconstruction and optimization of production-living-ecological space in rural homestay settlements from the perspective of self-organization theory: a case study of representative shared farms in Hainan Province. Southeast Academic Research, (03): 71–80.
Yu Kongjian. (2020). Work with and by nature: the essence of territorial spatial planning and ecological restoration. Landscape Architecture Frontiers, 8(01): 6–9+4–5.
Zhang Liang, & Yue Wenze. (2019). Integrated recognition of urban ecological space: a case study on Hangzhou. Acta Ecologica Sinica, 39 (17): 6460–6468.
Zhang Qifei, Wu Zhifeng, & Guo Guanhua. (2017). Thermal environment effect of urban ecological space in high density urban builtup areas: a case study of the central urban district of Guangzhou. Urban Insight, (03): 39–49.
Zhang Wenfang. (2018). *Discussion on fast-growing Sophora japonica and ecological poverty alleviation in Datong*. Soil and Water Conservation Science and Technology in Shanxi, (02): 4–6.
Zhang Xuemei. (2017). A study on the ecological spatial evolution in Shanghai. Shanghai: Tongji University.
Zhang Yun, & Chen Mei. (2014). Application research on Jinan native plants in the ecological urban construction.[C]//Chinese Society of Landscape Architecture. Proceedings of the 2014 annual conference of Chinese Society of Landscape Architecture (Volume II), 6. Chinese Society of Landscape Architecture: Chinese Society of Landscape Architecture.
Zhao JZ, Liu X, & Dong RC, et al. (2016). *Landsenses ecology and ecological planning toward sustainable development*. Intemational Joumal of Sustainable Development & World Ecology, 23(04):293–297.
Zhao Jing. (2007). Relationship between the change of geomorphic resources and urbanization process in Wuhan. Study Monthly, (22): 47–48.
Zhao Meng, Zhang Xueqi, & Zhang Yonglin, et al. (2020). Research on the improvement of urban ecological space service base on landsenses ecology: a case study in Shunyi District, Beijing. Acta Ecologica Sinica, 40(22): 8075–8084.
Zhou Chunshan. (2007). *Urban spatial structure and form*. Beijing: Science Press.
Zhu Junhua, Liu Jie, & Zhang Jianglin. (2012). *Low-carbon planning method based on the city's carbon sink assessment: a case study of Haizhu Eco-City planning in Guangzhou*//The Conference of China City Planning and Development in 2012 Paper .Collection Beijing: Chinese Society for Urban Studies.

Healthy City Planning from the Perspective of Territorial and Spatial Planning

Wang Lan, Wang Xinzhe, Jiang Xiji, and Jia Yinghui

With "Healthy China" becoming a significant development strategy in China, healthy city planning through space environment intervention can be an important way to achieve this goal. World Health Organization (WHO) fully recognizes and emphasizes the role of urban planning in public health, raising that "health must be the primary focus of urban planners."[1] Studies have shown that environmental risk factors, such as air and water pollution, chemical exposure, climate change, and ultraviolet radiation, can cause over 100 diseases and injuries. It is estimated that about 11.27 million people worldwide die each year due to the environmental risks, accounting for nearly one-third of global deaths (2019).[2] Unhealthy environment is a key source of global disease; among which the chronic non-communicable diseases are largely related to the improper planning, management and maintenance of urban environment.[3] With the spread of COVID-19, the great influence of spatial factors such as urban density, housing conditions and transport network on spreading infectious diseases has also been widely recognized[4]; Spatial interventions have re-entered the planner's vision for the prevention and control of infectious diseases. General Secretary Xi Jinping has clearly stated that health should be integrated into all policies, and the concept of life-cycle health management should run through all links

[1] WHO (2021).

[2] Murray et al. (2020).

[3] UN-HABITAT, WHO (2020).

[4] Hu et al. (2021).

W. Lan (✉) · J. Xiji · J. Yinghui
College of Architecture and Urban Planning, Tongji University, Shanghai, China
e-mail: wanglan@tongji.edu.cn

W. Xinzhe
Shanghai Tongji Urban Planning and Design Institute Co., Ltd., Shanghai, China
e-mail: wxinzhe@163.com

Z. Peng (ed.), *Theories and Frontiers in Territorial Spatial Planning*, Urban Sustainability, https://doi.org/10.1007/978-981-96-9825-7_11

and whole process of urban planning, construction and management.[5] In general, taking full account of spatial factors' effect on health in planning is the foundation for healthy city planning and construction.

At the same time, territorial and spatial development system in China has been initially established, and relevant planning practices are being actively carried out, which provides physical space carrier for integrating the concept of health into planning and promoting the health level of urban and rural residents. The establishment of territorial and spatial planning system aims to ensure our ecological civilization construction and enhance national governance capacities more comprehensively and systematically. In conclusion, two prominent characteristics are shown in territorial and spatial planning system:First, looking to the systematic governance of space resources at multiple levels comprehensively to achieve better quality, higher efficiency, fairer and more sustainable land development[6]; Second, breaking the barriers, between different departments, realizing "multi-spatial planning", and forming a policy mechanism of multi-sectoral collaborative planning and implementation.[7,8] The territorial and spatial planning system strives for scientific and efficient management of various spatial factors while offering a platform to integrate health concepts into planning.

Therefore, this chapter aims to introduce Healthy City Planning from the perspective of territorial and spatial planning systematically and makes a detailed discussion about how to integrate the concept of health into the current territorial and spatial planning system in China based on the theory and its characteristics, so as to obtain wider and long-term public health benefits. Finally, taking Gucun Town in Baoshan District, Shanghai as an example, the current practice and exploration of healthy city planning in China was introduced.

1 Healthy City and Healthy City Planning

1.1 Relationship Between the Two Concepts

In 1994, WHO put forward the definition of "healthy city": a city that is continually creating and improving those physical and social environments and expanding those community resources which enable people to mutually support each other in performing all the functions of life and in developing to their maximum potential. The conceptual subjects related to healthy city include city, people and planning, and the relationship among them under the goal of health value. Academician Wu Zhiqiang has pointed out that in planning whether a healthy city is "Health of City",

[5] En People.cn (2020).

[6] Zhang Shangwu (2019).

[7] Wang Xinzhe (2019).

[8] Wu Tinghai (2019).

"Health in City" or "City for Health" needs more accurate understanding. On the health of city, that is to say, whether the city itself develops in a healthy state, such as in terms of population and economy; on the other hand, health in city refers to the well-being of its urban residents; A city for health is a comprehensive concept including the former two.

It can be seen that a healthy city is the unity, which not only includes "Health of City", but also focus on "Health in City", then building "City for Health" through planning and design (Fig. 1). So in the field of urban science and urban-rural planning, healthy city is based on promoting the planning and construction of a healthy environment, economy and society, and takes the optimization of the built environment as its core to enhance residents' physical and mental health. Generally speaking, healthy city planning is a tool that takes urban planning for health as its core and uses urban planning for health, ultimately aiming to provide protection for people's health in cities (Fig. 2).

Fig. 1 Connotation of healthy city. *Source* Wang Lan (2018)

Fig. 2 Methodology and ontology of healthy city planning. *Sources* Wang Lan (2018)

1.2 The Health and Emergency Properties of Space

In territorial and spatial planning, as a starting point, space can promote the organic integration of the healthy city concept at multiple levels. For example, when carrying out "double evaluation", the possible impact of resources, environment and territorial space development on public health should be considered. In general, from the perspective of healthy city, urban space owns two significant properties named "health" and "emergency".

The "health" of space means that the whole urban space system has a certain role in promoting human health, including the prevention of infectious and chronic non-communicable diseases. In terms of infectious diseases, spatial pattern and environmental factors should play a preventive role in reducing the infesting of infectious sources, its contact with people, and improving the allocation of health facilities. While in chronic non-communicable diseases, the overall incidence of cardiovascular and respiratory diseases has accounted for 65% of the mortality rate of the world population. Meanwhile, the prevalence of obesity is also an key inducement for its occurrence which becoming a common problem for residents all over the world. Spatial planning should play a proactive role in reducing the risk of chronic non-communicable diseases; By the comprehensive optimization of land use, transportation organization, urban form and functional structure and open space and public system, healthy lifestyle is adopted, and the health risk exposure is reduced.

The spatial "emergency", which is a kind of urban emergency response ability under sudden public health events, has a certain correlation with resilience. It emphasizes the city's active and quick respond and recover in public health emergencies. Therefore the spatial "emergency" is mainly aimed at infectious diseases, which needs to be considered systematically in urban health care facilities layout, integration of basic medical facilities and community life circle, etc.

1.3 Theoretical Framework of Healthy City Planning

Building upon the understanding of the relationship between a healthy city and healthy city planning, as well as the two properties of space, the theoretical construction of healthy city planning needs to be further clarified: Identifying the key spatial environmental factors and their impact on health. There are significant differences in the causes of chronic non-communicable diseases and infectious diseases, so the theoretical framework of healthy city planning for both the diseases will be discussed respectively in this chapter.

- For chronic non-communicable disease

  The framework of healthy city planning for chronic non-communicable diseases can be categorized into "Four Factors" and "Three Paths". Thereinto, the four factors, which include land use, spatial form, road and transport, green

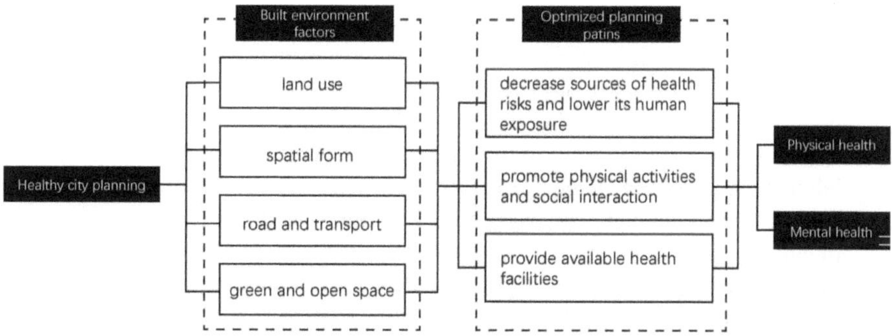

Fig. 3 Healthy city planning framework related to chronic non-communicable diseases. *Sources* Wang Lan (2018)

and open space, are the main types of built environment factors that can be intervened by planning; While the three main impact pathways are decrease sources of health risks and lower its human exposure, promote physical activities and social interaction, and provide available health facilities (Fig. 3). These impact paths have been adopted by the World Health Organization and UN-Habitat in their *Integrating health in urban and territorial planning: a sourcebook*[9].

From the three paths, the realization mechanism of space's effects on health is as follows:

(1) Decreasing sources of health risks and lower its human exposure. Health risks mainly include physical environment ones like air pollution, noise and heat exposure, and biological environment ones such as pathogen exposure, which are related to the morbidity and mortality of cardiovascular diseases, chronic respiratory infections, cancer and zoonotic infectious diseases. There are many polluting factors in urban and rural area (such as industrial facilities, urban transportation, etc.), whose layout, organization mode and spatial position relationship with other land (such as residential, working and recreational land) have a great impact on residents' exposure in polluted environment. On the one hand, the protective distance between these elements and residential areas, educational facilities and key working spaces should be considered in urban planning. On the other hand, it is necessary to fully analyze the potential impact of natural environmental factors such as ventilation, hydrology, and soil on the diffusion of pollutants, and to identify the potential affected areas and the degree of impact in order to develop appropriate planning and design strategies. In addition, centralized integration of pollution sources in planning can reduce decentralized and homogeneous layout, by which wider negative health impacts may be caused.

(2) Promoting physical activities and social interaction. Physical activity refers to "any physical exercise produced by skeletal muscle that consume energy",

[9] UN-HABITAT, WHO (2020).

including walking, cycling, other related entertainment and games, etc., which is helpful to prevent chronic diseases such as cardiovascular diseases and type 2 diabetes.[10] Promoting physical activity and social interaction by spatial planning and design can then reduce the incidence of chronic non-communicable diseases. As for physical activity, it includes professional work, housework, labor, transportation, leisure and entertainment, as well as physical exercise. Lack of physical activity leads to 3.2 million deaths every year.[11] Spatial planning and design can affect residents' lifestyle and working conditions. For example, promoting a work-life balance can encourage people to choose healthier modes of transportation for their commute, which can have a positive impact on reducing the incidence of chronic non-communicable diseases. WHO suggests that the average weekly physical activity time is 150 min per person. In terms of social interaction, urban life creates plenty of stress, like crowded housing and relative isolation from nature, which will have a negative impact on people's physical and mental health. High-quality public spaces (such as street corner gardens, woodland water systems and other blue-green spaces) have been proved to affect people's mental health. Through reasonable planning and design, these spaces will improve pro-natural activities and rich social interactions, which puts forward higher requirements on planning.

(3) Providing available health facilities. Health facilities mainly include medical, sports and health care facilities, which will provide residents health services. Its social justice is an important characteristic that affects health equity, which is mainly reflected in multiple aspects such as accessibility, availability, and spatial congruence of health facilities. First of all, basic facilities scale allocation and better spatial accessibility can provide basic health services, ensuring the physical and mental well-being of residents. Secondly, the spatial layout of health facilities that is compatible with social background factors such as age structure and income level will be more conducive to achieving precise health services, promoting health equity, and improving the efficiency of utilizing related facilities.

- For infectious diseases

  Apart from major infectious diseases like COVID-19, there are also various types of infectious diseases such as plague and avian influenza. They are closely related to the spatial environment in terms of their occurrence and spread, so it is necessary to focus on how to carry out prevention and emergency response in planning. According to epidemiology, the infectious diseases occur and spread in ecological process and social process.[12] Therefore, starting from the three basic links of infectious diseases spreading (isolating the source of infection, cutting off the route of transmission and protecting the vulnerable populations), we worked out the impact path of urban development on infectious diseases (Fig. 4), and

---

[10] WHO (2021).

[11] OMDL (2009).

[12] Eisenberg et al. (2007).

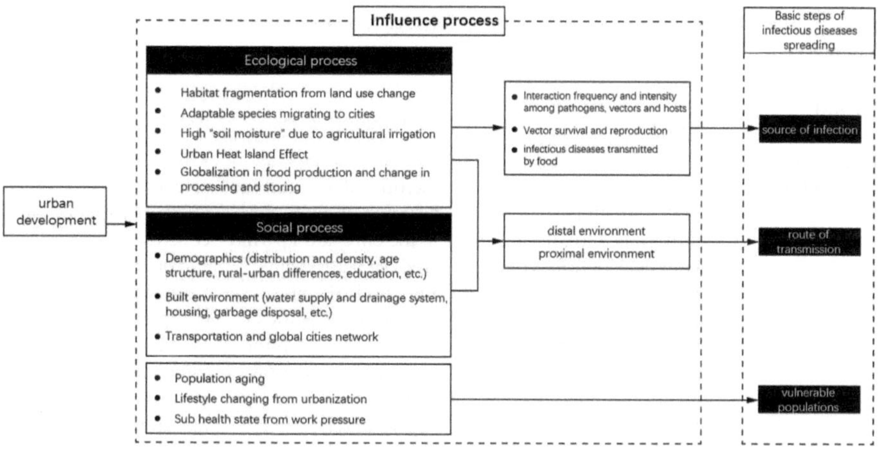

Fig. 4 Impact path of urban development on infectious diseases. *Sources* Wang Lan (2018)

based on which to, put forward specific spatial intervention strategies for different prevention and control links (Table 1).

2 Impact of Main Spatial Factors on Health in Territorial Spatial Master Planning

In terms of spatial factors, "*Several opinions on building a national territorial and spatial planning system and supervising the implementation by CPC Central Committee and State Council*" clarifies that territorial spatial planning must make overall arrangements for ecological space, agricultural space and urban space.

Based on the planning spatial factors and health-related research results as evidence-based practice, as well as the theoretical framework of healthy urban planning for chronic non-communicable diseases and infectious diseases, the path of spatial factors affecting public health is constructed in the national territorial spatial master planning (Fig. 5).

2.1 Ecological Space

Ecological space includes forests, grasslands, wetlands and water, etc., the total amount and layout of which affect public health. In terms of the total amount, the expansion of forest land, wetland and water area can effectively improve the

Table 1 Spatial intervention strategies for infectious disease prevention and control in urban planning

Prevent and control	Spatial intervention strategy		
	City level	Community level	Key facility
Isolate the source of infection	• Based on territorial spatial master planning, attach importance to all kinds of land distribution within the city, reduce ecological space fragmentation; • Set up epidemic prevention subsection and health insulated belt; • Allocate facilities such as integrated regional support centers,and make contingency plans for transport and materials	• Avoid placing facilities that could be a source of pathogens in the core and upwind areas of settlements; • Set up isolated supporting facilities and storage space for epidemic prevention materials, such as supermarkets, food markets and pharmacies within the 15-Minute Community Life Circle; • Set up fever clinics, temporary small-scale mild isolation areas, standby nursing spaces or emergency medical service centers based on community health facilities,to divert patients and avoid cross-infection	• Reasonable site selection of infectious disease hospitals and construction of surrounding supports; • Build a comprehensive medical city for infectious disease prevention and control research in mega cities; • **Consider both normal and pandemic periods** of key emergency facilities such as stadiums and exhibition centers to meet the need during an epidemic in terms of architectural design; • Improve the spatial adaptability of regular hospitals during the epidemic; • Equipped with facilities such as medical freezers, hospice care, palliative care or mortuaries, and designed to meet the needs in epidemic

(continued)

Table 1 (continued)

Prevent and control	Spatial intervention strategy		
	City level	Community level	Key facility
Cut off transmission route	• Close some public transportation, limit traffic and unnecessary activities; • Carry out time and space of residents behavior analysis and research based on big data, clarify the necessary traffic organization and protection means	• Guarantee emergency care, service and work-residential balance within the 15-Minute Community Life Circle, and help to form public health units; • Set up emergency medical service centers,antiviral drug supply points and divert patients; • Optimize the layout of mixed land and reduce mass communication through vehicles; • Strengthen the distribution of emergency care and life support facilities in rural areas	• For high risks but need to be opened places,like hospitals, supermarkets and so on, can be managed by controlling the flow of people in different periods to make concentrated spread impossible; • Central air-conditioning in public buildings should be redesigned, zoned or use natural ventilation more • Facilities such as emergency care and virus laboratory testing should be added in rural areas to protect susceptible people
Protect vulnerable populations	• Guarantee the accessibility of municipal infrastructure, and promote renovation for poor sanitary conditions, housing and crowded living areas; • Promote health equity and guarantee the accessibility of medical facilities and services for all groups of people; • Designing high-quality urban and rural environments plan that promote healthy lifestyles and work practices, reduce the prevalence of chronic non-communicable diseases and the population base of vulnerable populations	• Establishment of population health records in public health to track and provide information on vulnerable populations; • Health impact assessment of daily spaces in the community and updating of environmental risk factors; • Promoting healthy lifestyles through the configuration of community facilities and environmental design	• Establishment of health and epidemiological liaison points to provide home health care, outbreak response and long-term care services; • Ensure the spatial and service integration of facilities for specific population groups in places like elderly care services, residential/ nursing homes for the elderly, welfare homes, homeless shelters, etc

Source Wang Lan et al. (2020)

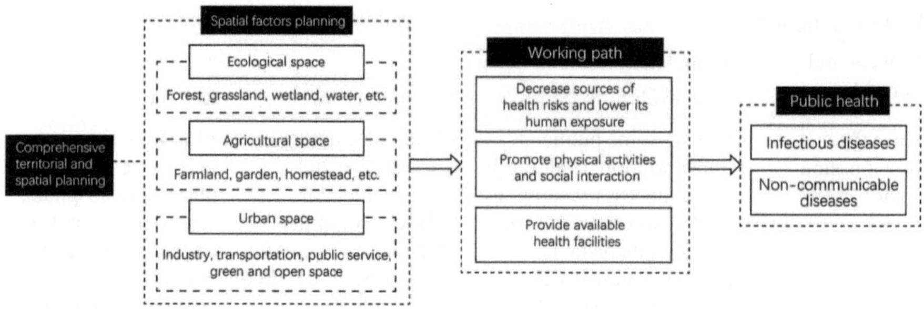

Fig. 5 Public health impact path of spatial factors in territorial spatial master planning. *Source* Wang Lan et al. (2021)

service value of ecosystem in regulating climate, absorbing air pollutants and purifying water,[13] so that to reduce the health risks of physical environment. In terms of layout, land patch size, shape and corridor setting of natural elements will affect biological environmental health risks. Specifically, the smaller the land patch, the lower the species diversity.[14] The loss of diversity will reduce the storage and dilution of pathogens by intermediate host species and increase the risk of pathogen spillover.[15] The more complex the patch shape is, the higher the edge ratio (the ratio of the edge area interacting with other ecosystems to the total patch area). The increase of edge ratio will intensify the interaction among pathogens, vectors and hosts,[16] and raise the risk of pathogen exposure; Corridor can connect multi-species habitats, enrich species in fine patches,[17] enhance dilution effect and reduce host infection rate, but at the same time, it will also cause pathogen host migration and increase the probability of contact with other species. Therefore, the health impacts of the total amount and layout should be fully considered when setting the ecological space in order to reduce the risk exposure of the population.

2.2 Agricultural Space

The agricultural space can be divided into agricultural production space (such as farm land and garden) and rural living space (including rural homestead, and public service facilities). The layout of these spaces can affect health through three paths. The overall layout of agricultural production space will affect "food miles", and long-distance food transportation will increase the health risks in physical environment

[13] Song and Deng (2017).

[14] Forman and Godron (1981).

[15] Schmidt and Ostfeld (2001).

[16] Patz et al. (2004).

[17] Jordán et al. (2003).

caused by transportation emissions.[18] Irrational distribution of agricultural land may also bring pollution to agricultural products. For example, near to the enterprises and high-grade road may lead to excessive heavy metals in agricultural products.[19]. In rural areas, the layout of homestead and the way of housing energy supply can affect residents' physical environmental health risks and their body exposure. For instance, villages located near high-grade road will aggravate people's exposure to traffic-related pollution, while the burning of household solid fuels will cause indoor air pollution exposure.[20] Rural medical facilities, educational facilities, sanitation facilities, etc. can provide daily health support[21]. Emergency nursing and living service facilities can respond to the needs[22,23] during the epidemic and provide residents with accessible health facilities. At the same time, the allocation of tourism service facilities to develop rural tourism can promote pro-natural activities and increase the total physical activities of urban and rural residents. In addition, agricultural space is related to the risk of occurrence and transmission of infectious diseases. The expansion of agricultural space will split ecological space, increase the contact between wild animals and people and livestock, and intensify the interaction between pathogens and hosts. The increase of soil moisture caused by agricultural irrigation will affect the abundance of mosquitoes and other vectors and increase the risk of pathogen transmission.[24] Therefore, the spatial layout of rural production and living space and their relationship with surrounding land should be considered in agricultural spatial planning.

2.3 Urban Space

Urban space includes industrial space, traffic space, public service facilities, green and open space, etc. Its total amount and layout can affect public health through three paths.

In terms of the total amount, the increase of urban development land will lead to a rise in pollutant emission,[25] heat island effect,[26] and physical environmental risks as well. In terms of layout, industrial spatial layout affects the distribution of pollutant

---

[18] Paxton (1994).

[19] Zhong Weike et al. (2001).

[20] Zhang and Smith (2007).

[21] Wang Lan et al. (2020).

[22] Wang Lan et al. (2020).

[23] Geng Hong et al. (2020).

[24] JORDÁN F, BÁLDI A, ORCI K-M, et al. Characterizing the Importance of Habitat Patches and Corridors in Maintaining the Landscape Connectivity of a Pholidoptera Transsylvanica (Orthoptera) Metapopulation [J]. Landscape Ecology, 2003, 18(01): 83–92.

[25] Feng et al. (2017).

[26] Li et al. (2017).

sources, commuting distances and lifestyles of residents,[27,28] which are related to health risk exposure and physical activity; road network layout and hierarchy affects the total amount of pollutants and their exposure to human beings, and the layout of Active transportation affects the total amount of physical activity and traffic pollution exposure to human beings.[29] The layout of public service facilities affects people's physical activities and interactions in daily life, and the accessibility of healthcare services,[30] which can also provide material support for home quarantine and effective patient triage and treatment during epidemics, while infectious disease hospitals may contribute to the spread of pathogens[31]; the layout of green and open spaces has an effect on the exposure risk to pollutants, the total amount of physical activity, and the ability to recover mentally,[32] and thus affects public health. Therefore, in the total amount and layout of urban space, it is necessary to fully consider its health impact to realize health promotion.

These spatial factors are the main carriers of influence from planning on health, which reflect the role of spatial planning from three paths. The specific planning content will produce health effects to varying degrees.

3 The Framework of Integrating Health into Territorial Spatial Master Planning

On the basis of identifying the health impact of spatial factors, this chapter proposes a framework for integrating health into territorial spatial master planning. The framework includes three stages: status diagonosis, proposal development and assessment, among which the first two stages can be incorporated into the basic work and proposal development of current planning. In the status diagonosis, the current health problems and planning optimization points are identified through recognizing the health characteristics of the population in the area, as well as diagnosing the health risks and health resources. In the proposal assessment, effective interventions are implemented through the configuration of various spatial factors, such as reducing risk exposure, providing health facilities, and promoting physical activities and interactions. The programme evaluation is a newly added part, which predicts and assesses the health impacts of the proposal. It identifies the possible health effects of the proposal implementation in terms of the health risks and health resources, and the health outcomes. This allows for the optimization and adjustment of the proposal to achieve the health-promoting effect (Fig. 6).

[27] Yang et al. (2004).
[28] Zhu Jing et al. (2014).
[29] Wang Lan et al. (2016).
[30] Tian Li et al. (2016).
[31] Wang Lan et al. (2020).
[32] Gan Liang et al. (2020).

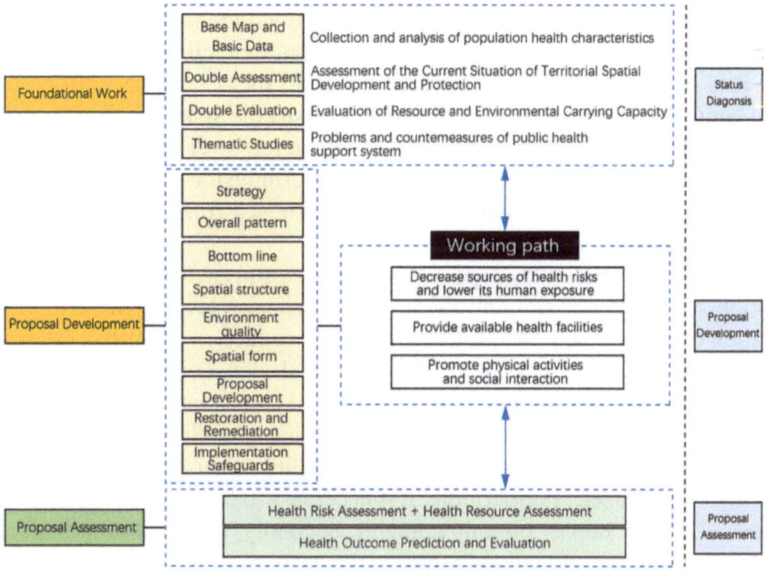

Fig. 6 Framework for Integrating Health into Territorial Spatial Master Planning. *Source* Wang Lan et al. (2021)

3.1 Status Diagonosis Stage

Status diagnosis can be integrated into the foundational stage of territorial spatial master planning. It involves adding health content to the base map and basic data, conducting "double assessment", "double evaluation" and thematic studies. This will enable the identification of population health characteristics, diagnosis of health risks and health resources in the planning area, evaluation of the capacity to address health considerations and research on health problems and countermeasures.

During the development of the base map and basic data, it is recommended to collect and analyze the health characteristics of the population (like incidence of major diseases, average life expectancy, occupational risk exposure, etc.), so as to understand the core needs and main issues related to health for local residents. This helps figure out the core work contents and methods of health integration planning, and provides a basis for subsequent work.

Within the "double assessment", the "assessment of the status quo in territorial spatial development and protection" is an important content of the preliminary analysis work; At present, it is required to complete the assessment of risks such as natural disasters, while the diagnosis of public health risks should be added. For example, combining with the wind environment simulation results, the built environment can be analyzed to calculate the physical environmental risks such as air pollution and noise and their spatial distribution; As for the specific high-incidence diseases of local residents, its potential environmental impacts are analyzed, so that to avoid it in subsequent planning. At the same time, the livability dimension of this part includes indicators such as accessibility of health-related public facilities facilities and green space. It is suggested to further increase the analysis of health facilities and their service equity to different groups of people, such as the spatial distribution and site reservation of 120 ambulances, and provide allocation level for aging communities; Gini coefficient and LISA analysis are used to determine the areas with high demand but low provision.[33,34] The status should be diagnosed from two aspects of health risks and health resources.

The "evaluation of carrying capacity of resources and environment" in the "double evaluation" requires determining the carrying capacity level under the functional orientation of ecological protection, agricultural production and urban construction. Firstly, the evaluation indicators for the ecological conservation function includes biodiversity conservation, water conservation, soil conservation, etc. It is suggested to enhance the analysis of the health benefits of various ecological elements, such as quantifying their contributions in reducing air pollution, noise, heat island effect, and enhancing biophilia, etc. Furthermore, it is necessary to clarify the current ecological problems and their impact on public health. Secondly, the evaluation indicators of agricultural production function includes land resources, water resources, etc. It is suggested to analyse the pollution risks of agricultural land including heavy metal pollution, air pollution, etc., so as to provide support for the delineation of edible crop planting areas and the division of village types (grouping and promotion, relocation and merger, etc.). Thirdly, the indicators for evaluating urban construction functions include natural resources such as land, water, atmosphere and risks of disaster. Health risk assessment is recommended to be added, for example, analyzing the health risks from large-scale transport infrastructure, and polluting industrial parks, etc., so as to identify the areas in which the risks are too high and unsuitable for urban construction. Integrating health-related indicators into the evaluation index system of resources and environment carrying capacity can provide support for the delineation of urban space, agricultural space and ecological space.

In addition, it is suggested to carry out "thematic studies on health" to discuss and analyze the problems and countermeasures of public health support system. For instance, after obtaining the data such as disease incidence and life expectancy per capita, a quantitative model of linking planning indicators with health outcomes can be developed to identify the built environment elements with significant influence, which provides an empirical basis for the selection of "double evaluation" elements, "double evaluation" indicators and the adjustment and optimization of subsequent planning schemes. Questionnaire survey and location data collection can also be employed to identify residents' spatial and temporal behaviors, analyzing the frequency, duration, metabolic equivalent and route choice of residents' physical activities. This helps to explore the location, site and quality characteristics of

[33] Song and Deng (2017).

[34] Tang Zilai and Gu Shu (2016).

public space where residents interact, and clarify the spatial factors and optimization methods for promoting physical activities and social interaction in the local environment, thereby promoting public health through spatial interventions.

3.2 Proposal Development Stage

According to the "*Guidelines for the Compilation of Territorial Spatial Master Planning (Trial)*" (2020), the compilation includes the overall pattern, bottom line constraints, spatial structure, environmental quality, spatial form, infrastructure etc.; Health can be integrated into multiple sectors through three optimized paths.

(1) Target strategy

"Healthy city" can be regarded as one of the development goals, and the control requirements of health risks and health resources can be added in the index system Planning indicators usually contains three categories: spatial bottom line, spatial structure and efficiency, and spatial quality. A suggested spatial bottom line index includes premature death rate from major chronic diseases, vector density control level, life expectancy and other indicators on the basis of constrainting the scale of forest land, wetland, farm land and development land; To reflect its fairness, the Gini coefficient of inter-distribution of specific health facilities is added to spatial structure and efficiency; Spatial quality index can include $PM_{2.5}$ concentration average, urban-rural temperature difference (Heat Island effect), the proportion of people who often take part in physical exercise and other indicators reflecting environmental quality.

(2) Overall pattern and bottom line constraints

The overall pattern requires to clarify the space for ecological protection, agricultural development and rural-urban development. The bottom line constraint involves delineating three control lines: ecological protection red line, permanent basic farmland, and urban-town development boundary. The total amount control and layout for health integration development can be taken into account from the two aspects: health risks and health resources. In terms of infectious diseases, the expansion of urban built areas reduces, breaks or changes the habitat of pathogen host species, which may lead to the rise in the contact rate between humans and livestock and increase the transmission risk of infectious diseases[35]; Therefore, as for total amount, it is necessary to strive for intensive and compact urban and rural development space, so that to avoid disorderly interweaving of three types of spaces in layout and fragmentation of ecological space, like excessively small ecological patches and low connectivity. Meanwhile, the more complex the boundary between ecological space and agricultural & urban space, the higher risk of zoonotic infectious diseases[36]; So it is

[35] Zhong Weike et al. (2001).

[36] Weinstein et al. (2020).

necessary to fully consider the species richness and density of pathogen hosts near the boundary, control the complexity of the boundary and set up buffer zones as appropriate. In chronic non-communicable diseases, the expansion of urban built areas may bring heat island effect, air, water and soil pollution, etc. Agricultural spatial layout affects local food production, and ecological space plays a role in regulating climate and air quality. Therefore, it is necessary to put forward avoidance and optimization planning strategies on the basis of corresponding health risk analysis. For example, the overlay analysis based on wind field and pollution source, catchment area and pollution source, the potential pollution areas of air and soil are defined, so that to put forward the scheme of relocating and merging, or define the spatial layout of agriculture and the areas where non-edible crops need to be planted.

(3) Spatial structure and infrastructure.

This part requires delineating planning zones, the structure and layout of development land, and the site selection of large-scale infrastructure, etc. It contains planning contents that affect public health through multiple paths. Among the contents and principles of routine planning, there are potential health promotion effects. For example, job-residence balance can reduce chronic non-communicable diseases caused by insufficient physical activity and obesity caused by motor commuting; Regional infrastructure corridor route selection and transportation hub site selection should try to avoid land with high species richness, reduce the loss of ecological benefits and the contact between people and pathogens, so as to reduce the risk of pathogen exposure. In order to better realize the health promotion of space, we can consider deepening the drawing of "urban microclimate zoning map", delineating infectious disease risk level areas, high temperature early warning areas, etc., so as to support the integration of health considerations into the layout of construction land. For example, in Hong Kong, China, the urban climate analysis map and the urban climate planning proposal map are drawn in the *Urban Climatic Map and Standards for Wind Environment*[37]. At the same time, the site selection and setting of large public service facilities should take the epidemic prevention and control into account. On the basis of the existing principles and indicators, the consideration and content of epidemic prevention and control should be added, such as gymnasiums, convention and exhibition centers and other facilities as alternative nursing places, and the surrounding life passages and special passages for epidemic prevention materials should be planned and set up.[38]

(4) Environmental quality and spatial form

This part focus on urban public service facilities, road network and Active Transportation System, green space and open space, and puts forward the requirements of shaping spatial form and style, which is closely related to many aspects of spatial health promotion. For example, in the layout of health facilities, it is important to stay away from the high-risk areas identified in health

[37] Planning Department and the Chinese University of Hong Kong (2012).

[38] Dai Shenzhi et al. (2020).

risk. It is necessary to set up health facilities that are safe and accessible at different levels, and these facilities should meet the requirements of epidemic prevention in terms of the layout and design.;Additionally, the characteristics of population distribution should be considered, and facilities should be allocated according to the needs of specific populations under the concept of health equity. Active Transportation System can rely on branch roads with high connectivity and safety, connect blue-green spaces and activity facilities, avoid overlapping with traffic arteries to reduce excessive pollution exposure to pedestrians, and encourage physical activities such as walking and cycling. The balance between economies of scale and accessibility should be fully considered in the layout of green space and open space, and the role of urban green space during the epidemic period should be considered at the same time. The spatial form section mainly considers the relationship between ventilation corridors, street canyon wind fields, and pollution source and source of infection, planning ventilation corridors and development intensity zoning, and guiding detailed planning through planning conduction.

In each section of proposal development, health can be integrated into its existing work and framework to varying degrees, either by optimizing health in the routine planning content or by adding health considerations. Specific principles and indicators can be supported by existing research, key issues and experiences from local research, innovative practices from interdisciplinary cooperation, and jointly support the detailed preparation of planning schemes.

3.3 Proposal Assessment Stage

After the completion of the proposal development, a health impact assessment can be carried out from three aspects: health risk, health resources and health outcomes.[39] This assessment aims to determine the positive effect, negative effect and overall effect on health. It also calculates the health gain compared with the status quo, or compares and selects multiple planning schemes.

The purpose of health risk assessment is to clarify the role of planning proposals in reducing resident health risk exposure. The assessment needs to identify the spatial factors in the programme that are likely to cause or exacerbate health risks (e.g., regionally significant transport and municipal facilities, urban arterial, on-ramps, industrial land, contaminated municipal land), and to assess the distribution of health risks in the planning area by systematically superimposing such elements on the urban context (e.g., wind farms, heat islands, hydrology), and socio-demographic factors (e.g., spatial distribution of the ageing population)[①]. Comparing the current health diagnosis results, we can judge the effect of the planning scheme in reducing health risks; If there are still high-value areas of health risk exposure, planning adjustment is needed.

---

[39] Wang Lan et al. (2018).

The Health Resource Assessment is designed to assess the positive health impacts of planning programme. The accessibility of health resources can be evaluated by minimum proximity distance method, two-step mobile search method, etc., and existing tools (health and economic assessment for riding and walking, green space assessment tool, etc.) can also be used to measure the health impact of the planning scheme in physical activity, green space, etc. The evaluation results can be compared with the current diagnostic results to get its health gain, and can also be used to judge its achievement and supple the corresponding health resources.

The aim of health outcome prediction and evaluation is to use mathematical models to predict the potential health outcomes, such as respiratory health, obesity, cardiovascular disease and depression, that may occur when the planning scheme is fully implemented. In terms of assessment methods, comparative assessment methods can be used, or local prediction models can be constructed. The former obtains Relative Risk (RR) based on empirical research results; Based on the local health data, the latter builds a model (such as regression model) to analyze the correlation of "specific health outcomes-built environment", and obtains the environmental factors and their degrees that significantly affect the health outcomes, which can be used as planning optimization clues and decision support.

The health impact assessment of the scheme follows the procedure of "assessment—scheme optimization—re-assessment—scheme finalization", improving the health benefit through continuous optimization of the scheme, and finally completing the planning scheme with health promotion effect.

4 Practice and Exploration of Gucun Town's Health Integration into Territorial Spatial Master Planning

Based on the practical framework of health integration into Territorial Spatial Master Planning, this chapter takes Gucun Town, Baoshan District, Shanghai as the case for the study. Gucun Town is located in the central and western part of Baoshan District, with an area of 41.65 square km^2, in the northwest of Shanghai's central city area, and has the spatial characteristics of both the central urban area and the suburbs. In view of the comprehensiveness of its spatial environment, Gucun was selected as a representative case of integrating health into the town-level territorial spatial master planning. The focus was on the impact of urban built environment on pollution exposure risk, health facilities availability and physical activity opportunity.

4.1 Status Diagnosis

In the status diagnosis stage, the current health risks and health resources were diagnosed in the "Assessment of the Status of Territorial Spatial Development"; At the same time, thematic studies were carried out to obtain residents' temporal and spatial behavior characteristics, which provided support for the preparation of the scheme.

In health risk diagnosis, the planning determined the location and type of pollution sources according to land use and road traffic information, and established a buffer zone to analyze the risk distribution of pollutants; The building model of Gucun Town was constructed, and the numerical simulation was carried out by software to analyze the distribution characteristics of wind field. In the planning, the distribution of pollution sources, wind environment simulation and residential land distribution were superimposed and analyzed to divide the areas that had high risk of pollution exposure and needed to be optimized.

In the diagnosis of health resources, the accessibility of its existing part and residents' needs were analyzed, and the Gini coefficient was used to calculate the spatial distribution fairness of health resources. The results showed that the spatial distribution of facilities for the elders was quite different, and it was urgent to make a supplement; The survey of residents' demand showed that the gap between supply and demand of health resources in various communities was different, for example, the gap between supply and demand of elderly care facilities for residents in commercial housing and affordable housing is large, and the medical facilities for farmers' resettlement houses were relatively insufficient, which needed to be supplemented by classification.

At the same time, the research on residents' spatio-temporal behavior was involved, and the basic scope of the community life circle was clarified by tracing routes on working days and weekends of the sample population based on the GPS positioning data. It was found that long-distance commuting was common among working people, and planning could consider increasing jobs in towns and increasing residence around industrial parks; Residents' daily non-commuting active travel (walking and cycling) usually does not cross the main road, so the allocation of public service facilities in the follow-up planning should meet the needs of the living circle unit.

The diagnosis of positive and negative health conditions and the analysis of residents' spatio-temporal behavior provided the basis for accomplishing the follow-up planning scheme, which could ascertain the high risk areas that need to be adjusted and optimized, the specific health facilities that need to be added, and the spatial scope and facility distribution that need to be considered to promote walking and cycling.

4.2 Proposal Development

The overall goal of building a healthy town is set in the target strategy of this plan, which draws the vision of healthy and livable development in public services and ecological space. Based on this, the overall pattern of the town is formed the four ecological corridors were controlled in the periphery. In accordance with the health risk diagnosis, high-risk villages were removed to reduce the risk of pollution exposure. In the spatial structure section, industrial land around rail transit stations and important industrial functional areas are expanded, the intensity of industrial land around stations are increased, and the proportion of long-distance commuting is reduced, the existing scattered industrial and logistics storage land to reduce health risks is vacated. For example, the storage industrial land on the north side is adjacent to sports and educational facilities, which poses a high risk of crowd exposure. Therefore, adjustments of the planning has been made. There might be pathogen exposure risks around municipal hospitals due to large population gatherings and heavy road traffic. As a result, the site selection of municipal cultural facilities has been changed.

In the environmental quality section, the facility allocation, road system and green space layout in townships are emphsized. According to the current diagnostic results, facilities with insufficient service coverage should be given priority in low-risk areas. In the dense road network planning of large residential communities, the grades of some branch roads and secondary roads are upgraded, and the traffic volume is guided to concentrate on high grade ones, thus reducing the pollutant concentration on branch roads around residential land; Combine green space, public service facilities and branch roads to plan active transportation system and encourage physical activity. For example, the health risk in the central part of the old town community is low, but the health resources are insufficient; therefore, the nursery and sports facilities were planned to be relocated into the center near the main road at the edge of the community, and add cultural and elderly care facilities to build a healthy and dynamic community center. The urban ventilation corridor of the "four corridors and six belts" are designed to improve the urban thermal and atmospheric environment. Additionally, in order to enhance the accessibility of park green space, various activities are incorporated, such as the addition of bicycle lanes and sports venues, thereby fully leveraging the health promotion function of green areas.

In addition, this plan demarcates the living circle according to the survey of spatiotempora behavior, configures corresponding facilities in the living circle, and build a health and safety unit as the basic unit for material reserve and epidemic isolation (Fig. 7).

4.3 Proposal Assessment

The program evaluation includes the assessment of health risk and health resource, as well as the calculation of health benefits compared with status quo. In terms of health risks, pollution sources are identified from the land use and road system of the planning schemes, and a three-dimensional spatial model is established to simulate the wind environment of the planning scheme. The results of superposition analysis of pollution distribution and wind environment show that, compared with the status quo, the maximum pollution value of the planning schemes is significantly reduced, and low-value health risk areas are significantly increased, which greatly

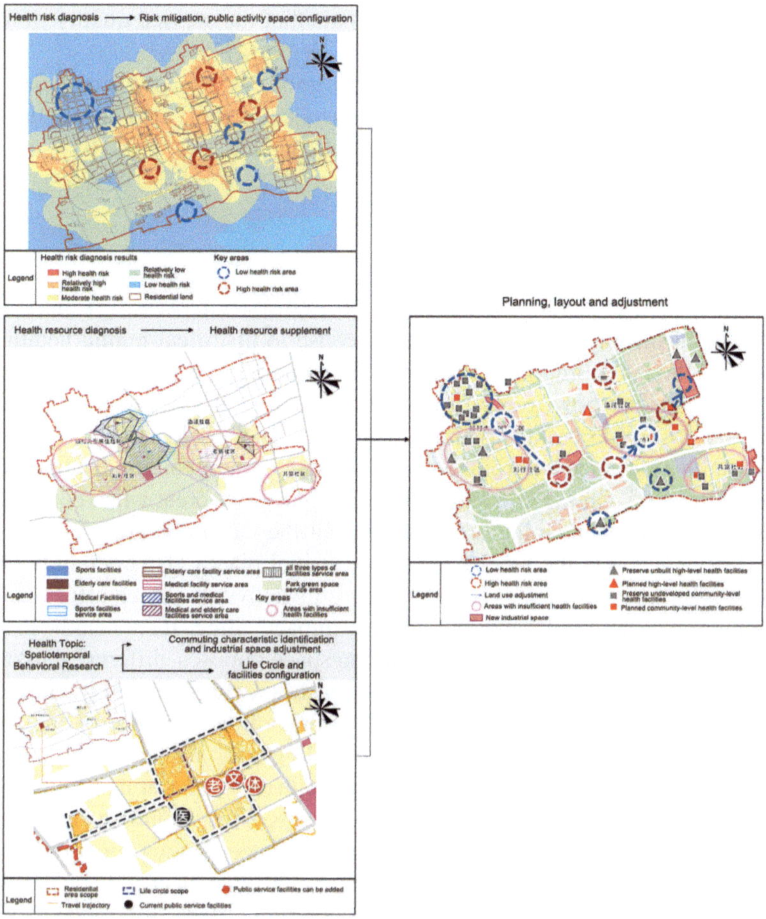

Fig. 7 Diagnostic results of status quo and schematic adjustment of land use layout. *Source* Wang Lan et al. (2021)

reduces the health risk exposure of residential land. In terms of health resources, the role of programs in physical activity promotion and health facilities supply should be evaluated. Through questionnaire survey, the changes of residents' walking and riding level before and after planning were obtained, and the health impact of physical activity changes was measured by health economic assessment tools. Result shows that the new green space and rail transit lines can increase the 25-min walking time and the 10-min riding per person each day; It can reduce 101 deaths related to physical activity every year. The accessibility analysis of health facilities and green space also shows that the service coverage rate has increased significantly. At the same time, the programme based on the concept of health equity, evaluates the accessibility differences of health resources in different types of communities by subdividing community types and facilities types, and finds that the service quality of various health resources has been improved to varying degrees.

The planning scenarios, overall, demonstrate good performance in reducing health risks, providing health facilities, promoting physical activity and advancing health equity, which are beneficial for promoting public health.

5 Conclusion

Modern urban planning originated as a response to historical public health crises in cities and the deterioration of residents' living conditions, and it is necessary for planners to consider how to return to their original goal and integrate health into the practice to enable better urban development. The World Health Organization pointed out that physical environment is a key factor among all the factors affecting individual health.[40] On the one hand, the urban places we creat when compiling the territorial and spatial planning are material, but they also contain social and even political meanings. On the other hand, these spatial places also contain human-to-human interactions and various social relationship activities, which affect residents' spatiotemporal behavioral patterns and thus impact affect their health. Professor Jason Corburn of Berkeley believes that urban space imposes different pressures on people. When the pressure on human body reaches a certain level, it can generate toxic factors, which will cause negative changes in hormone secretion of human body, and then affect health. Therefore, when compiling national territorial and spatial planning of various types and scales, reducing health risks, allocating health elements rationally and maintaining the balance of health resources should be emphasized.

However, it is worth emphasizing that significant uncertainties remain when translating research evidence into practical planning, more studies need to be carried out. These uncertainties mainly come from: (1) complex effects of spatial factors on health. Spatial elements may impact residents' health through various mechanisms, with substantial differences in research methods and conclusions[41]; (2) variation in influence effects brought by the background of research sites (such as health characteristics of different baseline population and the characteristics of built environment; (3) difference in health priorities across different time periods also contribute to variations in research focus. Therefore, the methodology constructed in this chapter is open and constantly updated, which can be continuously improved through the accumulation of various local health evidence to improve the effectiveness of planning intervention. Generally speaking, urban planners should take public health as an important value goal in specific planning practice; attach great importance to health demands in territorial spatial master planning, and incorporate health priorities identified through status diagnosis, proposal development and proposal assessment, which

[40] WHO (2021).

[41] Forsyth (2018).

helps reduce the occurrence of infectious diseases and chronic non-communicable diseases and improve the level of public health.

(Thanks to Gu Hao and Sun Wenyao, Ph.D. students of College of Architecture and Urban Planning; Li Xiaotian and Wang Zihan, master students. Thanks to Li Jijun and Wang Jinbai of Shanghai Tongji Planning and Design Institute for their contributions.)

References

Dai Shenzhi, Wang Jiangbo, & Liu Tingting. (2020).*Construction strategy of urban medical and epidemic prevention facilities system in both safe and disaster*. Urban Planning Forum, (03): 103–108.

Eisenberg, J. N., Desai, M. A., & Levy, K., et al. (2007). *Environmental determinants of infectious disease: a framework for tracking causal links and guiding public health research.* Environmental health perspectives, 115 (8), 1216–1223.

En.people.cn. (2020–06–03). *Xi stresses strong public health system to safeguard people's health* [EB/OL]. http://en.people.cn/n3/2020/0603/c90000-9696928.html.

Feng, H., Zou, B., & Tang, Y. (2017). *Scale-and region-dependence in landscape-PM2. 5 correlation: Implications for urban planning*. Remote Sensing, 9(09), 918.

Forman, R. T., & Godron, M. (1981). *Patches and structural components for a landscape ecology.* BioScience, 31 (10), 733–740.

Forsyth, A. (2018). *Evidence-based practice: challenges in a changing world* // Beatley, T., Jones, C. L., & Rainey, R. (Eds.). (2018). *Healthy environments, healing spaces: Practices and directions in health, planning, and design*. University of Virginia Press.

Gan Liang, Yang Weiguang, & Wang Lan. (2020). *Spatial characteristics of urban green space under different health impact paths*. Landscape Architecture, 27 (04): 95–100.

Geng Hong, Li Jiajia, & Lu Ningxing. (2020). Healthy development path of towns and villages with unconventional emergencies and normal response. Southern Architecture (04): 41–47.

Hu, M., Roberts, J. D., Azevedo, G. P., & Milner, D. (2021). *The role of built and social environmental factors in Covid-19 transmission: A look at America's capital city.* Sustainable Cities and Society, 65: 102580.

Jordán, F., Báldi, A., Orci, K. M., Racz, I., & Varga, Z. (2003). C*haracterizing the importance of habitat patches and corridors in maintaining the landscape connectivity of a Pholidoptera transsylvanica (Orthoptera) metapopulation*. Landscape Ecology, 18 (01): 83–92.

Wang, L., Jia, Y., Zhu, X., Wang, X., Li, J., & Gu, H. (2021). Integrating Health into Territorial Spatial Planning: Method Construction and Practical Exploration. *Urban Planning Forum*, No. 4: 81–87. https://doi.org/10.16361/j.upf.202104012

Li, X., Zhou, Y., & Asrar, G. R., et al. (2017). *The surface urban heat island response to urban expansion: A panel analysis for the conterminous United States*. Science of the Total Environment, 605, 426–435.

Murray, C. J., Aravkin, A. Y., & Zheng, P., et al. (2020). *Global burden of 87 risk factors in 204 countries and territories, 1990–2019: a systematic analysis for the Global Burden of Disease Study 2019*. The lancet, 396 (10258), 1223–1249.

OMDL S. (2009). *Global health risks*. Geneva: World Health Organization.

Patz, J. A., Daszak, P., & Tabor, G. M., et al. (2004). *Unhealthy landscapes: policy recommendations on land use change and infectious disease emergence*. Environmental health perspectives, 112 (10), 1092–1098.

Paxton, A. (1994). T*he food miles report: the dangers of long-distance food transport*. London: Safe Aliance.

Planning Department and the Chinese University of Hong Kong. (2012). *Urban climate map and wind environment assessment standard-feasibility study.* Hong Kong: Planning Department of the Hong Kong Special Administrative Region Government.

Schmidt, K. A., & Ostfeld, R. S. (2001). *Biodiversity and the dilution effect in disease ecology.* Ecology, 82 (3), 609–619.

Song, W., & Deng, X. (2017). *Land-use/land-cover change and ecosystem service provision in China.* Science of the Total Environment, 576: 705–719.

Tang Zilai, & Gu Shu. (2016). Re-discussion on social performance evaluation of public green space distribution in downtown Shanghai:From Social Fairness to Social Justice. Urban Planning Forum, (01): 15–21.

Tian Li, Li Jingwei, & Ouyang Wei. (2016). Relationship between urban and rural planning and public health and conception of interdisciplinary research framework. Urban Planning Forum, (02): 111–116.

UN-HABITAT, WHO. (2020). *Integrating health in urban and territorial planning: a sourcebook.* World Health Organization.

Wang Lan, Jia Yinghui, & Li Xiaotian. (2020). *Urban spatial intervention strategies to prevent and control of infectious diseases.* Urban Planning Review, 44 (08): 13–20 +32.

Wang Lan, Liao Shuwen, & Zhao Xiaojing. (2016). *Analysis on the Path and Elements of healthy city planning.* Urban Planning International, 31 (04): 4–9.

Wang Lan, Sun Wenyao, & Gu Jiayu. (2018). Method construction and practical exploration of health-oriented urban design;Take Huangpu District of Shanghai as an example. Urban Planning Forum, (05): 71–79.

Wang Lan, Zhang Surong, & Yang Xiu. (2020).*Influence Analysis of satisfaction with built environment on residents'self-accessed health in new rural communities:Taken 4 Communities in the Outer Suburbs of Chengdu as Examples.* Landscape Architecture, 27 (09): 57–62.

Wang Xinzhe. (2019). *Position and function of territorial spatial master planning in prefecture-level cities.* Urban Planning Forum, (04): 31–36.

Weinstein, J. S., Leslie, T. F., & von Fricken, M. E. (2020). *Spatial associations between land use and infectious disease: Zika Virus in Colombia.* International Journal of Environmental Research and Public Health, 17 (04): 1127.

WHO. (2021-05-19). *Health must be the number one priority for urban planners* [EB/OL]. http://www.who.int/zh/news-room/comments/details/health-must-be-the-number-one-priority-for-urban-planers.

WHO. (2021-5-19). *Physical activity* [EB/OL]. https://www.who.int/news-room/fact-sheets/detail/physical-activity.

WHO. (2021-5-19). *Determinants of Health* [EB/OL]. https://www.who.int/news-room/q-a-detail/determinants-of-health.

Wu Tinghai. (2019). Preliminary study on urban planning in the territorial spatial planning system. Urban Planning Review, 43 (08): 9–17.

Yang, C. Y., Chang, C. C., & Chuang, H. Y., et al. (2004). *Increased risk of preterm delivery among people living near the three oil refineries in Taiwan.* Environment International, 30(3), 337–342.

Zhang Shangwu. (2019). *Topics and prospects of spatial planning reform: thoughts on planning compilation and discipline development.* Urban Planning Forum, (04), 24–30.

Zhang, J., & Smith, K. R. (2007). Household Air Resolution from Coal and Biomass Fuels in China: measurements, health impacts, and interventions Environmental health perspectives, 115 (06): 848–855.

Zhong Weike, Fan Yaobo, & Wang Minjian. (2001). *Pollution and Prevention Measures of Heavy Metal to Crops in China.* Journal of Agro-Environment Science, (04): 270–272.

Zhu Jing, Gao Penghua, & Wu Xiao. (2014). *Study on health effects of commuting behavior of residents in big cities:Take Xi'an as an example.* Urban Planning Forum(06): 46–51.

Urban Organic Renewal and Fine Management

Wu Jiang

Abstract Currently, urban development in China is undergoing a significant transformation. The large-scale old town reconstruction and rapid urban expansion are only short-term and exceptional phenomena under special circumstances in the urban development history. After more than 30 years of rapid and large-scale construction, it is inevitable for urban development to progress from an extraordinary state to a more regular mode, that is, from extensive incremental development to refined stock development. It is the law of urban development.

A city can be compared to an organism, exhibiting vitality through its continuous metabolism, or, the ability of continuously renew itself. However, the typical expression of vitality in urban areas should be the ongoing vitality of cellular renewal, which occurs on a smaller scale of urban spatial units. This renewal consistently follow and develop the unique spatial and cultural genes formed by different factors such as geography, climate, natural ecological environment, economic production mode and human history development track. In urban construction, urban organic renewal is characterized by small scale, gradual progress and persistence, rather than large scale, fracture and operation. This development model is called urban organic renewal.

Cities are influenced by a variety of enduring and encompassing factors, including the external and internal natural environments and factors such as economic, social, and cultural activities. Different cities are bound to have different external features and internal operation logic, thus forming unique spatial and cultural genes of each city. The need for continuous renewal in cities arises due to both the aging and decline of their built environmen caused by time, as well as the unsuitability of their original functional structure and spatial layout for development purposes.. So urban organic renewal is far from just constantly transforming and repairing the material environment, but also needs to continuously stimulate social vitality, maintain social network, repair urban ecology and protect cultural heritage through renewal

W. Jiang (✉)
College of Architecture and Urban Planning of Tongji University, Shanghai, China
e-mail: wujiang@tongji.edu.cn

Z. Peng (ed.), *Theories and Frontiers in Territorial Spatial Planning*, Urban Sustainability, https://doi.org/10.1007/978-981-96-9825-7_12

behavior.[1] Only when such organic renewal activities become a normal state, the city as an organism will last forever. Therefore, urban organic renewal is also an ongoing process that promotes urban vitality.

The goals of urban organic renewal are diverse and complex, and it is a continuous process. The diversity and complexity as well as the continuity of the state necessitate pluralistic and dynamic updates in terms of content, subject, and mode. It involves not only the continuous material repairing and space quality improvement, but also the promotion of intangible function. The conventional planning and design work done by planners and architects is only the starting point of creating high-quality urban space, which cannot run through the whole process of urban development. Therefore, in the new era of stock development, the scope of urban renewal needs to be expanded from "one-time sale" to "lifelong service", that is, extending the traditional design work before construction to the whole life cycle of the designed object. Urban builders and decision makers must accurately predict the entire life cycle of urban development, from planning and development to management, operation, and maintenance. They also need to carry out continuous dynamic monitoring and maintaining urban development process, and respond to various problems in the process timely and accurately. Refinement of urban governance is an important way to achieve these goals. At present, relevant research has noticed the necessity and urgency of urban renewal from material transformation to comprehensive social governance transformation.[2] However, the existing research predominantly concentrates on the "top-level design"with regards to system construction, and there is little systematic discussion on the specific realization path and implementation means of fine management in urban renewal. This paper will sort out the transformation characteristics facing China at present, summarize the main characteristics of organic renewal, and put forward the realization path of organic renewal based on the refinement of urban construction management.

1 From Increment Expansion to Stock Upgrading:Transformation of Urban Development Model

In the early days following the foundation of the People's Republic of China, urban construction primarily focused on rapid basic industrial construction and improving the workers' living conditions. For the original city, it is mainly to make full use of the existing built-up areas and greatly improve its spatial use intensity. As a result, many cities' historical spaces became overcrowded and the original infrastructure is overwhelmed. In the limited urban reconstruction activities, small-scale physical renewal, such as local reconstruction of dilapidated buildings, renovation of old urban roads and municipal infrastructure construction, has been the main content of

[1] Zheng Shiling (2019).

[2] Tang Yan (2020).

China's urban construction until the early 1990s.[3] The tendency of "emphasizing on production and neglecting life" didn't change until the end of 1990s.[4]

The reform and opening up, especially the rapid development of the real estate sector, has triggered the urban construction upsurge in China. At the national level, an unprecedented urban reconstruction and construction movement has been fully rolled out nationwide, marking a shift from the "active and steady" policy mindset. Its scale and speed are unique in the history of urban construction. Big demolition and construction became the normal state of urban development during this period.

During the 1990s, urban renovation and construction projects in China mainly concentrated on improving urban housing and urban infrastructure. Although Mr. Wu Liangyong proposed the concept of "organic renewal" as early as 1994, criticizing the phenomenon of large-scale demolition and construction that had already appeared at that time,[5] it still emphasizes the gradual improvement of the material living environment, and lacking in extensive discussion on the related economic, social and urban vitality. Although "increasing public open space" has become one of the main goals of our urban construction in 1990s,[6] it was mainly to beautify the urban landscape and enhance the image of the city in that period, and even was mistakenly used as political achievements in some places, where a large and improper "municipal square" that far away from citizens' lives was built. However, the social benefits of public spaces, their contribution to economic vitality, and their positive impact on the quality of life are frequently disregarded.

Since the twenty-first century, the perception of urban development in China has changed. The research on urban renewal has gradually gained popularity, leading to a diversified and comprehensive understanding of the topic.[7] Urban renewal is regarded as "multi-objective rather than a single material renewal objective".[8] In recent years, more and more attention has been paid to the improvement of urban renewal on spatial quality. " Providing high-quality public services and facilities and nurturing the quality and vitality of urban built environments have become the "basic responsibilities" of urban renewal for China in the new era.[9] In Shanghai, the focus of urban renewal work has shifted from solely improving living conditions to comprehensive development that encompasses commercial and business sectors, as well as public services. Efforts are dedicated to enhancing the overall quality of urban spaces.[10] In 2014, at the Sixth Planning Land Work Conference in Shanghai, it was clearly stated that the scale of planned construction land should achieve "negative growth", and the transformation of land utilization mode would force the transformation and

[3] Ding Fan and Wu Jiang (2017).

[4] Li Jianbo and Zhang Jingxiang (2003).

[5] Wu Liangyong. (1994).

[6] Wu Binghuai (1999).

[7] Zhang Gengli (2004).

[8] Dong Mali et al. (2009).

[9] Wang Shifu and Shen Shuangting (2015).

[10] Zhuang Shaoqin (2015).

development of the city.[11] In 2015, the promulgation of the *Measures for the Implementation of Urban Renewal in Shanghai* marked that urban renewal with the goal of improving stock quality officially became the core work of Shanghai's urban development. The first Shanghai Urban Space Art Season in 2015 took urban renewal as its theme, bringing professional attention to this issue and promoting it to the public's attention. In 2017, Shanghai completed the "*Shanghai Urban Master Plan (2017–2035)*" (referred to as "Shanghai 2035"), which identified "Innovative City, Humanistic City, Ecological City" as the three major goals for urban development. The concept of urban renewal was further clarified in the official document.

2 Five Bottom Lines Abided for Urban Development Under the Stock Model

The city is a large and intricate system, and ongoing academic discussions regarding urban renewal exist. However, it is crucial to establish a general consensus on certain matters promptly. One such matter is the need for urban development to comply with specific constraints and standards during the course of urban renewal.

2.1 Bottom Line of Land Resources

The bottom line on land resources is still in dispute within the academic circles. However, when societies experience rapid growth and population expansion, spatial capacity is significantly affected. In the process of rapid industrialization and urbanization, most of the western countries have paid a huge price for breaking through the constraints of space. Their urbanization process is closely related to spatial constraints. However, the urban development speed in China today is much faster than that in western developed countries (Fig. 12.1). The rapid consuming of land and other natural resources in this process has greatly impacted the safety of farm land and ecological security. In Shanghai, the total area for construction land and the city's population size had already neared or even surpassed the targeted goals for 2020 as early as 2014, which has raised concerns and debates among academics.[12] Resource limitation is one of the critical reasons for the transformation of Chinese urban development. Urban development must confront the of resources limitation and make efficient use of existing built environments. This approach allows for a greater utilization of land to support urban development.

[11] Zhou Jian et al. (2019).

[12] Zhuang Shaoqin (2015).

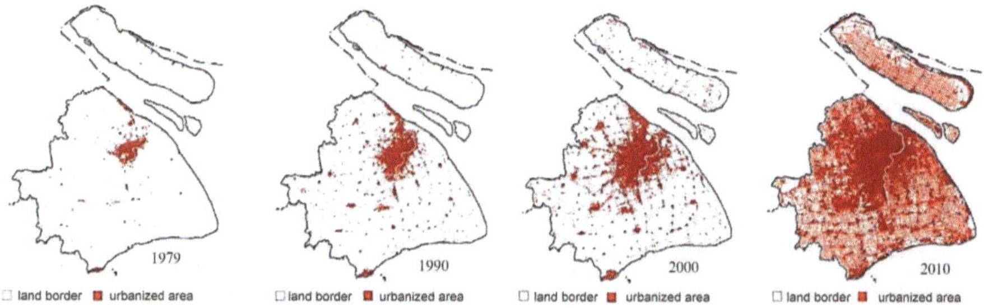

Fig. 12.1 The scope of Built-up Areas in Shanghai in 1979, 1990, 2000 and 2010. *Source* Author's self-drawn

2.2 Ecological Livable Bottom Line

The goal of a city is to establish an optimal man-made setting for human life and production activities. Rapid urban development leads to a series of environmental problems, including air pollution, waste of resources and ecological degradation. The degradation of ecological conditions will not only destroy the ideal living environment of human beings, but also reduce the livability of cities, weaken the attraction of investment, increase the probability and frequency of disasters, and then inhibit the economic and social development of cities. In this context, the 2013 Central Urbanization Work Conference put forward the specific requirements of "appreciating natural scenery and recalling memories of home" for China's current urban development. The 18th National Congress of the Communist Party of China pointed out that we should vigorously promote ecological civilization construction. In the report of the 19th National Congress of the Communist Party of China, the goal of improving the development quality of new urbanization and taking the high-quality development path of green, intensive, efficient, low-carbon, innovative and intelligent urbanization appeared continuously. These measures indicate that China's urban construction is transitioning from an outstretched and over-consumed model characterized by production factors and investment to a new model of coordinated development of economy, society and environment, which is more suitable for the stock model and characterized by innovation and wealth.[13] In such a trend, urban renewal work needs to consider the existing urban space as an integral part of the urban ecosystem. The harmonious development of cities and ecosystems is the primary task and bottom-line requirement for archieving urban sustainable development.

[13] Duan Jin (2021).

2.3 Historical Humanistic Bottom Line

City serves as a crucial spatial carrier of human civilization, originating from its continuous practicing and accumulation. The form of civilization takes time, but it is also easy to be destroyed. While in the process of urban renewal, historical areas with long-term accumulation are more likely to become the initial renewal target and be destroyed. Although the revised "*Cultural Relics Protection Law of the People's Republic of China*" in 2002 clearly stipulates the protection of historical and cultural districts as a whole. However, there are still many ordinary historical buildings and historical districts in cities that do not fall under the category of cultural relics but hold rich historical and cultural information. The conservation value of the built environment depends on its embodiment of the overall social and cultural identity and its tangible role in representing historical and geographical dimensions.[14] Therefore, many commonplace structures in the historic district, serving as the carrier of material spatial form in a specific period of the city, also have the value of overall protection. Due to insufficient understanding of the importance of a large number of ordinary buildings in the old town, also driven by economic interests and guided by the wrong concept of cultural heritage protection, some local governments tend to pay one-sided attention to the economic benefits of land, and carry out destructive construction behaviors such as large-scale demolition and construction, demolition of old buildings and new buildings in the old town (Fig. 12.2). Shanghai has suffered significant consequences from demolishing about 30 million m^2 traditional buildings.[15] The 2013 Central Urbanization Work Conference clearly pointed out that the current urban development should prioritize historical and cultural preservation, natural and regional characteristics, contemporary style and Chinese traditions. At the 2015 Central Urban Work Conference, the central government also put forward the requirement of protecting "regional environment, cultural characteristics and architectural style genes". On May 21, 2021, General secretary Xi Jinping presided over the 19th meeting of the Central Committee for Comprehensively Deepening Reform. He pointed out that efforts should be made to solve prominent problems such as repeated destruction and demolition of historical and cultural heritage in urban and rural construction, strengthening the top-level design of the system, make overall plans for protection, utilization and inheritance, and adhering to systematic and complete protection, so as to protect not only single buildings, but also streets and towns, as well as historical areas, natural landscapes and humanistic environment. In this context, the preserving urban historical and cultural heritage and achieving a balance between future urban development and past history are fundamental requirements for current urban renewal work.

[14] Chang Qing (2018).

[15] Zheng Shiling (2017).

Fig. 12.2 Large-scale demolition and construction projects can indeed disrupt the overall historical context of urban areas. *Source* Author's self-photoed

2.4 The Bottom Line of Social Justice

After long-term accumulation, the old town has often formed a stable "social network" and "social ecological chain". Urban renewal differs significantly from new district construction as it has the potential to modify formed social networks and social ecological chains. As these social networks contain complex property rights and interests, urban renewal is actually a process of redistributing spatial interests.[16] Urban renewal should pursue the balance of interests among government, society, developers and residents, which is also the core work of land administration reform.[17] The report of the 19th National Congress of the Communist Party of China pointed out that "it is the true meaning of people's democracy to have something and someone to discuss with". Sustainable and organic urban renewal needs to rationally adjust the balance and game of interests in the process of renewal, and realize the fairness and legalization of the public interest guarantee mechanism. In particular, shantytown renovation, old residential area renovation and other livelihood projects closely related to the vital interests of citizens. However, the process of urban renewal through large-scale demolition and construction has led to the removal of longstanding living spaces and social networks within old districts. But these living spaces and social networks play a crucial role in providing low-income and socially disadvantaged groups with the opportunity to reside in the city center

[16] Tang Yan (2020).

[17] Zhuang Shaoqin (2015).

and have fair access to urban public resources. They serve as an important guarantee for social equity and justice. Urban renewal must not only pay attention to the economic interests of capital investment and local finance, but also pay attention to the overall public interests of cities and keep the bottom line of fairness and justice.

2.5 The Bottom Line of Public Safety

At last, it is necessary to emphasize the bottom line of public safety in the process of urban renewal. Security is integral to any urban development, as without it, everything is at risk. When faced with various disasters and security challenges from nature and human, cities should become shelters rather than boosters. The high concentration of urban population requires extremely high management wisdom and management mechanism to cope with various security challenges, and also needs to make adequate preparations in terms of space resources and hardware facilities. Failure to do so would result in severe consequences. To realize the sustainability development of the city, we must keep these bottom lines.

3 Normalized Urban Renewal: Organic Renewal

Under the constraint of the five bottom lines, China's urban renewal activities face increasing transformation pressure. How to realize the sustainable urban development while taking into account resources, ecology, history, society and security, and without damaging any dimension? How to improve residents' living standards while safeguarding the history and culture? How to achieve economic gains while promoting community development and protecting the ecology?? The responses to these inquiries lead to an "organic" urban renewal model. How to define weather one urban renewal model is "organic"?Let's follow the three aspects below.

3.1 Small-Scale

At first glance, small-scale refers to architectural projects or spatial scales, but in fact it is discussing urban land administration. The purpose of small-scale is to preserve urban spatial texture, which contains urban spatial genes. In the past, land administration for urban redevelopment often took the whole neighborhood or even multiple neighborhoods as the basic unit of unified development. This development mode played an clear role in enhancing the capital's effective for development and operation, but the city's spatial texture was completely wiped out in this process. Small-scale means more detailed and dynamic management. If the scale of land use can be reduced and the nature of land use can be diversified at the same time, it become possible to preserve urban spatial texture, sustain collective memory and release block vitality.

3.2 Progressive

Progressive is achieved through small-scale developments. In fact, whether it is a traditional natural village, a Chinese residential building in the old Shanghai or a lane in the concession, construction takes place gradually. Taking Lane as an example, except for the super-large Lane projects such as Sven Lane, which occupy the whole block, the most Lanes collage from several projects charged by different developers, and the changeable and natural texture also appears. Nostalgia is not an end, and new buildings are not unacceptable. A constructive outlook is necessary for urban renewal and construction with the caveat of keeping limited scale in mind. The maintenance of urban spatial texture will have the opportunity to go hand in hand with the renewal of buildings, that is, single units can be rebuilt and built, but Key information like lanes, orientations, and density are preserved.

3.3 High Quality

After achieving the first two aspects, high quality will surely become a vital element of organic renewal standards. Whether it is for the residents living in it or the city's overall style, it requires long-term continuous improvement. In addition to cleaning old toilets, installing stair handrails and lighting, new technologies in the Internet age will also provide solutions that are more suitable for the needs of contemporary life and society. However, achieving high-quality outcomes relies heavily on the architects, planners, social workers, volunteers and other groups involved. And the efficiency of stakeholder communication and the final effect of implementation may become an impact factor in achieving it.

4 The Main Bottleneck of Realizing Organic Renewal

Organic renewal mode is the guide of future urban renewal in China. However, there are currently three bottlenecks in the construction and engineering technology of the existing system that need to be addressed to achieve organic renewal.

4.1 Lack Considering on Existing Spatial Elements in Planning Technical Method

The current urban planning technical of China, particularly regulatory detailed planning with legal effect, only have a control effect on new incremental construction, and lack of attention to spatial elements with protection and preservation value. For urban central areas, especially the high-density urban central areas of megacities like Shanghai, there are often a large number of historical and cultural heritage resources and a large number of natural and reasonable spatial elements. How to break through the existing technical mode of urban planning and construction, resolve the contradiction between construction development and historical preservation, fully consider all kinds of existing spatial environment elements of cities and put them into the urban construction management system to shift from a simplistic towards a complex system of management. and transform urban construction management into urban operation and maintenance management, which is a major bottleneck for Chinese cities to realize organic renewal at present.

4.2 Lack of Accurate Dynamic Information Support in Decision-Making and Management

After over 30 years of large-scale construction activities, Chinese cities generally show the characteristics of high building density, numerous elements and complex spatial relations due to the superposition and changes of built-up environment in different ages. Meanwhile, the urban management department lacks sound and systematic basic data, and the old information collection and processing technology can neither provide accurate, complete and dynamic three-dimensional basic information of existing built environment for reconstruction and construction activities, nor provide flexible and intuitive decision support for the decision-making and management of new planning and construction schemes, which can cope with complex time and space environment. Integrating diverse geographic information through digital technology to realize fine collection, dynamic monitoring, intelligent extraction and management, and visual interactive display of high-density built-up environment three-dimensional information and health status, and provide basic information support for urban planning and construction, is the main bottleneck in the basic data acquisition level of urban organic renewal in China at present.

4.3 *Difficult to Cope with the Multiple Constraints' Existing Environment for Project Implementation*

Urban organic renewal tends to activate and reuse existing buildings through small-scale development and construction, and mostly adopts the construction methods of finding opportunities and integrating old and new. In the past, the technical methods of statutory planning put too much emphasis on the effective control of urban construction in terms of "quantity", such as density, height, floor area ratio, etc. However micro-scale "small moves" lacked refined control means. At the same time, the construction under the complex conditions of high-density environment is also faced with many problems, such as high safety risks, narrow and low operation space, and the influence of operation process on the daily use of existing buildings, which are subject to multiple time and space constraints. It is the major bottleneck in engineering implementation to realize Urban organic renewal as to develop new plans and engineering technology that can deal with the development and construction of small-scale space under sensitive environment and multi-constraint conditions, promoting the renovation and reusing of existing buildings, and improving the overall quality of the built environment while minimizing the impact and intervention on existing urban areas.

5 Important Path to Realize Organic Renewal: Fine Management

Improving urban management is vital to overcoming the aforementioned obstacles. The system is a comprehensive systematic project. It extensively involves the transformation design of urban material space, the accurate monitoring and simulation to existing built environment, and the fine control and management of construction process.

5.1 *Full-Cycle Multi-factor Collaborative Spatial Planning*

As a statutory planning, the technical index system set by regulatory detailed planning has shaped the basic urban spatial form, which is the most critical breakthrough in the transformation of urban planning technical system in China. The index system proposed by the traditional regulatory detailed planning can only control and guide the incremental-oriented new development, failing to adapt to the new normal, which emphasizes urban maintenance while considering protection and renewal under a stock-oriented approach. On the basis of years practice summary, the team put forward the technical method of fine planning of all elements of built environment,

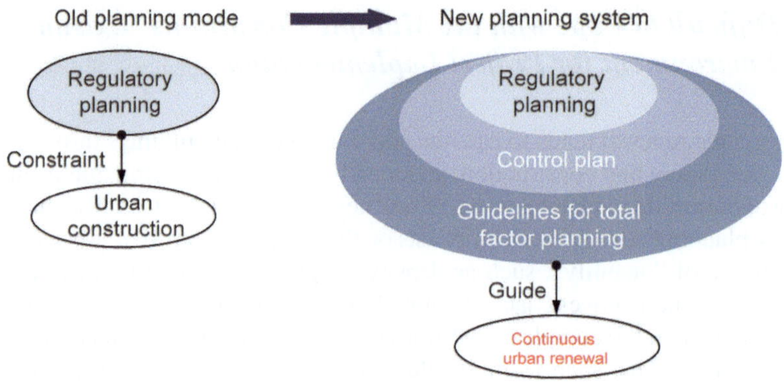

Fig. 12.3 Compilation system of organic renewal planning composed of control planning, control drawing and total factor planning guidelines. *Source* Author's original illustration

which integrates protection and renewal. At the level of regulatory detailed planning, the problem of incompatibility between heritage protection and renewal and construction in existing urban areas is solved. Then a multi-level refined planning methodology is proposed that incorporates regulatory planning, control charts, and total factor planning guidelines. It aims to break through the bottleneck that the existing planning can only control large-scale construction but cannot take small and micro-scale stock transformation into account (Fig. 12.3).

First of all, the regulatory detailed planning under the guidance of organic renewal takes the control orientation under the background of integrity, authenticity, sustainability and classified protection as the planning principle, and on this basis, a set of planning index system covering different style types and landscape scales was established (points, lines and areas), and integrating protection and renewal into one. The system incorporates numerous planning and control elements for historical preservation, which including specific conservation requirements for historical buildings and planning and construction requirements for each existing building, as well as the spatial texture of the land parcel.. In addtion, the system consist of planning and control indexes, such as building capacity, scope of construction, building density, building height, plot ratio, green space ratio, and public facilities, etc. and establishes detailed zoning, classification and categorization standards for the organic renewal of the existing urban areas. Based on this, computer technology can assist in establishing a method and software to automatically generate a three-dimensional city model using index parameters. to quantify the relationship between design and regulation of city style quality and renewal and construction intensity.

Secondly, based on multi-department collaborative management, a set of technical methods of all factor fine planning was put forward. It takes the street space network as the carrier and focuses on the "pedestrian space" of the city. In order to break through the limitation of urban road planning method with motor vehicle traffic as the core, all the built-up environmental factors are divided into 3 categories (namely building elements, public spatial elements and environmental elements) and 18 subcategories,

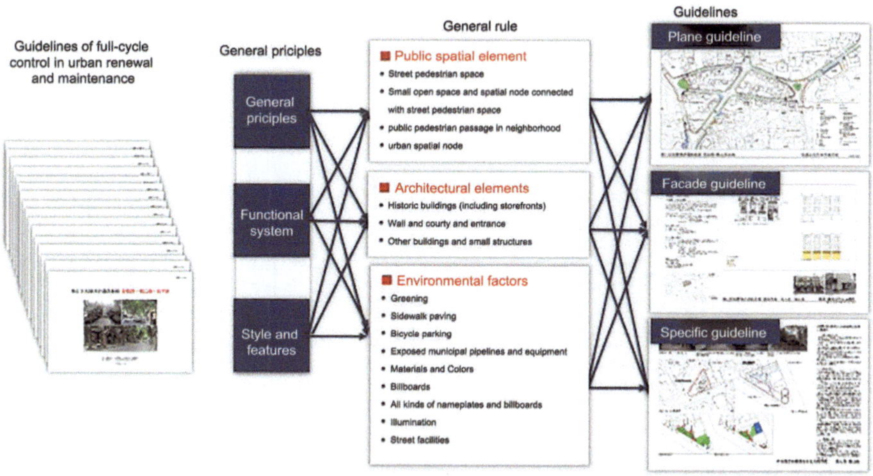

Fig. 12.4 Based on General Principles-General Principles1Multi-level refined planning framework of guidelines. *Source* Author's original illustration

so as to establish a systematic analysis model. Plans are adopted to achieve effective management and control based on the framework of general-principles-guidelines (Fig. 12.4). This set of technique also includes a multi-department collaborative management technical platform composed of basic information database and planning information database, so as to overcome the problems of lack of systematic basic data support and action basis when different government functional management departments formulate urban governance projects and implement remediation measures in a compartmentalized state, as well as the lack of necessity and rationality, without combining, repeat investment and so on. Using a refined set of planning techniques, the team conducted an exploratory study of roads protected by statutory landscape regulations in Shanghai.

5.2 *Information Acquisition and Decision Support of Total Factor Digital Twin*

Mastering the detailed information of urban built environment is the premise and foundation of fine planning and construction. Facing the more and more complex existing urban areas, the traditional static spatial information collection and management methods of manual means and two-dimensional drawings are inefficient, cumbersome and inaccurate, which can not meet the requirements of dynamic urban planning, construction and management decision-making. Cooperating with Shanghai Construction Engineering Second Construction Group,

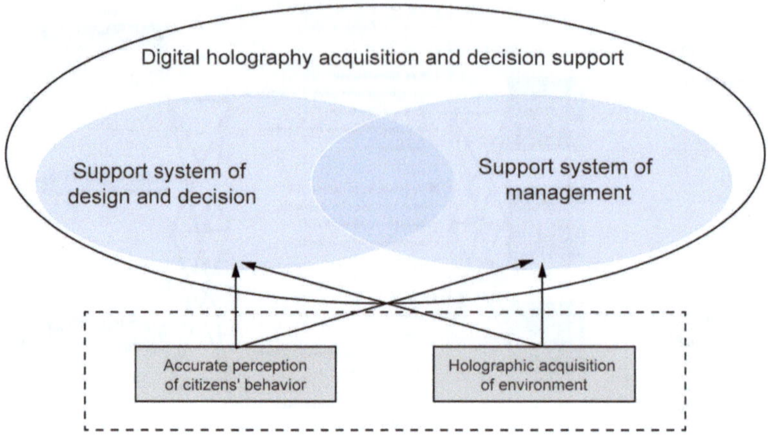

Fig. 12.5 Decision support system based on accurate perception of citizen behavior and holographic acquisition technology of physical eenvironment. *Source* Self-painted by the author

Shanghai Surveying and Mapping Institute, Shanghai Geology and Mineral Engineering Survey Institute and other units, the team established a Comprehensive application system. Through the comprehensive utilization and integration innovation of surveying and mapping technology and digital technology, this system can get high-precision collection of three-dimensional information of built environment, accurate perception of citizen behavior. Combine with intelligent design decision support system (DDSS) and management support system (MSS), it provides information support forUrban organic renewal (Fig. 12.5).

With the help of the multi-machine collaborative network, the system can collect and dynamically monitor the built environment information with high precision, so as to build a refined three-dimensional model and high-precision digital archives with millimeter accuracy in high-density complex urban areas. Meanwhile, various methods such as infrared thermal imaging and structural nondestructive testing are adopted comprehensively to carry out fine analysis and emergency mapping on the deformation of existing buildings, so as to realize the investigation of the current diseases of buildings, and realize all-round CT visual intelligent detection and dynamic monitoring on their health status. In addition, the Web GIS monitoring system provides two dimensions of spatial and temporal monitoring and early warning and decision support model for design and management decision-making. Using high-precision dynamic monitoring and early warning platform, it overcomes the problems that conventional system can't meet the requirements of large data volume, real-time fast matching and emergency decision-making service, then provides support for a new urban renewal planning.

Complex environmental information requires intelligent management and visual decision support. Based on 3D dynamic layer technology, the team, Shanghai Institute of Surveying and Mapping and other cooperative units, accurately integrate building information model (BIM) and geographic information system (GIS), and carry out multi-source, multi-scale and multi-temporal refined system management for the whole life cycle of urban 3D model through semantic-based BIM information screening and extraction tools and OGC standards and metadata, so as to realize intelligent extraction and efficient management of urban 3D information. In addition, with the help of interactive augmented reality and digital twinning technology, this decision support system can also realize collaborative interaction and visual display between urban 3D solid model and built environment performance data, which obtained by virtual scene calculation, and provide highly intuitive decision basis for decision management of urban planning and construction (Fig. 12.6).

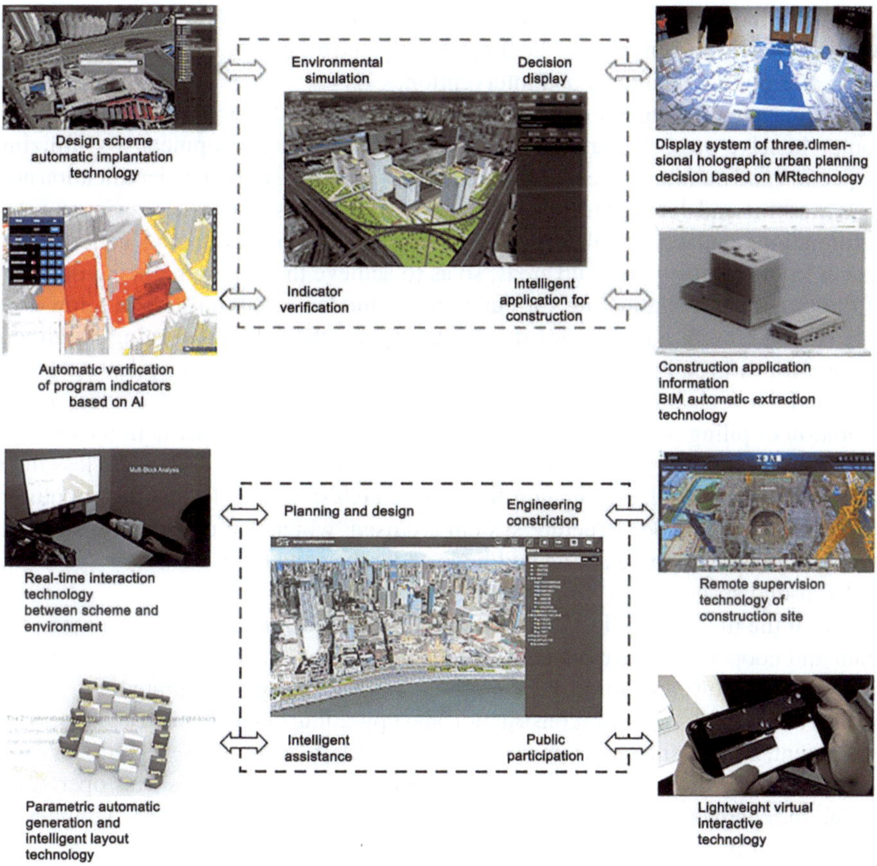

Fig. 12.6 Intelligent Platform for Decision Support of Urban Construction (Part I) and Management (Part II). *Source* Self-drawn by the author (the photo in the picture is provided by Shanghai Surveying and Mapping Institute and Shanghai Construction Engineering Second Construction Group Co., Ltd.)

5.3 *Refinement Engineering Implementation of Multi-Dimensional Tight Constraint*

The existing areas of the central city are more sensitive to urban renewal and construction activities. Thus, in the process of regeneration, urban construction and development face a number of constraints, such as height and width limitations on the space for construction operations, contradictions between construction activities and the daily operation of the city, and the destruction of the cultural heritage and ecological quality of high-density existing urban areas due to rough regeneration works. To this, the team and the cooperative units have developed a comprehensive and refined construction and environmental remediation, with the goal of realizing organic renewal and enhancing the functionality and quality of the built environment while minimizing its impact and intervention.

To renovate existing buildings constrained by space, the technical system makes comprehensive use of a variety of non-destructive testing equipment. By utilizing inspection and monitoring technical methods, as well as risk source identification and early warning and disposal mechanisms, the system precisely predicts environmental deformations and structural settlement in the process of developing underground spaces in high-density built-up areas, so as to achieve the goal of reducing the risks of the project and eliminating various types of hidden safety hazards in a low-cost way (Fig. 12.7). In addition, for the limited space clearance in existing buildings, the pile-forming equipment of common pile types such as steel pipe piles and bored piles under low clearance operation is used to realize the highly refined construction operation of piling in existing buildings. Simultaneously, the construction technologies including top-down construction, precast column construction, beam less floor construction of self-supporting non-linear basement structure, Tian Zi beam-column cap and beam less floor structure system are used, which can realize fine, safe and efficient construction of existing buildings under the constraints of narrow space and underground space (Fig. 12.8).

As for the time constraints imposed by urban renewal in existing urban areas, the team and cooperators have adopted various techniques, including safety-controlled multi-person flow and multi-zone segregation technology, integrated fire safety control technology, non-stop construction and upgrading of mechanical and electrical equipment, and on-site replacement of existing curtain walls. These techniques offer solutions to the huge economic loss due to unable to maintain business operations during the renovation and upgrading of commercial facilities with high volumes of pedestrian traffic in existing urban areas. By adopting this approach, it becomes possible to minimize such significant economic losses and enable continuous renovation and construction without any disruptions to existing operations.

In urban renewal, polluted water is among the existing environmental challenges. To address this issue, refined technologies such as microbial solidification and ultra-fine air flotation, along with other equipment, have been deployed. The adopted method involves using ultra-high pressure air–water mixing. The generation of large

Fig. 12.7 Accurate Monitoring of Ground Deformation in Existing Urban Areas. *Source* Provided by Shanghai Geology and Mineral Engineering Survey Co., Ltd

Fig. 12.8 Refined urban renewal construction method under strong spatial constraints. *Source* Povided by Shanghai Construction Engineering Second Construction Group Co., Ltd.)

numbers of bubbles at the micron, submicron, and nanometer levels in an over-saturated condition provides inexpensive and highly efficient removal of microbial contamination and enables environmental remediation during urban redevelopment, contributing to the protection and restoration of urban ecology.

6 Conclusion

A city functions like an organism with an inevitable birth and eventual demise. But cities should have a life cycle much longer than that of people. Urban renewal activities are the normal state throughout the whole life cycle of urban development. Urban organic renewal should adhere to the fundamental principles of urban development and respect the inherent characteristics of urban living. In its process, the normalized vitality of urban vitality is mostly manifested in the living body at the cellular level, which is always present and uninterrupted, but imperceptible. In this normalized process of Urban organic renewal, alongside the spatial structure adjustment and important leading construction activities aimed at improving major urban functions and upgrading urban economic energy level, a significant number of normalized "micro-renewal" activities and acupuncture-style rectification and improvement are required on a periodic basis.. This kind of renewal activity focus not only on the quality of urban material space, but also on the special attention and continuous maintenance of resources and environment, ecology, society, history and humanities. In addition to the foreseeable urban renewal activities under the guidance of planning and policies, a large number of unpredictable and spontaneous natural renewal processes often occur in cities. The refined mechanism design of urban management should also fully respect this kind of natural renewal activities. The vitality of cities can only be maintained through compliance with urban development regulations and the provision of sufficient flexible space for natural regeneration activities.

(Thanks to all colleagues of Tongji University Institute of Fine Management of Megacities (International), Shanghai Construction Engineering Second Construction Group, Shanghai Institute of Surveying and Mapping, Shanghai Institute of Geology and Mineral Engineering Survey and other units for their joint efforts in long-term cooperation.)

References

Wu Binghuai. (1999) *Retrospective analysis and development suggestions of urban renewal theory and practice in China.* Urban Research, (05): 46-48.

Ding Fan, & Wu Jiang. (2017). *Evolution of concepts related to urban renewal and its practical significance today.* Urban Planning Forum, (06): 87-95.

Zhang Gengli. (2004). *Towards a tripartite partnership:The Evolution of Western Urban Renewal Policies and Its Enlightenment to China.* Urban Development Studies, 11 (04): 26-32.

Zhou Jian, Yan Shuxin, & Wan Zhiying. (2019). *Thoughts on Perfecting Shanghai's Urban Renewal System.* Urban Planning Forum, (01): 20-26.

Li Jianbo, & Zhang Jingxiang. (2003). *A comparative study on urban renewal and evolution between China and the West.* Urban Problems, (05): 68-71.

Duan Jin. (2021–4–26). *Urban design, shaping the territorial space with high artistic quality.* China Natural Resources News (001).

Wu Liangyong. (1994). *Beijing Old Town and Juer Hutong.* Beijing:China Architecture Industry Press.

Dong Mali, Chen Tian, & Wang Liyan. (2009). *Development process and policy evolution of urban renewal in western.* Human Geography, (05): 42-46.

Chang Qing. (2018). *The future of the past:Critical cognition and practice on the problem of built heritage.* Architectural Journal, (04): 8-12.

Zhuang Shaoqin. (2015) *Intensive and economical use of land in Shanghai under the "new normal".* Shanghai Land & Resources, 36 (03): 18.

Zhuang Shaoqin. (2015). *Intensive and economical land use in Shanghai under the "new normal".* Shanghai Land & Resources, 36 (03): 1-8.

Wang Shifu, & Shen Shuangting. (2015). *From Renovating Three Old to Urban Renewal; Thoughts on the establishment of urban renewal bureau in Guangzhou.* Urban Planning Forum, (03): 22–27.

Zheng Shiling. (2017). *Urban Renewal and Historic Building Protection in Shanghai.* Bulletin of Chinese Academy of Sciencess, 32 (07): 690-695.

Zheng Shiling. (2019). *Thoughts on urban renewal in Shanghai.* Architectural Practice, (07): 8-11.

Tang Yan. (2020). *Strengthen system construction and promote urban renewal;From simple material transformation to comprehensive social governance.* Environmental Economy, *(13)*:39-43.

Frontiers of Modern Urban Design Theory

Tong Ming

1 The Cutting Edge Trends in Contemporary Urban Development

1.1 Trends in Urban Development

The cutting-edge areas of modern urban design are related to current urban development trends.

Since the 1960s and 1970s, cities and social spaces worldwide have been undergoing a major structural transformation. This structural shift has coincided with the fastest and most extensive urbanization in history. By 2008, it was the first time in history that more than half of the world's population resided in urban areas. By 2030, the urban population is projected to reach 5 billion, accounting for 60% of the global population. From a global perspective, future population growth will be concentrated in cities, and almost exclusively in developing countries, particularly China and India.

Essentially, at the core of this structural transformation is a new technological revolution organized around information technology. Just as the profound impact of the industrial technological revolution of the seventeenth and eighteenth centuries on traditional cities (Fig. 13.1), this new technological revolution relies on the infrastructure and economic mode formed by information technology, and the resulting globalization process, which is changing the way of production, consumption, management, communication and the thinking way of human society.

Furthermore, China's current urbanization process is not only a natural result of the rapid socio-economic development, but also an inevitable manifestation of the transformation of socio-economic structures in the information age. It is just

T. Ming (✉)
School of Architecture, Southeast University, Nanjing, China
e-mail: tongming@seu.edu.cn

Z. Peng (ed.), *Theories and Frontiers in Territorial Spatial Planning*, Urban Sustainability, https://doi.org/10.1007/978-981-96-9825-7_13

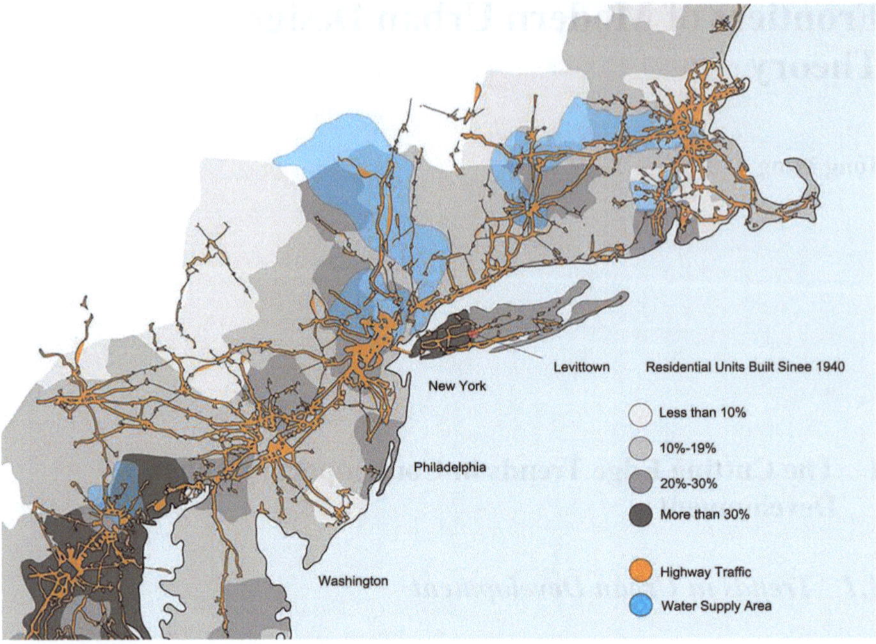

Fig. 13.1 Jean Gottmann, Metropolis, 1961. This map depicts the urban development trend that emerged in the northeastern United States in the second half of the twentieth century and spread all over the world. In the graph, the system composed of expressway, suburban expansion, and water supply intertwined into one, bringing individual cities together into a holistic urban community. *Sources* Repaint from GOTTMANN J. Megalopolis (1961). *The Urbanized Northern Seaboard of the United States*. New York: Twentieth Century Fund, 1961: 233

because the technological innovation led by information technology has brought revolutionary new ideas and methods to urban infrastructure, leading to significant transformations in many geographical regions and urban spaces within the process of globalization. It not only rapidly changes the pattern between cities and villages in terms of quantity, but also adjusts the traditional regional spatial model in terms of quality. It has profoundly affected the social and economic system of cities, and gave rise to new types and qualities of urban space as well.

From a macro perspective, the level and quality of urbanization in China have shown regional variations. Although the pace of urbanization in the central and western regions has been accelerating in recent years, metropolitan areas or urban sprawl zones have gradually emerged in the eastern coastal areas. The advanced transportation and information systems have facilitated the connection of these metropolitan areas across vast territories and also linked them with similar regions around the world, resulting in a continuous improvement in internationalization and an increasingly significant process of globalization.

From a city-level perspective, great changes have taken place in the spatial structure and appearance of many cities with the development of the social economy, but at the ground level, there are profound transformations in industrial structure, professional structure, consumption structure, and lifestyle. With the rapid popularization of information technology, social development is rapidly networked, showing a multi-level and multi-dimensional deductive process. During this process, many cities have shattered the original concept of urban areas, breaking through the original physical space, and turned to suburban expansion. The flow space formed by information networks is gradually replacing the traditional and stable urban space.

Simultaneously, new trends in urban development which are more and more different from the past, are constantly emerging, mainly reflected as follows: macro-market-driven fluctuations in industrial structure, the liquidity and speculation of investment and home ownership are gradually increasing, and the restrictive role of urban space is generally reduced; With the continuous improvement of social networking, citizens' autonomy and mobility are increasing day by day; Infrastructure is insufficient; Urban social composition is continuously diversifying, with increasing segregation. The more frequent population mobility leads to the difficulty of urban public service and management, and the government departments are increasingly lacking sufficient resources and capabilities to solve urban problems, which has brought great influence and impact to the original urban social structure and planning management.

These development trends are often perceived as negative consequences of urban development due to their frequent departure from the conventional scope of understanding. But from another point of view, they also represent a profound transformation of information and globalization that is taking place in many cities and will become more and more prominent in the future of social development and economic life. To a certain extent, the arrival of the information age is making the original urban social space lose their existing significance. Information technology is a technological revolution with historical significance, which not only significantly improves the efficiency of economic production, but also gradually eliminates the spatial distance between various fields of social life, thus challenging the structural change of urban social space.

Some new urban places become popular since the beginning of the twenty-first century, such as large-scale comprehensive shopping centers, suburban business groups, characteristic high-tech towns, etc., have gradually occupied a prominent position in urban life, thus announcing the birth of a kind of "placeless". The experience of being in these commercial centers or in the crisscrossing transportation system is similar to that of the rest of the world. The resulting urban spatial experience is one of places and locations that are free from hierarchy and differentiation, dotting the extensive network of expressways that connect in all directions.

Fig. 13.2 Tamashita Kenzo, Tokyo Bay Project, 1960. The city is evolving towards a colossal structure. *Sources* LIN Z (2010). *Kenzo Tange and the Metabolist Movement, urbanutopias of modern Japan*. New York: Routledge, 2010: 137

1.2 Trends in Urban Design

Compared to the traditional urban design, the current urban development reality poses a more profound challenge to urban space and urban design. Since the twenty-first century, a key issue in urban development has been the intersection of economic globalization and social networking, which not only ushers in a new era different from the past for urban research but also makes most topics related to urban design lose their original background and significance (Fig. 13.2).

In such a brand-new historical context and development environment, some topics in traditional urban design theory, mainly focusing on city image and space quality, are undergoing a major turning point in their practical context.

(1) Issues regarding functional attributes

In the discourse on functional attributes, the consistency traditionally espoused in urban spatial planning is disintegrating, and the corresponding between economic activities and spatial carriers is becoming more detached. The new technology based on the dual revolution of communication systems and microelectronics has transformed space and place into flows and channels.[1] Simultaneously, with the rapid development of fast traffic and transportation system, the economic trade and its organization form are becoming more and more global and flexible, and the rapid flow of investment and capital leads to the great reduction of the limiting role of geographical location.

[1] Custer, Wang Zhihong (2006).

In this new trend of industrial economy networking, urban consumption is also gradually getting rid of the constraints of physical space factors. Non-spatial methods such as the Internet, television, telephone and so on are becoming people's habitual shopping methods, and convenient logistics systems are reducing the frequency of people entering urban street space. Technically speaking, large shopping malls have fallen behind, not to mention the traditional business environment. Therefore, urban planning contents such as "spatial location", "integration of production and city" and "layout research" in the traditional sense will lose their efficiency, and the design method that pursues stable spatial correspondence will face the challenge from a fragmented and fractured urban landscape, and the relationship between architecture and spatial layout in the traditional sense will be gradually replaced by the invisible network constructed by information technology.

(2) Issues related to social dimensions

In the discourse of social dimensions, the organizational relationship between capital and labor has undergone structural changes in a reconstructed spatial economy. In this spatial economy, the geographical imbalances of the original socio-economic development are being reconstituted in a new and often unpredictable way, and the urban population will continue to migrate and reorganize according to the increasingly specialized spatial adjustment. A significant influx of rural people who have just been released from the rapid urbanization process are flocking to developed coastal cities, while the eastern region is accordingly relocating those outdated factories and industries to the land-locked western regions. The frequency of population mobility in cities is far higher than before, and social isolation is becoming increasingly obvious. The public space, which needs to be considered emphatically, can no longer become a quiet and pleasant environmental container, and the planning and layout involving spatial fairness, such as community supporting facilities and resource allocation, which need to be considered emphatically, are also difficult to implement.

At the micro level, the separation between urban affairs and spatial forms has led to a clearer distinction between citizens' lives and the meanings of the city. The classic models such as social security space and defensible space, which need to be considered emphatically in urban design, have given way to the camera monitoring system all over the streets, and the community space is constantly degenerating due to the loss of substantive public communication.

(3) Issues related to cultural topics

In discussions on cultural topics, the appearance of space of flows replaces the basic meaning of space of places, and what is about to disappear is the place meaning that exists because of citizens. Urban space is more and more closely linked with the development of a "No Place" environment.[2] Every city and every place will acquire its own social value by turning to its hierarchical position in the social network, and the functioning and rhythm of this social network will also be detached from these places, even from the people in these places.

---

[2] Su Jia (2004).

In such an environment, the discussion of historical and cultural preservation has lacked substantive significance, and the new urban development is ruthlessly sweeping the traditional community. Even for the strictly preserved urban historical environment, the new urban significance is leading people to break with the historicity and culture of their space. The "homeland" no longer exists, and the identity characteristics attached to the place will disappear. The rapid and chaotic movement of people and material gradually replaces the original place, the original context and the original cultural implication, which makes this socially alienated and individual closed space become more and more undifferentiated. The living beings are transformed into an abstract concept, and the real environment becomes an illusory shadow.

(4) Issues related to public affairs

In the discourse of public affairs, global capital flow and the rise of network society have become the main themes of the current era. More and more abstract capital can come to a certain piece of land at will, which stimulates more violent speculative urban development. There is no longer a need for a close correlation between urban development projects and their places, and there is no need for coordination and unity between one urban space and another in theory. As a consequence, the aggregation of public resources is weakened, social consensus is diminished, the reality of urban space is often beyond the control of governments, and the leverage of traditional economic policy is weakened. The development reality of urban space often exceeds the control scope of the government, and the leverage of traditional economic policies has become extremely weak. It is in such an increasingly popular contemporary urban scene that the current urban space has begun to be filled with a non-integral and irregular atmosphere, which has spread widely with the increasing globalization process, and its operation mechanism is difficult to really explain.

The problems brought by this trend to current urban design are: is it still possible to have a spatially oriented urban design in this context? If it still exists, what would it be like? What common demands can it still have with previous urban designs? Who will be responsible for them, how will they be implemented, what issues will they target, and what kind of response will they make?

2 The New Meaning of Modern Urban Design

In the context of the current information-based global economy and network society, the traditional meaning of urban space has gradually lost its original meaning because of the ever-changing social and economic system, and spatial integration through traditional urban design methods has become more difficult to implement. At the same time, however, this trend does not indicate that the effectiveness of urban design in the current urban environment will decline sharply, on the contrary, its role may be far more effective than ever before. This is because the form of a city, that is, the main work area of urban design (material organization and design), will play an increasingly important role in the world economy. As a visual system, the value of urban form in a society will never be lost. However, in the context of the new era, the work on urban form will acquire new significance.

2.1 The New Engine of Economic Development

From a global perspective, the importance of shaping a beautiful urban form in a city pattern has not declined but has improved. Large-scale projects such as West Kowloon in Hong Kong, Binhai in Dubai and Port Area in Hamburg are being carried out in large numbers all over the world, and there are also small and medium-sized projects that are more difficult to count and easier to operate. Through proper space creation and environmental improvement, many cities have ushered in a strong recovery of the service industry, retail industry, entertainment industry and tourism after the recession stage caused by technological upgrading, industrial transformation and population aging. Although information economy and network-based communication can distribute population and resources to every corner of the city, the spatial environment full of cultural temperament and emotional appeal is still irreplaceable, which makes the agglomeration effect of some urban centers continuously enhanced.

Meanwhile, urban spatial factors such as public safety, ecological environment and livable characteristics have become more important in an increasingly horizontal labor market. The mobile economic mode will also allow people to choose freely according to the required lifestyle, thus affecting the production mode of urban space. In such circumstances, the main goal of building production is to create a landscape with changing materials, and the same is true of urban design, which has been closely related to the driving force of a city's development and has become an important factor to directly mobilize economic power.

2.2 New Tools for the Governance of Urban Space

The increasing mobility of the population and the increasing mixing of social structures are strengthening the trend of space fragmentation. At the same time, this trend also stimulates people's demand for safety protection and close monitoring of the public environment, and leads to the separation of space in the field of life. A large number of access control communities have appeared all over the world since the 1980s, which shows that creating exclusive space through appropriate spatial layout has become an important consideration in urban design, which will contribute to form a substantive tool of spatial governance, so as to isolate social marginalization.

The phenomenon of spatial differentiation in cities is not only evident in residential areas but also gradually emerging in bustling commercial spaces. They are more and more inclined to choose to stay away from the city center with a complex environment, not only to avoid the expensive land rent and traffic cost, but also to avoid the possible negative impact on urban public safety through private transportation. In this sense, suburbanized mega-malls have also had an important impact on the division of contemporary urban space and society, and urban design, as a spatial policy tool, usually plays a role in fueling the flames through construction and packaging in this process.

2.3 The New Foundation of Cities Competition

In this world of scarce resources, economic fluctuations and intense competition, city governments of all kinds are actively seeking opportunities for independent development and the support of entrepreneurs and developers, which leads to almost all cities being passively in a state of crisis and being stimulated. They hope to attract more working professionals through more beautiful and pleasant environmental characteristics, and at the same time further enhance their cultural taste. Only then can they achieve better prospects on the world stage, and then may get better economic benefits.

This phenomenon reflects the interaction between the great wave of globalization and the rapidly changing urban space. The information capital flow constantly redivides high-level talents, symbolic economy and virtual time and space, which depends on people's changing realistic feelings about the built environment. To a great extent, a city's ability to change also depends on its ability to readjust and shape its own spatial environment. In this context, successful cities survive and prosper, while failing cities need to think more about space creation, all of which needs to be promoted by active urban design that can effectively produce a promotional image.

2.4 The Significance of Urban Design

With these phenomena, urban design today seems to have acquired a different important meaning from the past. From the unique contribution of urban design, it is distinguished from the past material form design which only emphasizes integrity. Creating places, developing tourism and attracting commerce, the beautiful urban space has become an essential component of urban amenities. Urban design is more like running a theater, by constantly changing the attractive content, so as to attract customers and keep them enjoying.

In line with this, the current popular urban design is losing its characteristics as public goods but becoming a more specific and targeted space marketing tool. If there is anything that can be understood as an essential change, then we can summarize it as follows: the current urban design is changing from "public institution" to the "production of spaces".

On the one hand, this trend is dismantling the previously unassailable views on urban space design, which claimed that a harmonious urban environment would naturally lead to a better society. Facing the current pattern of the times, most of the views of traditional urban design theory have lost their explanatory effect, at least this situation has been reflected in various controversies such as modernism and postmodernism, industrialization and post-industrialization.

On the other hand, it is reassembling other new factors, and some previously unrelated fields are connected by a brand-new network relationship. New urban spatial phenomena have emerged constantly over the past decade or so, such as novelty, exaggeration, retro, and worship of foreign countries…These spatial phenomena that cannot be understood according to the routine in the process of rapid urbanization need to be explained from a convincing point of view.

Despite the increasing emphasis on incorporating disciplinary perspectives from sociology, politics, economics, human geography and other fields in the theoretical research of urban design, these viewpoints have not been effectively integrated into the theoretical explanation of contemporary urban space construction. The inevitable result of such circumstances is that more and more research paradigms about urban spatial form and composition do not necessarily come from outside the discipline, but just come from the basic viewpoints within the discipline of urban design. The old paradigm has died out and the new one has not yet been established. This situation requires us to constantly reflect on what the urban design theory that can cope with the changes of the times will be (Fig. 13.3).

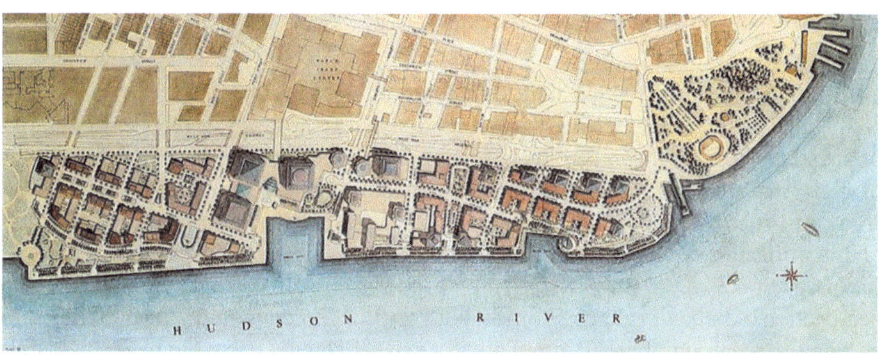

Fig. 13.3 Cooper & Axto Office, Bartley Park City, New York, 1978. This urban design exemplifies how urban design can spearhead the development of urban renewal. *Sources* YOST C. (2011). *Battery Park City: it's a wrap.* Architectural Record, 199 (09): 29

3 Urban Design in the Context of Contemporary Social Transformations

In a dynamic and historical context, what other reasonable explanations can we get about urban design? What kind of urban design can we have tomorrow? There may not be a definitive answer to this question, as we cannot possibly have all the answers in advance in the face of an ever-changing reality. We can only say that it is impossible to presuppose an urban design method, and it is impossible to have an urban design that does not face the need for change. There can be no urban design without concrete objects, nor can there be a universally applicable abstract model. There can be no presupposition standard, and there can be no urban design without decision-making.

Urban design should not be treated as a formal theory, thus assuming concepts and methods to be effective for all societies and cities. The background environment of every society in every era possesses a distinctiveness, which precludes a general framework that can be universally applied to all situations.

However, this does not mean that we are unable to propose a general perspective on the interaction among different cities, societies and historical changes, and develop a fruitful and expansive research perspective. In such an expansive domain, there can be a fresh interpretation of urban morphology in design.

3.1 Urban Design Represents the Comprehension and Research of Cities

Urban design is an extended field that explores how the material form of the city forms and evolves. In such a vision, urban space is not only a physical container of social process, but also a collective memory and association with the future, full of glory and disrepair, humdrum and passionate, praise or conflict. Therefore, the study of urban form will be the core of urban design. In Costov's view, the core of urban design is to pay attention to "how and why cities form what they are". In this perspective, "I am not studying abstract forms or forms explained from the perspective of behavioral possibilities, but I am concerned with the form of meaning carrier, the meaning of architecture always exists in the relationship between history and culture."[3]

Urban design is the manifestation of urban meaning existing in a specific urban form. In order to fully understand why a city is a city, we must go from abstract social science into the field of human experience and creative process.

This involves not only the functional problem of how the city works, but also the problem of how to achieve social consensus. Urban design is not only the skill of designing a city but also the knowledge of how a city grows and changes. It investigates how civilization is presented in spatial form and how specific urban form

[3] Costov. (2005).

is produced. Thus we need a theory that can interpret how urban form is produced, and also need a theory of vision with sufficient flexibility to explain the production and operation of urban functions and forms under different backgrounds.

The patterns of urban design cannot only exist in the work of architects or urban planners, and can be generated, realized and accepted through their evaluation. If we fail to uncover how a city evolved into the secret of historical life in a specific social sense, the work of urban design will still be only a technical revision, and the urban form will only be a subjective taste problem.

3.2 *Urban Design is a Form of Practice*

Urban design is a deliberate process of shaping the physical form of a city within its unique urban context. Throughout the ages, human society has always been a spatial existence, and people actively participate in the ubiquitous spatial social construction around them. If we regard a city as a spatial organization composed of human life, it not only refers to all kinds of static and permanent material entities in the city, but also means all ways, ways of using and trading physical space, and means people's distribution in space, the layout of economic activities, the circulation relationship between regions (Fig. 13.4).

There is no doubt that the objective of urban design is to make cities aesthetically pleasing. However, as a social practice, urban design needs to try its best to adapt to the changes of urban life, and provide applicable and different solutions for each change and activity as much as possible.

Fig. 13.4 Paley Clark Paley, San Francisco Trans-Bay Station Complex, 2007. Such projects also include the second phase of the International Finance Centre Tower, which is essentially a transportation hub, connecting the subway system, regional railway network and local public transport system Based on this, it forms an urban community. *Source* SHANE D G. (2011). *Urban design since 1945, a global perspective*. New York: Wiley, 2011:298

It is within this perspective that we can delve into the unique "professional field" that urban design possesses. Compared with its specific and independent construction activities, the uniqueness of urban design as a whole lie in the fact that:

(1) It is social, and the urban design in particular is to try to introduce the materiality of spatial form from social economy, cultural system, and political conflict in the analysis of social changes. Therefore, we need to make efforts to understand how the significance of cities is formed, thus opening up the path to uncover the historical process.
(2) It is dynamic, urban development is always composed of a series of continuous fragments. Local control can only act on its development and form, and there is no final result. Urban design is always confronting a dynamic field, urban design is to deal with this dynamic change, adjusting the picture of moving.

Therefore, as an independent professional field, urban design needs to give up a relatively stable and static attitude paradoxically, to resist a seemingly complete and standardized "conventional design", a solidified theoretical rhetoric and stereotyped method. It must not be determined by a single thinking, or the result determined by a molding thinking, but should be embodied as a structural and time-consuming process.

3.3 *Change and Transformation in Urban Design*

In the process of alternating transformation of urban patterns, we can see that when urban designers conceive urban life patterns, they need to constantly re-examine the social and economic changes that have taken place in the process of urban development, and create opportunities for innovative urban development from the process of mutual learning. In this global context, we can see the change and transformation of urban design.

An earlier case study comes from British architect John F. Turner. At the first UN Habitat Conference in Vancouver in 1976, Turner thought that self-built housing in slums in Latin America, with its bottom-up action process and small but flexible operation mode, provided a better solution for economically underdeveloped countries. In fact, Otto Konigsberger, Turner's teacher at AA School, began to persuade Indian Prime Minister Nehru in 1949 to allow 7 million homeless refugees caused by India's split with Pakistan in the process of independence to build their own houses.

Turner's early work began in the barriadas of Lima, Peru. In the late 1960s, he organized an international slum upgrading competition, and architects such as Jim Sterling and Christopher Alexander also put forward solutions, some of which were implemented in the Previ District of Lima, including various schemes of mixed housing types.[4]

[4] Turner J F. (1976).

In Asia, Africa and Latin America, an increasing number of urban designers are learning how to communicate with inhabitants of self-organizing systems and use remote satellite sensing systems (such as Google Earth and handheld GPS devices) to depict these areas. David Satterthwaite, who works for UN-HABITAT in Vancouver and the International Institute for Environment and Development in London, pointed out that although the population of mega-cities only accounts for 8% of the global urban population, this is the hope for the future, because local governments and non-governmental organizations are better able to improve the urban conditions of small urban agglomerations with a population of 1 million on such a scale, feedback and participation from residents can be more effective.

The constraints of limited resources faced by underdeveloped countries urges people to think about how to improve the complex efficiency of cities in practice. In the 1960s and 1970s, cities in the United States, Canada and Latin America faced difficulties in their economic development driven by cheap oil, which threatened their decentralized spatial logic. As energy supply becomes more and more expensive, urban transportation will become uneconomical. During the tenure of two visionary mayors, Bogota, Colombia, developed a low-energy consuming megastructure, providing an ecological blueprint for low-cost construction for slum upgrading; Curitiba, Brazil, a relatively wealthy country, developed a large-scale mode of transportation based on bus lines in the 1980s. In Bogota, urban designers adopted the early ideas of Le Corbusier and J.L. Sert, and created a new linear transportation system for parks in 1990s, connecting pedestrian areas, fast bus lanes and local communities in the city center, and providing these local communities with fresh water, hospitals with Internet, schools and parks to upgrade them.

In Africa and Asia, over half of the population still resides in rural settlements, connected to cities through mobile phone services. Due to the proximity of agriculture in the surrounding areas of these small urban zones, it means that food resources are readily available nearby. This model has prompted many European urban scholars to rethink the relationship between man-made and nature from the perspective of urban complexity.

In the book *La Citt à Elementare*, Paola Viganó made a similar analysis of Italian urban communities based on the Veneto region, and envisioned a kind of "reverse city", in which the spacing between urban elements in the landscape becomes the focal point of urban design. Early regional studies emphasized the ecological aspect of landscape design with Bernardo Secchi. Like many Asian big cities, Sage and Vigan ó emphasized a positive and valuable force, and called it the "reverse city" code, which was embedded in the outline of a productive agricultural landscape in the fabric spreading through the city as part of the overall design.[5]

Other European theorists, such as Thomas Sieverts from Germany, as an organizer of the Rhine-Ruhr corridor project, conceived a new kind of "in-between-city" with half city and half township in *Cities without Cities*.[6]

---

[5] Viganó P (1999).

[6] Sieverts T (2003).

These theorists, originating from affluent regions within the European Union, are confronted with complex landscapes that bear some resemblance to the situations in rural areas of low-income Indonesia and the urban–rural interface in Bangkok, both of which are a mixture of urban–rural interfaces in the outskirts of Bangkok. In response to this reality, some American theorists have proposed a concept of the urban–rural junction and organic urban agriculture, which provides a new concept for urban development with localized food supply. For example, Brian McGrath and Victoria Marshall of Parsons New College in New York put forward the idea for Baltimore Ecosystem Research (BES).[7] Dickson Despommier presents a more representative model for metropolises: Develop innovative green skyscrapers and cultivate crops on the "air farm" in the center of the city; In Italy, home to the Slow Food movement, designer Carlo Petrini envisions a concept called "slow city" (città slow), where citizens can grow and produce their own food in the city and surrounding vacant areas.

Another phenomenon worth reflecting in Latin American slums is the flexible and adaptive design system they present when dealing with their own complex problems, thus combining the urban village system with the self-construction process of residents using waste so that large shanty towns can repair themselves over time. Teddy Cruz's project in Tijuana, Mexico, was carried out in the absence of industrial support, which correspondingly suggests that the suburban housing of single families in the United States can also learn from the temporary buildings in the shanty towns in Mexico, and the suburban temporary housing in the United States can do better because it can be supported by industrial production.

The issue of complexity also raises the questions about how to promote the compatibility and inclusiveness of urban development. In the book *Phylogenesis*, FOA Architecture put forward the idea of FOA Ark,[8] that is, folding and classifying architectural forms and their scripts, and endowing each building with DNA and code (Fig. 13.5). Reiser + Umemoto demonstrates how to get inspiration from these emerging systems, build creative profile forms in cities, use undulating ground and inclined parks, and provide more choices for citizens.

For contemporary cities, there is another issue that poses a significant threat from the ecological perspective. From a global perspective, many megacities and human settlements in the world are built in delta areas where rivers enter the sea. Sea level rise caused by climate change and the greenhouse effect will seriously threaten the future living conditions of these areas.

The design thinking regarding urban ecological problems began in the United States, but was subsequently neglected because of its extensive urban development mode. Europeans have further developed eco-city design methods and tried to learn from the low-carbon footprint of Asian cities. Stockholm and London, for example, have introduced urban centre pricing from Singapore to tackle congestion. Meanwhile, European designers export eco-city models to other big cities in the world, such as Shanghai Dongtan Eco-city, which became famous a few years ago.

---

[7] Mccrath B, Marshall V (2005).

[8] Foreign Office Architects (2003).

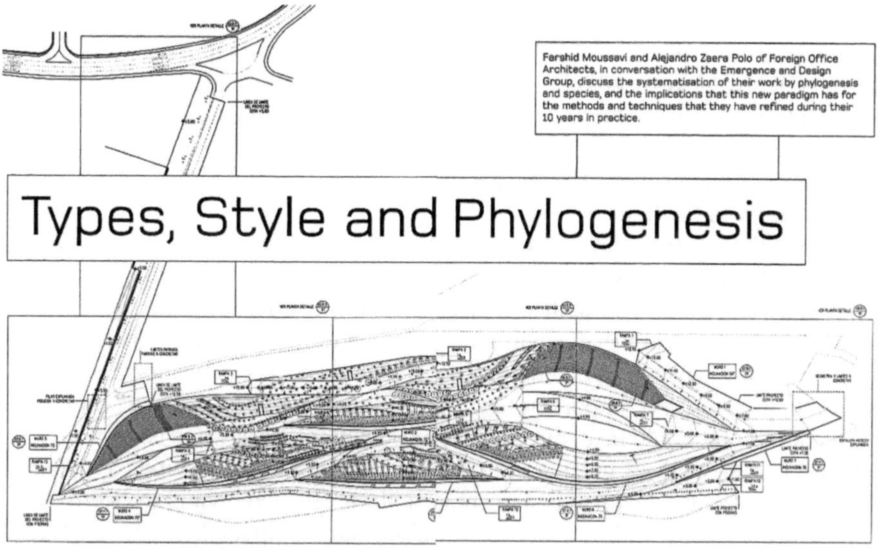

Fig. 13.5 Master Plan for Parks and Auditorium on the Barcelona Southeast Coastal Park by FOA Architects. The tissue prototype is based on coastal sand dunes, which are a material organization with few internal structures and are affected by wind. Program allocation is based on modeling sports and leisure activities as circuit networks. *Sources*: Foreign Office Architects (2003). *Phylogenesis*. New York: Actar

However, experience from underdeveloped regions may be more efficient. For example, in Vietnam, where the Mekong Delta is particularly vulnerable, designers are planning a new type of elevated urban island and urban archipelago to provide people with safe areas; A team from the School of Architecture and Urban Planning (UN-ASRO) at the University of Leuven in Belgium prepared a preliminary plan for Vinh in Vietnam to continue the tradition of desa-kota in Asian cities and integrate agriculture and irrigation systems with urban structures. As a result, the relevant teams from Leuven University in Belgium and Hanoi University in Vietnam invented a new waterfront, which can allow floods to pass through, but at the same time ensure that the houses on the shore can safely exceed the predicted water level.

4 Theoretical Research on Urban Design Theory with Respect of Urban Complexity

As a man-made environment, the urban real world comes from countless urban operators, among which architects, urban designers and planners try their best to think and conceive the future of the city. At the same time, countless and ever-changing types of owners stand behind them: from the central government, local governments, communities, non-governmental organizations (NGOs) and various economic institutions, to all kinds of settled residents and floating people.

The effectiveness of urban design depends on operator's ability to think about the people they are working with, identify key elements and adapt them to local environment and conditions. This ability to reflect, adapt, discuss and change is very important for the sustainable creation of new urban content and the renewal and utilization of old facilities.

Despite theoretical and computational innovations make the traditional typological method adapt to a dynamic relationship and change with situations, problems, actors and time, it is still difficult to closely relate to the self-constructed and informal phenomena in the real urban environment. Urban designers need to recognize the interconnection, diversity, and complexity of cities. This collective form of individuals and groups is very important for how to grasp the dynamic situation of cities. Urban designers should take into account those "inter" areas when constructing the analytical framework of cities and integrating the type units of cities and include the strength of individual builders to create a pluralistic and organic collective form (Fig. 13.6). To cope with the chaotic variables and compound values involved in the future development of cities, so as to be able to work under various unstable situations and respond to urban problems in an indirect and nonlinear way.

Fig. 13.6 Overall Planning for Bonavista Community in Singapore by Toyo Ito Architects. The strategy employed here is to integrate urban complexes with agricultural landscapes to form an ecological-driven future urban development. Sources: ITO T. (2011). *Singapore buona vista masterplan competition, Singapore.* Architectural Design, 81(01):90–93

Therefore, from a global perspective, the dissemination of models and mutual learning of patterns between different regions, different structures and different backgrounds and the learning between them do not happen naturally, but are the result of the joint efforts of countless participants. From the various cases of experience exchange and interaction between European and American metropolitan model transformation and Asian and Latin American cities, we can see that the current urban design thought has been like a giant tapestry shuttle, which has been woven into one without distinction under the series connection of some themes. Due to the unprecedented complexity in the current process of urban transformation, urban design continues to play and should play an even more crucial role.

First, the various changes in cities are derived from the will and actions of the actors involved, that is, the subjects, organizations, associations involved in urban construction and the ideas they use to build cities. Although cities often seem to lose control during the transformation process, urban actors and their ideas still exert tremendous energy and organizational skills. Even the most randomly built cities are the result of human intervention. This seemingly chaotic agglomeration is actually the comprehensive result of many personal choices and actions made by urban actors when they pursue their own interests. These behaviors may bring many unexpected consequences after being repeated on a large scale. In the process of urban construction, as well as in the process of urban renewal, adjustment and regeneration, urban actors need to cooperate with each other. Cities are the product of people living together, which requires organization and skills to manage the affairs of local communities and larger cities.

Secondly, as cities expand in size and become more complex, any attempt to exercise complete control has become completely ineffective. This calls for research on urban construction to pay attention to the self-organization behavior in cities and construct a model for it. With the development of cybernetics in the twentieth century, that is, the research on management information flow and the principle of analyzing urban self-organization through ranking and feedback mechanism, more and more urban actors supported by computer technology can master and process more information than before, and analyze the complexity of cities by establishing corresponding models.

Finally, real urban outcomes depend on urban actors' ability to reflect on their work, to reorganize urban elements, and to shift models to suit local circumstances. This ability of reflect, adapt, adjust and change is crucial for continuously enhancing the adaptability of traditional cities, promoting the revitalization of declining urban areas, and advocating for innovation in the urban environment.

References

Costov. (2005). *The formation of cities: Urban model and urban significance in the historical process*. Shan Hao, translated. Beijing: China Construction Industry Press, 2005:9.

Custer, Wang Zhihong. (2006). *Flow space*. Foreign Urban Planning, 21 (05): 70.

Foreign Office Architects. (2003). *Phylogeresis*. New York: Actar.
Su Jia. (2004). *Postmodern geography*. Wang Wenbin, translated. Beijing: Commercial Press, 2004:367.
Mccrath B, Marshall V. (2005). *Operationalizing Patch Dynamics*. Landscape Architecture, Architectural Design, March/April, 77(02):52-59.
Sieverts T. (2003). *Cities without cities, an interpretation of the Zwischenstadt*. London: Routledge.
Turner J F. (1976). *Housing by the people: towards autonomy in building environments*. London: Marion Boyars.
Vigano P. (1999). *La città elementare*. Milan Geneva: Skira.

MIX
Papier aus verantwortungsvollen Quellen
Paper from responsible sources
FSC® C105338

If you have any concerns about our products,
you can contact us on
ProductSafety@springernature.com

In case Publisher is established outside the EU,
the EU authorized representative is:
Springer Nature Customer Service Center GmbH
Europaplatz 3, 69115 Heidelberg, Germany

Printed by Libri Plureos GmbH
in Hamburg, Germany